Soliton

(孤立子)

Guo Boling   Yao Yuqin   Zhao Lichen

(郭柏灵   姚玉芹   赵立臣)

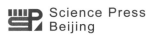

Science Press
Beijing

Responsible Editor: Li Xin

Copyright © 2025 by Science Press
Published by Science Press
16 Donghuangchenggen North Street
Beijing 100717, P. R. China

Printed in Beijing

ISBN 978-7-03-082514-8

Preface

With the development of modern physics and mathematics, the phenomenon of "solitary waves" discovered by the famous British scientist Russell as early as 1834 has attracted great attention and interest in the past decade. It has been confirmed from numerical calculations, theoretical analysis, and physical experiments that a large number of nonlinear evolutionary equations have soliton solutions. The interaction between these solitons exhibits a very peculiar phenomenon, and solitons have stable waveforms. These new phenomena are being utilized to explain some of the new problems that arise in physics. In mathematics, new methods such as scattering inversion, prolongation structure, and Bäcklund transformation have emerged to accurately solve special solutions of nonlinear evolution equations, and have gradually formed a relatively systematic mathematical theory related to soliton problems.

The main purpose of writing this book is to provide a brief introduction to the fundamental problems of solitons, along with their mathematical and physical methods. In addition to the most fundamental knowledge and concepts related to soliton problems, some recent developments and achievements in these research fields were also introduced. We hope that these will help readers sort out some rough but clear threads from the vast works and literatures, so that when readers are interested in a certain aspect of the problem, they can carry out research work based on consulting relevant references.

Below is a brief introduction to the background of the formation of this book. In 1978, one of the authors wrote a delivered notes *KdV Equations and Solitons*. In 1979, at the National Finite Element and Soliton Conference, colleagues such as Feng Kang, Zhang Xueming, Tu Guizhang, and Li Yishen from the conference leadership group believed that there was a need for a book introducing solitons, and entrusted the task of writing this book to the author. This matter also received active support from colleagues from the Science Press. Therefore, based on the "lecture notes", the author began writing, further consulted relevant materials, added content to the original "lecture notes", supplemented the latest literature, and made a new attempt by combining mathematical methods with physical content. However, due to the extensive knowledge involved in theoretical issues related to solitons and limited for the author's level, there were inevitably inappropriate or erroneous aspects in the book. Readers were kindly requested to correct them.

Finally, we would like to express our special gratitude to colleague Tu Guizhang for his consistent care and enthusiastic assistance in the compilation of this book, as well as for providing many valuable suggestions.

In the process of translation, we strive to maintain the academic value of the original work while taking into account the reading habits and thinking patterns of Chinese readers, and have added a chapter on rogue waves and wave turbulence.

Guo Boling

November 9, 2024

Contents

Preface

Chapter 1   Introduction $\cdots\cdots\cdots\cdots\cdots\cdots\cdots\cdots\cdots\cdots\cdots\cdots\cdots\cdots\cdots\cdots\cdots\cdots$ 1

1.1   The Origin of Solitons $\cdots\cdots\cdots\cdots\cdots\cdots\cdots\cdots\cdots\cdots\cdots\cdots\cdots\cdots\cdots\cdots$ 1

1.2   KdV Equation and Its Soliton Solutions $\cdots\cdots\cdots\cdots\cdots\cdots\cdots\cdots$ 4

1.3   Soliton Solutions for Nonlinear Schrödinger Equations and Other Nonlinear Evolutionary Equations $\cdots\cdots\cdots\cdots\cdots\cdots\cdots\cdots\cdots\cdots\cdots\cdots\cdots$ 6

1.4   Experimental Observation and Application of Solitons $\cdots\cdots\cdots\cdots\cdots$ 10

1.5   Research on the Problem of Soliton Theory $\cdots\cdots\cdots\cdots\cdots\cdots\cdots\cdots$ 10

References $\cdots\cdots\cdots\cdots\cdots\cdots\cdots\cdots\cdots\cdots\cdots\cdots\cdots\cdots\cdots\cdots\cdots\cdots\cdots\cdots\cdots\cdots$ 11

Chapter 2   Inverse Scattering Method $\cdots\cdots\cdots\cdots\cdots\cdots\cdots\cdots\cdots\cdots\cdots$ 12

2.1   Introduction $\cdots\cdots\cdots\cdots\cdots\cdots\cdots\cdots\cdots\cdots\cdots\cdots\cdots\cdots\cdots\cdots\cdots\cdots\cdots$ 12

2.2   The KdV Equation and Inverse Scattering Method $\cdots\cdots\cdots\cdots\cdots\cdots$ 12

2.3   Lax Operator and Generalization of Zakharov, Shabat, AKNS $\cdots\cdots\cdots\cdots$ 21

2.4   More General Evolutionary Equation (AKNS Equation) $\cdots\cdots\cdots\cdots\cdots$ 28

2.5   Solution of the Inverse Scattering Problem for AKNS Equation $\cdots\cdots\cdots\cdots$ 35

2.6   Asymptotic Solution of the Evolution Equation $(t \to \infty)$ $\cdots\cdots\cdots\cdots\cdots$ 46

    2.6.1   Discrete spectrum $\cdots\cdots\cdots\cdots\cdots\cdots\cdots\cdots\cdots\cdots\cdots\cdots\cdots\cdots\cdots$ 46

    2.6.2   Continuous spectrum $\cdots\cdots\cdots\cdots\cdots\cdots\cdots\cdots\cdots\cdots\cdots\cdots\cdots$ 49

    2.6.3   Estimation of discrete spectrum $\cdots\cdots\cdots\cdots\cdots\cdots\cdots\cdots\cdots\cdots$ 52

2.7   Mathematical Theory Basis of Inverse Scattering Method $\cdots\cdots\cdots\cdots\cdots$ 56

2.8   High-Order and Multidimensional Scattering Inversion Problems $\cdots\cdots\cdots$ 74

References $\cdots\cdots\cdots\cdots\cdots\cdots\cdots\cdots\cdots\cdots\cdots\cdots\cdots\cdots\cdots\cdots\cdots\cdots\cdots\cdots\cdots\cdots$ 83

Chapter 3   Interaction of Solitons and Its Asymptotic Properties $\cdots\cdots\cdots\cdots$ 85

3.1   Interaction of Solitons and Asymptotic Properties of $t \to \infty$ $\cdots\cdots\cdots\cdots$ 85

3.2   Behaviour State of the Solution to KdV Equation Under Weak Dispersion and WKB Method $\cdots\cdots\cdots\cdots\cdots\cdots\cdots\cdots\cdots\cdots\cdots\cdots\cdots\cdots\cdots$ 94

3.3   Stability Problem of Soliton $\cdots\cdots\cdots\cdots\cdots\cdots\cdots\cdots\cdots\cdots\cdots\cdots\cdots\cdots$ 100

3.4   Wave Equation under Water Wave and Weak Nonlinear Effect $\cdots\cdots\cdots\cdots$ 102

References $\cdots\cdots\cdots\cdots\cdots\cdots\cdots\cdots\cdots\cdots\cdots\cdots\cdots\cdots\cdots\cdots\cdots\cdots\cdots\cdots\cdots$ 109

Chapter 4 Hirota Method $\cdots\cdots\cdots\cdots\cdots\cdots\cdots\cdots\cdots\cdots\cdots\cdots\cdots\cdots\cdots\cdots\cdots\cdots\cdots$ 111

  4.1 Introduction $\cdots\cdots\cdots\cdots\cdots\cdots\cdots\cdots\cdots\cdots\cdots\cdots\cdots\cdots\cdots\cdots\cdots\cdots\cdots\cdots$ 111

  4.2 Some Properties of the D Operator $\cdots\cdots\cdots\cdots\cdots\cdots\cdots\cdots\cdots\cdots\cdots$ 113

  4.3 Solutions to Bilinear Differential Equations $\cdots\cdots\cdots\cdots\cdots\cdots\cdots\cdots$ 115

  4.4 Applications in Sine-Gordon Equation and MKdV Equation $\cdots\cdots\cdots\cdots$ 117

  4.5 Bäcklund Transform in Bilinear Form $\cdots\cdots\cdots\cdots\cdots\cdots\cdots\cdots\cdots\cdots$ 125

  References $\cdots\cdots\cdots\cdots\cdots\cdots\cdots\cdots\cdots\cdots\cdots\cdots\cdots\cdots\cdots\cdots\cdots\cdots\cdots\cdots\cdots$ 127

Chapter 5 Bäcklund Transformation and Infinite Conservation Law $\cdots\cdots\cdots\cdots$ 129

  5.1 Sine-Gordon Equation and Bäcklund Transformation $\cdots\cdots\cdots\cdots\cdots\cdots$ 129

  5.2 Bäcklund Transformation of a Class of Nonlinear Evolution Equation $\cdots$ 134

  5.3 B Transformation Commutability of the KdV Equation $\cdots\cdots\cdots\cdots\cdots$ 141

  5.4 Bäcklund Transformations for High-Order KdV Equation and

       High-Dimensional Sine-Gordon Equation $\cdots\cdots\cdots\cdots\cdots\cdots\cdots\cdots\cdots$ 143

  5.5 Bäcklund Transformation of Benjamin-Ono Equation $\cdots\cdots\cdots\cdots\cdots\cdots$ 145

  5.6 Infinite Conservation Laws for the KdV Equation $\cdots\cdots\cdots\cdots\cdots\cdots$ 151

  5.7 Infinite Conserved Quantities of AKNS Equation $\cdots\cdots\cdots\cdots\cdots\cdots\cdots$ 154

  References $\cdots\cdots\cdots\cdots\cdots\cdots\cdots\cdots\cdots\cdots\cdots\cdots\cdots\cdots\cdots\cdots\cdots\cdots\cdots\cdots\cdots$ 157

Chapter 6 Multidimensional Solitons and Their Stability $\cdots\cdots\cdots\cdots\cdots\cdots$ 159

  6.1 Introduction $\cdots\cdots\cdots\cdots\cdots\cdots\cdots\cdots\cdots\cdots\cdots\cdots\cdots\cdots\cdots\cdots\cdots\cdots\cdots$ 159

  6.2 The Existence Problem of Multidimensional Solitons $\cdots\cdots\cdots\cdots\cdots\cdots$ 160

  6.3 Stability and Collapse of Multidimensional Solitons $\cdots\cdots\cdots\cdots\cdots\cdots$ 174

  References $\cdots\cdots\cdots\cdots\cdots\cdots\cdots\cdots\cdots\cdots\cdots\cdots\cdots\cdots\cdots\cdots\cdots\cdots\cdots\cdots\cdots$ 180

Chapter 7 Numerical Calculation Methods for Some Nonlinear Evolution

          Equations $\cdots\cdots\cdots\cdots\cdots\cdots\cdots\cdots\cdots\cdots\cdots\cdots\cdots\cdots\cdots\cdots\cdots\cdots$ 182

  7.1 Introduction $\cdots\cdots\cdots\cdots\cdots\cdots\cdots\cdots\cdots\cdots\cdots\cdots\cdots\cdots\cdots\cdots\cdots\cdots\cdots$ 182

  7.2 The Finite Difference Method and Galerkin Finite Element Method for

       the KdV Equations $\cdots\cdots\cdots\cdots\cdots\cdots\cdots\cdots\cdots\cdots\cdots\cdots\cdots\cdots\cdots\cdots$ 184

  7.3 The Finite Difference Method for Nonlinear Schrödinger Equations $\cdots\cdots$ 189

  7.4 Numerical Calculation of the RLW Equation $\cdots\cdots\cdots\cdots\cdots\cdots\cdots\cdots$ 194

  7.5 Numerical Computation of the Nonlinear Klein–Gordon Equation $\cdots\cdots$ 195

  7.6 Numerical Computation of a Class of Nonlinear Wave

       Stability Problems $\cdots\cdots\cdots\cdots\cdots\cdots\cdots\cdots\cdots\cdots\cdots\cdots\cdots\cdots\cdots\cdots$ 197

  References $\cdots\cdots\cdots\cdots\cdots\cdots\cdots\cdots\cdots\cdots\cdots\cdots\cdots\cdots\cdots\cdots\cdots\cdots\cdots\cdots\cdots$ 202

Chapter 8 The Geometric Theory of Solitons $\cdots\cdots\cdots\cdots\cdots\cdots\cdots\cdots\cdots\cdots$ 204

  8.1 Bäcklund Transform and Surface with Total Curvature $K = -1$ $\cdots\cdots\cdots$ 204

  8.2 Lie Group and Nonlinear Evolution Equations $\cdots\cdots\cdots\cdots\cdots\cdots\cdots\cdots$ 207

  8.3 The Prolongation Structure of Nonlinear Equations $\cdots\cdots\cdots\cdots\cdots\cdots$ 211

  References $\cdots\cdots\cdots\cdots\cdots\cdots\cdots\cdots\cdots\cdots\cdots\cdots\cdots\cdots\cdots\cdots\cdots\cdots\cdots\cdots\cdots$ 217

Chapter 9   The Global Solution and "Blow up" Problem of Nonlinear Evolution Equations ··· 219

  9.1   Nonlinear Evolutionary Equations and the Integral Estimation Method ··· 219

  9.2   The Periodic Initial Value Problem and Initial Value Problem of the KdV Equation ··· 221

  9.3   Periodic Initial Value Problem for a Class of Nonlinear Schrödinger Equations ··· 229

  9.4   Initial Value Problem of Nonlinear Klein-Gordon Equation ············ 235

  9.5   The RLW Equation and the Galerkin Method ························· 243

  9.6   The Asymptotic Behavior of Solutions and "Blow up" Problem for $t \to \infty$ ··· 251

  9.7   Well-Posedness Problems for the Zakharov System and Other Coupled Nonlinear Evolutionary Systems ··· 256

  References ··· 258

Chapter 10   Topological Solitons and Non-topological Solitons ·············· 261

  10.1   Solitons and Elementary Particles ······························· 261

  10.2   Preliminary Topological and Homotopy Theory ·················· 265

  10.3   Topological Solitons in One-Dimensional Space ················· 270

  10.4   Topological Solitons in Two-Dimensional ······················· 276

  10.5   Three-Dimensional Magnetic Monopole Solution ················ 282

  10.6   Topological Solitons in Four-Dimensional Space—Instantons ·········· 288

  10.7   Nontopological Solitons ··· 292

  10.8   Quantization of Solitons ··· 296

  References ··· 301

Chapter 11   Solitons in Condensed Matter Physics ························· 303

  11.1   Soliton Motion in Superconductors ······························· 304

  11.2   Soliton Motion in Ferroelectrics ································· 315

  11.3   Solitons of Coupled Systems in Solids ··························· 318

  11.4   Statistical Mechanics of Toda Lattice Solitons ·················· 322

  References ··· 327

Chapter 12   Rogue Wave and Wave Turbulence ························· 329

  12.1   Rogue Wave ··· 329

  12.2   Formation of Rogue Wave ······································· 329

  12.3   Wave Turbulence ··· 333

  12.4   Soliton and Quasi Soliton ······································· 336

    12.4.1   The Instability and Blow-up of Solitons ················· 338

    12.4.2   The Case of Quasi-Solitons ··························· 339

  References ··· 341

Chapter 1
Introduction

1.1 The Origin of Solitons

In 1834, British scientist Scott Russell accidentally observed a wonderful water wave. In 1844, he vividly described this phenomenon in his article *On Waves* published in the *Report of the* 14*th Conference of the British Association for the Advancement of Science*: "I observed the movement of a ship, which was pulled by two horses and rapidly advanced along a narrow canal. Suddenly, the ship came to a stop, but the large amount of water pushed by the ship did not stop. They accumulated around the bow of the ship and violently disturbed. Then, the waves suddenly appeared as a round, smooth, and well-defined huge isolated peak, rolling forward at a huge speed and rapidly leaving the bow. Its shape and speed did not change significantly during the journey. I rode on a horse and followed closely to observe that it rolled forward at a speed of about eight to nine miles per hour and maintained its original shape of about 30 feet long and 1 to 1.5 feet high. Gradually, its height decreased. When I tracked for 1 to 2 miles, it finally disappeared into the meandering river channel." This was a peculiar phenomenon observed by Russell, who then believed that this isolated wave was a stable solution to fluid motion and called it "solitary wave". Russell was unable to successfully prove and convince physicists of his argument at the time, thus blaming mathematicians for not being able to predict this phenomenon from known fluid motion equations. Subsequently, the issue of solitary waves sparked widespread debate among many physicists of the time. Until 1895, 60 years later, Korteweg de Vries studied the motion of shallow water waves and established the following shallow water wave motion equation for unidirectional motion under the assumption of long wave approximation and small amplitude

$$\frac{\partial \eta}{\partial t} = \frac{3}{2}\sqrt{\frac{g}{l}}\frac{\partial}{\partial x}\left(\frac{1}{2}\eta^2 + \frac{2}{3}\alpha\eta + \frac{1}{3}\sigma\frac{\partial^2 \eta}{\partial x^2}\right) \tag{1.1}$$

here, η is the wave height, l is the water depth, g is the gravitational acceleration, α and σ are constants. They conducted a relatively complete analysis of the solitary wave phenomenon and derived a pulse like solitary wave solution with shape invariance from Eq.(1.1), which is consistent with Russell's description, thus theoretically proving the existence of solitary waves. However, is this wave stable? Can two solitary waves deform after collision? These questions have not been answered yet. Some people even suspect that Eq.(1.1) is a nonlinear partial differential equation, and the superposition principle of solutions does not satisfy it. After collision, the shape of the two solitary waves may be completely destroyed. This viewpoint has led many people to believe that this type of wave is "unstable", and solitary waves remain buried for a long time until new discoveries are made.

Another question is, do solitary waves like what Russell said appear in other physical fields besides fluid mechanics? In the early 20th century, this was an elusive issue. It was not until the 1950s that a new situation emerged due to the work of Fermi, Pasta and Ulam. They connected 64 particles with nonlinear springs to form a nonlinear vibrating string. Initially, all the energy of these resonators was concentrated in one, while the initial energy of the other 63 was zero. According to classical theory, as long as nonlinear effects exist, there will be phenomena such as energy equalization and ergodicity of states, that is, any weak nonlinear interaction can cause the system to transition from a non-equilibrium state to an equilibrium state. But the actual calculation results surprised them greatly, that the concept of achieving energy balance mentioned above was incorrect. In fact, from Fig.1-1, it can be seen that after a long period of time, almost all the energy returns to its original initial distribution, which is the famous FPU problem. At that time, due to their only examining the frequency space, solitary wave solutions could not be found, so the problem was not properly explained. Later, people regarded the crystal as a chain

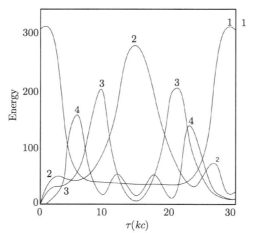

Figure 1-1 Energy curve of FPU problem. The unit is arbitrary, and the initial form of the string is a single sine wave

of elastic tubes with a mass scene and approximately simulated this situation. Toda studied the nonlinear vibration of this mode and obtained solitary wave solutions, which further aroused people's interest in solitary wave research.

Subsequently, in 1962, Perring and Skyrme applied the sine-Gordon equation to the study of elementary particles, and numerical calculations revealed that such solitary waves did not disperse, retaining their original shape and velocity even after a collision.

In 1965, the renowned American scientists Zabusky and Kruskal used numerical simulation methods to investigate in detail the nonlinear interaction process of soliton collisions in plasma, obtaining relatively complete and rich results, and further confirming the theory that soliton interactions do not change the waveform, which surprised people.

Due to the above results and the fact that stable solitary waves with unchanged waveforms after collisions have been discovered in many physical models, many physicists and mathematicians have shown great interest and pay attention in this field.   Research on soliton problems boomed and gradually formed a relatively complete soliton theory.

So, what exactly is a "soliton"? Usually, we refer to the local traveling wave solutions of nonlinear evolution equations as "solitary waves". The term "local" denotes that the solutions of differential equations approach zero or a constant as one approaches spatial infinity. We refer to these stable solitary waves, which are solitary waves that do not disappear after colliding with each other and whose waveforms and velocities do not change or only slightly change (just like the common collision of two particles), as "solitons". But there are also several articles and books that confuse solitons with solitary waves.

In physics, solitons are also defined as stable, finite-energy and non-dispersive solutions to classical field equations. That is, if we denote the energy density of a soliton as $\rho(x,t)$, then

$$0 < H = \int \rho(x,t)d^m x < +\infty \ (m \text{ is the dimension of the space})$$

and

$$\lim_{t \to \infty} \max \rho(x,t) \neq 0 \text{ (for certain } x)$$

That is to say, an soliton can be seen as a finite stable "mass" with non diffusive field energy, which is not destroyed even in motion or collision. For a large number of nonlinear wave equations, their solitons generally take on four shapes, as depicted in Fig.1-2 (a), (b), (c) and (d), namely bell-shape (or wave packet-shape), vortex-shape (anti-bell-shape), kink-shape (knot-shape) and anti-kink-shape (anti-knot-shape).

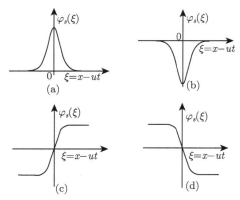

Figure 1-2 Types of solitons

$\varphi_s(\xi)$ represents the traveling wave solution, and $\xi = x - ut$ (u is a constant wave velocity)

Based on the research on elementary particles, Li Zhengdao and others further divided existing solitons into two categories: topological solitons and non topological solitons. We will specifically elaborate on their definitions and detailed information in Chapter 10, and provide a brief introduction to the research work of Li Zhengdao and others on non topological solitons.

1.2   KdV Equation and Its Soliton Solutions

As mentioned above, in 1895, Korteweg and de Vries established the shallow water wave equation (1.1), which we slightly modified to obtain the following form

$$u_t + uu_x + \mu u_{xxx} = 0 \text{ (Add punctuation marks after each formula)} \qquad (1.2)$$

here, the constant μ can be positive or negative. If $\mu < 0$, apply the transformation $u \to -u$, $x \to -x$, $t \to t$, then Eq.(1.2) becomes

$$u_t + uu_x - \mu u_{xxx} = 0 \qquad (1.3)$$

Thus, it can be assumed that $\mu > 0$. Eq.(1.2) is referred to as the KdV equation.

Let $u(x,t) = u(\xi)$, $\xi = x - Dt$, $D = $ const, substituting them into Eq.(1.2) and integrating with respect ξ twice, yields

$$3\mu \left(\frac{du}{d\xi} \right)^2 = -u^3 + 3Du^2 + 6Au + 6B = f(u) \qquad (1.4)$$

here, A, B are integral constants. The solution of Eq.(1.4) can only be real if $f \geqslant 0$ ($\mu > 0$). If $f(u)$ has only one real root, then it is unbounded. Now let's assume that function $f(u)$ has three real roots, namely $f(u) = -(u - c_1)(u - c_2)(u - c_3)$,

$c_1 < c_2 < c_3$. From this, we can obtain $D = \frac{1}{3}(c_1 + c_2 + c_3)$, $A = \frac{1}{6}(c_1c_2 + c_2c_3 + c_3c_1)$, $B = \frac{1}{6}c_1c_2c_3$. The general form of function $f(u)$ is shown by the curve in Fig.1-3.

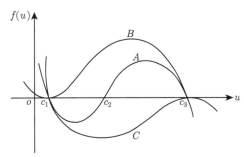

Figure 1-3

The exact solution of Eq.(1.4) can be expressed as a Jacobian elliptic function

$$u = u(x,t) = c_2 + (c_3 - c_2)c_n^2\left[\sqrt{\frac{c_3 - c_2}{12\mu}}\left\{x - \frac{1}{3}(c_1 + c_2 + c_3)t\right\}, k\right] \qquad (1.5)$$

here, $k^2 = (c_3 - c_2)/(c_3 - c_1)$. The periodic wave train Eq.(1.5) is usually referred to as the "Cnoidal wave". Since the real period of the function c_n is $2K$ and K is the first type of elliptic integral, the period of the "Cnoidal" wave is $T_p = 4K\sqrt{\frac{3\mu}{c_3 - c_1}}$.

If $K = 0$, $c_n(\xi, 0) = \cos\xi$. Eq.(1.4) has an oscillatory solution

$$u = \bar{c} + a\cos\left[2\sqrt{\frac{c_3 - c_1}{12\mu}}\left\{x - \frac{1}{3}(c_1 + c_2 + c_3)t\right\}\right] \qquad (1.6)$$

here $\bar{c} = \frac{c_2 + c_3}{2}$, $a = \frac{c_3 - c_2}{2}$.

If $K = 1$, $c_n(\xi, 1) = \text{sech}\xi$. At this time, the period becomes infinite when $c_2 \to c_1$, as shown in curve B in Fig.1-3. Soliton solution of KdV Eq.(1.2) is obtained

$$u = c_1 + (c_3 - c_1)\text{sech}^2\left[\sqrt{\frac{c_3 - c_1}{12\mu}}\left\{x - \frac{1}{3}(2c_1 + c_3)t\right\}\right] \qquad (1.7)$$

Setting $c_1 = u_\infty$, $c_3 - c_1 = a$, then (1.7) becomes

$$u = u_\infty + a\,\text{sech}^2\left[\sqrt{\frac{a}{12\mu}}\left\{x - \left(u_\infty + \frac{a}{3}\right)t\right\}\right] \qquad (1.8)$$

Here, u_∞ represents a uniform state at infinity, and a represents the amplitude of the soliton. From (1.8), it can be seen that the velocity of this solitary wave relative to the uniform state is proportional to the amplitude, while the width of the wave is inversely proportional to the square root of the amplitude, and the amplitude is independent of the uniform state. If $u_\infty = 0$, $\mu = 1$, it can be inferred from (1.8) that

$$u(x, t) = 3D\operatorname{sech}^2 \sqrt{\frac{D}{2}}(x - D_t) \tag{1.9}$$

as shown in Fig.1-4.

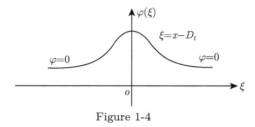

Figure 1-4

Nowadays, it is constantly discovered that a considerable number of wave equations and systems describing weak nonlinear effects can be reduced to the KdV equations under the long wave approximation and small and finite amplitude assumptions. For example: (1) The motion of magnetohydrodynamic waves in cold plasma; (2) Vibration of non resonant lattice; (3) Ionic acoustic waves of plasma; (4) Longitudinal dispersion fluctuations in elastic rods; (5) Pressure wave motion in two mixed states of liquid and gas; (6) The rotation of fluid at the bottom of a pipe; (7) Thermal excitation of phonon wave packets in nonlinear lattices at low temperatures.

1.3 Soliton Solutions for Nonlinear Schrödinger Equations and Other Nonlinear Evolutionary Equations

Cubic nonlinear Schrödinger equations

$$iu_t + u_{xx} + \nu|u|^2 u = 0 \tag{1.10}$$

or in a more general form

$$u_t - \gamma u_{xx} = \chi u - \beta|u|^2 u \tag{1.11}$$

here, $\beta = \beta_0 + i\beta_1, \gamma = \gamma_0 + i\gamma_1, i = \sqrt{-1}, \beta_0, \beta_1, \gamma_0, \gamma_1, \chi, \nu$ all are real constants. Such equations have been found in many physics problems, for example, in fine beam flow (nonlinear optics), there is the following equation

$$2ik\frac{\partial\Psi}{\partial x} + \nabla_\perp^2\Psi + \frac{n_2}{n_0}k^2|\Psi|^2\Psi = 0 \tag{1.12}$$

here, $\nabla_\perp^2 = \dfrac{\partial^2}{\partial r^2} + \dfrac{m}{r}\dfrac{\partial}{\partial r}$, and $m = 0$ is a plane. $m = 1$ is cylindrical symmetry. $\Psi = ae^{ik\theta}, \theta = kx - wt + ks(x,r), k$ is wave number, and $n = \dfrac{c_0 k}{w} = n_0 + \dfrac{1}{2}n_2 a^2$. For two-dimensional flow, there is an equation

$$\frac{n_2}{n_0}k^2|\Psi|^2\Psi = -2ki\Psi_x - \Psi_{xx} - \Psi_{yy} \tag{1.13}$$

Other examples include Langmuir waves in plasma, self-modulation of one-dimensional monochromatic waves, self-focusing of two-dimensional steady-state plane waves, and the Ginzbug-Landau equation for the motion of superconducting electron pairs in electromagnetic fields can be described using the nonlinear Schrödinger equation.

Now consider the traveling wave solution of Eq.(1.10). Set

$$u(x,t) = e^{irx-ist}v(\xi), \quad \xi = x - Dt$$

where r, s are all undetermined constants, and $D = $ const. Substituting the expression for u into (1.10), the ordinary differential equation for v can be obtained

$$v'' + i(2r - D)v' + (s - r^2)v + \nu|v|^2v = 0 \tag{1.14}$$

Now taking $r = \dfrac{D}{2}, s = \dfrac{D^2}{4} - \alpha \ (\alpha > 0)$, and eliminating v' (v is real) gives

$$v'' - \alpha v - \nu v^3 = 0 \tag{1.15}$$

After integrating, we obtain

$$v'^2 = A + \alpha v^2 - \frac{\nu}{2}v^4 \tag{1.16}$$

When $\nu > 0$, $A = 0$, we have

$$v(x,t) = \left(\frac{2a}{\nu}\right)^{\frac{1}{2}} \text{sech}\,\alpha(x - Dt) \tag{1.17}$$

Obviously, $|u|^2 \propto \text{sech}^2\,\alpha(x - Dt)$, and $v(x,t)$ is called an envelope solitary wave. Now consider a more general form of solution for Eq.(1.10)

$$u(x,t) = \Phi(x,t)e^{i\theta(x,t)} \tag{1.18}$$

where Φ, θ are all real functions and θ is called carrier wave, Φ is called an envelope wave. Substituting Eq.(1.18) into Eq.(1.10) and separating the real part and imaginary part, we can obtain

$$\Phi_{xx} - \Phi\theta_t - \Phi\theta_x^2 - \nu\Phi^3 = 0, \nu > 0 \tag{1.19}$$

$$\Phi\theta_{xx} + 2\Phi_x\theta_x + \Phi_t = 0$$

Setting $\theta = \theta\,(x - D_1 t)$, $\Phi = \Phi\,(x - D_2 t)$ and substituting them into (1.19) gives

$$\Phi_{xx} + D_1\Phi\theta_x - \Phi\,(\theta_x)^2 + \nu\Phi^3 = 0 \tag{1.20}$$

$$\Phi\theta_{xx} + 2\Phi_x\theta_x - D_2\Phi_x = 0 \tag{1.21}$$

For fixed t, regarding (1.21) as a function of x and integrating (1.21) yields

$$\Phi^2\,(2\theta_x - D_2) = \varphi(t)$$

Taking $\varphi(t) = 0$ yields $\theta_x = \dfrac{D_2}{2}$. Substituting it into (1.20) and integrating (1.20) with respect to x, we obtain

$$\int_{\Phi_0}^{\Phi} \frac{d\Phi}{\sqrt{P(\Phi)}} = x - D_2 t$$

where

$$P(\Phi) = -\frac{\nu}{2}\Phi^4 + \frac{1}{4}\left(D_2^2 - 2D_1 D_2\right)\Phi^2 + C$$

If $C = 0, D_2^2 - 2D_1 D_2 > 0$, then $\Phi = 0$ is the double root of equation $P(\Phi) = 0$ and there are still two roots $\Phi = \pm\Phi_0$, where $\Phi_0 = \sqrt{\dfrac{D_2^2 - 2D_1 D_2}{2\nu}}$. At this point

$$\Phi = \Phi_0\mathrm{sech}\left[\sqrt{\frac{\nu}{2}}\Phi_0\,(x - D_2 t)\right]$$

as shown in Fig.1-5.

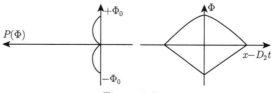

Figure 1-5

If $P(\Phi) < 0$, there cannot be a traveling wave solution. If $C \neq 0$, and if

$$\left[\frac{1}{4} \left(D_2^2 - 2D_1 D_2 \right) \right]^2 + 2\nu C \geqslant 0 \text{ or}$$

$$C > -\frac{1}{8\nu} \left(\frac{D_2^2}{2} - 2D_1 D_2 \right)$$

and $C < 0$, then $P(\Phi)$ has single roots $\pm\Phi_1, \pm\Phi_2$, as shown in Fig.1-6.

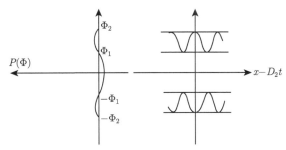

Figure 1-6

Here, the integral is an elliptic function

$$\Phi = \Phi_1 \left[1 - \left\{ \left(1 - \frac{\Phi_1^2}{\Phi_2^2} \right) \text{sn}^2 \left[\frac{\sqrt{\nu}}{2} \left(x - D_2 t \right) \right] \right\} \right]^{-\frac{1}{2}}$$

where elliptic function sn has the modulus $\text{mod} r = 1 - \dfrac{\Phi_1^2}{\Phi_2^2}$.

Other well-known equations such as the sine-Gordon equation

$$u_{tt} - u_{xx} + \sin u = 0 \qquad\qquad (1.22)$$

has soliton solution

$$u = 4 \tan^{-1} \pm \left\{ \pm \left(1 - D^2 \right)^{-\frac{1}{2}} \left(x - Dt \right) \right\}$$

When both inside and outside the curly braces take a positive sign, it represents a kink from $\Phi = 0(x = -\infty)$ to $\Phi = 2\pi(x = +\infty)$; When both inside and outside the curly braces take a negative sign, it indicates a kink from $\Phi = -2\pi(x = -\infty)$ to $\Phi = 0(x = +\infty)$; When different signs are used inside and outside the curly braces, it indicates an anti-kink, as shown in Fig.1-7.

positive   negative

Figure 1-7

There are many other nonlinear evolution equations with soliton solutions, such as the nonlinear Klein-Gordon equation, Toda nonlinear lattice equation, ferromagnetic chain equation, nonlinear electronic filtering equation, Boussinesq equation, Hirota equation, Born-Infeld equation, etc.

1.4   Experimental Observation and Application of Solitons

The objective existence of solitons has long been confirmed by Russell's observations of water waves. In addition to the soliton image obtained through numerical calculations, it has also been continuously observed in experiments. For example, in the early 1970s, Ikezi, Taylor, Baker, and others witnessed the propagation of KdV type solitons in shallow water waves during water tank experiments. And it is consistent with the results calculated by the computer. In laser shooting, we also observed the propagation of vortex solitary waves caused by collapse and the generation of solitons when the laser beam self focuses in nonlinear media. Recently, L. F. Mollenauer, R. H. Stolen, and J. P. Gordon from the Holmdel Bell Telephone Laboratory in New Jersey, USA, observed the propagation of light pulse type solitons in quartz core fiber materials, which is consistent with the numerical results calculated by J. Satsuma and N. Yajima. In addition, the Josephson effect of superconduction is one of the most important topics in contemporary physics and electronic technology. In the two superconducting materials that make up the Josephson junction, the phase difference φ between the superconducting wave functions satisfies the sine-Gordon equation. The existence of soliton solutions has been confirmed using a superconducting transmission line with a Josephson tunnel junction branching.

The theory of solitons has successfully explained the density pit problem and infrared outward shift problem that have not been explained by classical theory for many years in laser shooting. Recently, Bell Laboratories in the United States have been researching the use of solitons to improve signal transmission systems and increase their transmission rates, which means that they have the advantages of no loss of waveform, no change in speed, high fidelity, and good confidentiality during transmission. They claimed to have made remarkable progress in their experiments on soliton transmission of optical fibers, which undoubtedly promotes the development of research work on soliton theory and provides a more solid foundation.

1.5   Research on the Problem of Soliton Theory

Due to the occurrence of soliton problems in a large number of nonlinear physics fields and their shared characteristics, they have aroused the interest of many physicists. They hope to use soliton theory to study some difficult problems in plasma physics, fundamental quantum physics, and condensed matter physics, as well as the motion laws of matter under nonlinear effects. From a mathematical perspective, it has been discovered that a large class of nonlinear evolutionary equations has soliton solutions, and these nonlinear evolutionary equations with soliton solutions share a series of important and common characteristics. They have infinite conservation laws. There is an analytical method for solving linear equations — inverse scattering method. There is a Bäcklund transformation. They are completely integrable, etc. Some research has also been conducted on the interrelationships between these important characteristics. There have emerged numerous distinctive branches within mathematical methods. The inverse scattering method, formulated through solving boundary value problems of ordinary differential equations and Gelfand-Marchenko integral equations, has evolved into a useful approach for solving a broad class of nonlinear evolutionary equations, developed by Lax, Zakharov, Shabat and AKNS. Utilizing specific function transformations, including the Bäcklund transform and Hirota transform, a multitude of soliton solutions have been identified. The prolongation structure method, formulated using exterior differential forms and Lie groups, offers a robust tool for solving nonlinear evolutionary equations. To explore the stability and interactions of solitons, a plethora of numerical methods for nonlinear evolutionary equations have been proposed and have achieved significant development. It is no exaggeration to say that the study of soliton theory has emerged as a vibrant field within applied mathematics, attracting numerous mathematicians. Consequently, this research is poised to significantly advance fields like differential equations, functional analysis, group theory, homotopy theory, and topology.Current research on soliton theory has achieved significant advancements, particularly in one-dimensional problems, which are progressing towards greater depth and essentiality. The current development trend is towards multidimensional aspects, with some achievements in multidimensional inverse scattering methods, multidimensional Bäcklund transformations, multidimensional Fourier transforms, multidimensional analytical solutions, and numerical calculations. The second is to explore the development of nonlinear coupled evolutionary equation systems, such as investigating solitons in non fully integrable systems in plasma physics. Although these issues are relatively difficult and there have been few achievements so far, there is still great potential.

References

[1] Scott A C, Chu F Y F, McLaughlin D W. The soliton: a new concept in applied science[J]. Proceedings of the IEEE, 1973, 61(10): 1443-1483.

[2] Bullough R K, Caudrey P J. The soliton and its history[M]//Solitons. Berlin, Heidelberg: Springer, 1980: 1-64.

[3] 张学铭. 关于最用量原理与拉格朗日场论 [J]. 应用数学和计算数学, 1979,(3): 50.

[4] 屠规彰, 秦孟兆. 波和孤立子 [J]. 数学的实践与认识, 1980, (03): 46-52.

[5] 郭柏灵. 非线性波和孤立子 [J]. 力学与实践, 1982, 4(2): 8-16.

Chapter 2
Inverse Scattering Method

2.1   Introduction

With the formation and development of soliton theory, the inverse scattering method has become an extremely important method for exact solutions of solitons and a rather wide range of a large number of nonlinear evolution equations, and it occupies an important place in soliton theory. The most important feature and advantage of this method is that this rather complex group of nonlinear equations can be solved exactly by combining several linear equations. In the last decade, this method has been mutually promoted and progressively improved with the consequent development of other methods for finding soliton solutions, such as Bäcklund transform, Hirota method, prolongation structure method, etc. This method was first discovered by GGKM in the KdV equation, and later generalized by Lax, Zakharov, Shabat, AKNS and others to a large and very wide range of nonlinear evolutionary equations (which include systems of equations and higher dimensional cases), making them a more general method for linear exact solutions. More recently, good work has been done linking it to the qualitative theory of differential equations, opening up a new avenue for research in the theory of differential equations. In this chapter, we will introduce the important concepts of the inverse scattering method, its most basic elements as well as some recent research results, and present some open problems.

2.2   The KdV Equation and Inverse Scattering Method

We know that using the well-known Hopf-Cole transformation for the Burgers equation

$$u_t + uu_x - \alpha u_{xx} = 0 \ (\alpha > 0) \tag{2.1}$$

i.e., by making

$$u = -2\alpha \frac{w_x}{w} \tag{2.2}$$

it can be reduced to a linear heat conduction equation

$$w_t = \alpha w_{xx} \tag{2.3}$$

for the unknown function w, and the solution of the original equation has the expression

$$
u(x,t) = \int_{-\infty}^{\infty} \frac{x - \xi}{t} \exp\left[-\frac{(x - \xi)^2}{4\alpha t} - \frac{1}{2\alpha} \int_0^{\xi} u_0(\xi')d\xi' \right] d\xi
$$
$$
\bigg/ \int_{-\infty}^{\infty} \exp\left[-\frac{(x - \xi)^2}{4\alpha t} - \frac{1}{2\alpha} \int_0^{\xi} u_0(\xi')d\xi' \right] d\xi
\tag{2.4}
$$

where $u_0(x)$ is the initial condition, and $u|_{t=0} = u_0(x)$. When $\alpha \to 0$, it can be shown that it converges to a generalized solution of the quasilinear hyperbolic equation

$$
u_t + uu_x = 0 \tag{2.5}
$$

Is there a transformation similar to (2.2) for the KdV equation? Now consider the KdV equation of the following form

$$
u_t - 6uu_x + u_{xxx} = 0 \tag{2.6}
$$

Let $u = v^2 + v_x + \lambda$. If $u(x,t)$ is considered as a known function in this equation, it is the Riccati equation for the unknown function $v(x,t)$. Making the transformation $v = \psi_x/\psi$ gives the one-dimensional Schrödinger equation

$$
\psi_{xx} - (u - \lambda)\psi = 0 \tag{2.7}
$$

where ψ is the wave function, u is the potential and λ corresponds to the energy spectrum. We note that the function u depends not only on x but also on the parameter t. Thus, in general, ψ and λ also depend on the parameter t. We know that if the solution of the KdV equation is smooth, bounded, and tends to zero when $|x| \to \infty$, then in Schrödinger equation (2.7), there are finite discrete spectrums when $\lambda < 0$, given by $\lambda_m = -k_m^2$ $(m = 1, 2, \cdots, N)$; $\lambda > 0$ indicates the continuous spectrum $\lambda = k^2$ $(-\infty < k < \infty$, k is real). For fixed t, we define the solution of the scattering problem (2.7) to satisfy the boundary conditions

$$
\begin{aligned}
\psi(x,k,t) &\sim e^{-ikx} + b(k,t)e^{ikx}, \quad x \to +\infty \\
\psi(x,k,t) &\sim a(k,t)e^{-ikx}, \qquad\qquad x \to -\infty
\end{aligned}
\tag{2.8}
$$

and the solution of its bounded state satisfies the boundary conditions

$$
\begin{cases}
\psi_m(x, k_m(t), t) \sim c_m(k_m(t), t)e^{-k_m x}, & x \to +\infty \\
\psi_m(x, k_m(t), t) \sim e^{k_m x}, & x \to -\infty
\end{cases}
\tag{2.9}
$$

where $b(k,t)$ is called the reflection coefficient, $a(k,t)$ is the penetration coefficient, c_m is the attenuation factor, and

$$\int_{-\infty}^{\infty} \psi_m^2 dx = 1, \quad |a|^2 + |b|^2 = 1 \tag{2.10}$$

as shown in Figure 2-1.

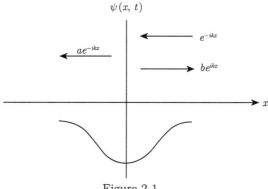

Figure 2-1

The "direct" formulation of the scattering problem is: given the potential u, seek the scattering parameters k_m, c_m, $a(k)$, $b(k)$, and determine the wave function ψ. The inverse scattering problem is now presented as follows: given the scattering parameters k_m, c_m, $b(k)$, determine the potential u. Regarding the connection between the potential u, energy spectrum λ and the scattering parameters in Schrödinger equation (2.7), the following results are obtained:

The potential u of the inverse scattering problem is

$$u(x,t) = -2\frac{d}{dx}K(x,x,t) \tag{2.11}$$

where K satisfies the following GLM integral equation:

$$K(x,y,t) + B(x+y,t) + \int_x^{\infty} B(y+z,t) \cdot K(x,z,t)dz = 0 \tag{2.12}$$

$$y > x, \quad K(x,z,t) \to 0, \ z \to \infty$$

The kernel of the integral equation (2.12) is

$$B(x,t) = \sum_{m=1}^{N} c_m^2(t)e^{-k_m x} + \frac{1}{2\pi}\int_{-\infty}^{\infty} b(k,t)e^{ikx}dk \tag{2.13}$$

where summation denotes to the discrete spectrum and integration denotes to the continuous spectrum. Obviously, the connection between the potential and the scattering parameters pointed out above does not really address the solution of the inverse scattering problem. Because, in order to determine the potential u, K must be determined; in order to determine K, one must determine the scattering parameters c_m, k_m, $b(k)$, etc., which, in turn, depends on the potential u, creating an endless loop. In order to break this loop and actually solve the potential u from the inverse scattering problem (where u is the solution to the KdV equation we require), an important fact is revealed about the KdV equation in relation to the corresponding Schrödinger equation. We have the following theorem:

Theorem 2.1   *Consider the Schrödinger equation*

$$\psi_{xx} - (u - \lambda)\psi = 0, \quad -\infty < x < +\infty$$

If $u(x,t)$ is a solution to the KdV equation and converges sharply to zero when $|x| \to \infty$, then the discrete eigenvalues $\lambda_1, \lambda_2, \cdots, \lambda_N$ of (2.7) are constants (independent of t).

Proof   Substituting $u = \dfrac{\psi_{xx}}{\psi} + \lambda$ from (2.7) into the KdV equation (2.6) and multiplying by ψ^2 gives

$$\lambda_t \psi^2 + [\psi R_x - \psi_x R]_x = 0 \tag{2.14}$$

where

$$R \equiv \psi_t + \psi_{xxx} - 3(u + \lambda)\psi_x$$

When $|x| \to \infty$, the eigenfunctions corresponding to the eigenvalues λ_n and their derivatives converge to zero. Integrating (2.14) yields

$$\lambda_{nt} \int_{-\infty}^{\infty} \psi_n^2 dx = 0$$

From the normalization condition $\int_{-\infty}^{\infty} \psi_n^2 dx = 1$, we have

$$\lambda_{nt} = 0, \text{ i.e. } \lambda_n = \text{const} \qquad \square$$

The result that λ_n is constant is very important, it means that the inverse scattering problem of the Schrödinger equation only needs to start from the initial value $u_0(x)$ of the KdV equation to determine its discrete eigenvalues and all scattering parameters. Using $\lambda_t = 0$, (2.14) becomes

$$\psi R_{xx} - R\psi_{xx} = 0$$

i.e. $R_{xx} - (u - \lambda)R = 0$. It has the exact same form as (2.7), so R can be represented by a linear combination of the eigenfunctions of (2.7), i.e.

$$R \equiv \psi_t + \psi_{xxx} - 3(u + \lambda)\psi_x = C\psi + D\varphi \qquad (2.15)$$

where C, D depend on t and φ is the solution to the equation (2.7) that is linearly independent of ψ. If take $\varphi = \psi \int_0^x \frac{dx}{\psi^2}$, we have

Theorem 2.2   *Under the condition of Theorem 2.1, the scattering parameters of scattering problem (2.7) are*

$$c_n(t) = c_n(0)e^{4k_n^3 t}$$
$$b(k, t) = b(k, 0)e^{8ik^3 t} \qquad (2.16)$$
$$a(k, t) = a(k, 0)$$

where $c_n(0), b(k, 0), a(k, 0)$ are determined by the initial value $u_0(x)$ of the KdV equation.

Proof   If ψ_n is the eigenfunction, $\varphi_n = \psi_n \int_0^x \frac{dx}{\psi_n^2}$, then φ_n is exponentially unbounded when $x \to +\infty$, so $D(t) = 0$ in (2.15). Multiply (2.15) by ψ_n and integrate over infinite intervals to get

$$\int_{-\infty}^{\infty} \left(\frac{1}{2}\psi_n^2\right)_t dx + \int_{-\infty}^{\infty} \left(\psi_n \psi_{nxx} - \frac{3}{2}\psi_{nx}^2 - 3\lambda\psi_n^2\right)_x dx = c \int_{-\infty}^{\infty} \psi_n^2 dx$$

It can be seen from $\int_{-\infty}^{\infty} \psi_n^2 dx = 1$ and the boundary conditions that the left integrals of the above equation are both zero, so $c(t) \equiv 0$. And since $x \to +\infty$, $\psi \sim c_n(t)e^{-k_n x}$, from (2.15) and $u \to 0$ $(x \to +\infty)$, we can get

$$c_n'(t) - 4k_n^3 c_n(t) = 0$$

Therefore, $c_n(t) \equiv c_n(0)e^{4k_n^3 t}$.

Secondly, for the continuous spectrum, it can be considered that λ is independent of t, so ψ satisfies (2.15), then using the steady radiation condition of the plane wave, that is

$$\psi \sim a(k, t)e^{-ikx}, \quad x \to -\infty$$

substituting it into (2.15) yields

$$(a_t + ik^3 a + 3k^3 a)e^{-ikx} = Ca(k, t)e^{-ikx} + \frac{D}{a}e^{-ikx} \int_0^x e^{2ikx} dx,$$

$$a_t + 4ik^3 a = Ca + \frac{D}{a} \int_0^x e^{2ikx} dx$$

Then we have $D = 0$, and $a_t + (4ik^3 - C)a = 0$. When $x \to +\infty$, substituting $\psi \sim e^{-ikx} + b(k, t)e^{ikx}$ into (2.15), given that the coefficients of the linearly independent function $e^{\pm ikx}$ is zero, we get

$$C = 4ik^3, \quad b_t - 8ik^3 b = 0, \quad b(k, t) = b(k, 0)e^{8ik^3 t}$$

From $C = 4ik^3$, we have $a_t = 0$, i.e., $a(k, t) = a(k, 0)$. □

Theorem 2.1 and Theorem 2.2 provide a concrete way to solve the initial value problem of KdV equation

$$\begin{cases} u_t - 6uu_x + u_{xxx} = 0, \quad -\infty < x < \infty, \ t > 0 \\ u(x, 0) = u_0(x) \end{cases}$$

by Schrödinger equation and scattering problem. First, we solve the eigenvalue problem

$$\psi_{xx} - [u_0(x) - \lambda]\psi = 0 \tag{2.17}$$

For this problem, the scattering parameters k_n, $c_n(0)$, $b(k, 0)$ are determined, then $c_n(t)$, $b(k, t)$ is obtained by (2.16), which leads to

$$B(x + y, t) = \sum_{n=1}^N c_n^2(t)e^{-k_n(x+y)} + \frac{1}{2\pi} \int_{-\infty}^{\infty} b(k, t)e^{ik(x+y)} dk$$

$$= \sum_{n=1}^N c_n^2(0)e^{8k_n^3 t - k_n(x+y)} + \frac{1}{2\pi} \int_{-\infty}^{\infty} b(k, 0)e^{i[8k^3 t + k(x+y)]} dk$$

Then $K(x, y, t)$ is determined by GLM equation

$$K(x, y, t) + B(x + y, t) + \int_x^{\infty} B(y + z, t)K(x, z, t)dz = 0 \ (y > x)$$

Thus, we obtain the solution

$$u(x, t) = -2\frac{d}{dx}K(x, x; t)$$

The solution steps are shown in Figure 2-2.

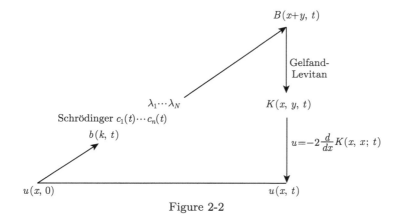

Figure 2-2

In this way, we turn an initial value problem for solving a nonlinear equation (KdV equation) into a problem for solving two linear equations, one is the Sturm-Liouville problem for second-order ordinary differential equation, and the other is solving a linear integral equation. Let's take two examples to illustrate this.

Let $u_0(x) = -2\operatorname{sech}^2 x$. The eigenvalue problem corresponding to (2.17) which we can solve precisely with hypergeometric functions, has a discrete value $k_1 = 1$, the corresponding normalization constants $c_1(0) = \sqrt{2}$, $b(k,0) = 0$, and therefore $b(k,t) \equiv 0$, $\forall t \geqslant 0$. The corresponding Gelfand-Levitan equation is

$$K(x,y,t) + 2e^{8t-x-y} + 2e^{8t-y}\int_x^\infty K(x,z,t)e^{-z}dz = 0$$

Let $K(x,y,t)$ be a separable variable[①], i.e. $K(x,y,t) = L(x,t)e^{-y}$, then we obtain that

$$L(x,t) + 2e^{8t-x} + 2e^{8t}L(x,t)\int_x^\infty e^{-2z}dz = 0$$

$$L(x,t) = \frac{-2e^x}{1+e^{2x-8t}}, \quad K(x,y,t) = \frac{-2e^{x-y}}{1+e^{8x-8t}}$$

It is easy to verify that $K(x,y,t)$ determined in this way does satisfy the Gelfand-Levitan equation and is the unique solution. Therefore, the solution of KdV equation is

$$u(x,t) = \frac{8e^{2x-8t}}{(1+e^{2x-8t})^2} = -2\operatorname{sech}^2(x-4t)$$

---

① This is generally not possible for non separable variables, where $K(x,y,t)$ should be extended as a series.

which, as can be seen from the travelling wave solution, is indeed the exact solution of the initial value problem of KdV equation.

Another example is taking $u(x,0) = u_0(x) = -6\,\mathrm{sech}^2 x$, then two different eigenvalues $k_1 = 2$ and $k_2 = 1$ can be solved. From $b(k,0) = 0$, the solution of KdV equation can be obtained as

$$u(x,t) = -12\,\frac{3 + 4\cosh(2x - 8t) + \cosh(4x - 64t)}{[3\cosh(x - 28t) + \cosh(3x - 36t)]^2}$$

Next, we consider the problem of solving N solitons by inverse scattering method. In this case, the reflection coefficient $b(k,t) \equiv 0$, which is only the case of the discrete spectrum. Then the Gelfand-Levitan equation is

$$K(x,y,t) + \sum_{m=1}^{N} c_m^2(t)e^{-k_m(x+y)} + \sum_{m=1}^{N} c_m^2 e^{-k_m y} \int_x^{\infty} e^{-k_m t} K(x,z,t)dz = 0 \quad (2.18)$$

where $c_m = c_m(t) = c_m(0)e^{4k_m^3 t}$, and $k_m > 0$ are different. Let $K(x,y,t)$ have the form

$$K(x,y,t) = -\sum_{m=1}^{N} c_m \psi_m(x)e^{-k_m y} \quad (2.19)$$

where ψ_m is the undetermined function and c_m is the normalized factor. Substituting (2.19) into (2.18) and taking the coefficient of $e^{k_m y}$ as 0, we get that $\psi_m(x)$ should satisfy the following linear algebraic equation

$$\psi_m(x) + \sum_{m=1}^{N} c_m c_n \frac{e^{-(k_m+k_n)x}}{k_m + k_n}\psi_n(x) = c_m e^{-k_m x} \quad (m = 1, 2, \cdots, N) \quad (2.20)$$

If we use the matrix representation

$$I \equiv (\delta_{mn}), \ \ C \equiv \left(c_m c_n \frac{e^{-(k_m+k_n)x}}{k_m + k_n} \right), \ \ \psi = (\psi_1, \psi_2, \cdots, \psi_N)^{\mathrm{T}}$$

$$E = (c_1 e^{-k_1 x}, c_2 e^{-k_2 x}, \cdots, c_N e^{-k_N x})^{\mathrm{T}}$$

then (2.20) can be written as

$$(I + C)\psi = E \quad (2.21)$$

For the solution ψ in equation (2.20) to exist, it is sufficient to prove that matrix C is positive definite. In fact, there is

$$\sum_{m=1}^{N}\sum_{n=1}^{N} p_m p_n c_m c_n \frac{e^{-(k_m+k_n)x}}{k_m + k_n} = \int_x^{\infty} \left[\sum_{m=1}^{N} p_m c_m e^{-k_m z} \right]^2 dz > 0$$

So we know that $I + C$ is positive definite, and thus (2.20) has a unique solution. We can solve for ψ using Cramer's rule. Let Q_{mn} be the algebraic cofactor of an element a_{mn} in matrix $I + C$, and expand on the nth row to get

$$\Delta \equiv \det(I + C) = \sum_m \left(\delta_{mn} + c_m c_n \frac{e^{-(k_m + k_n)x}}{k_m + k_n} \right) Q_{mn}$$

$$\psi_n(x) = \Delta^{-1} \sum_m c_m e^{-k_m x} Q_{mn}$$

From (2.19), $y = x$, there is

$$K(x, y, t) = -\sum c_n \psi_n(x) e^{-k_n x}$$

$$= -\Delta^{-1} \sum_m \sum_n c_m c_n e^{-(k_m + k_n)x} Q_{mn}$$

$$= \Delta^{-1} \frac{d}{dx} \Delta$$

Therefore, the potential without reflection coefficient is

$$u(x, t) = -2 \frac{d^2}{dx^2} \log \det(I + C) \tag{2.22}$$

which is the solution to the KdV equation.

In the Gelfand-Levitan integral equation, due to the symmetric position of K and B, the method of determining B by the scattering parameter first and then solving K may not be adopted, instead, the solution K satisfying the linear hyperbolic equation is used to solve B by the Gelfand-Levitan equation, thus forming another solution to the scattering inversion problem. The solution block diagram is shown in Figure 2-3.

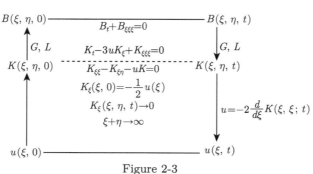

Figure 2-3

2.3 Lax Operator and Generalization of Zakharov, Shabat, AKNS

Consider the general nonlinear evolution equation

$$u_t = K(u) \tag{2.23}$$

where $K(u)$ represents a nonlinear operator on some appropriate function space. If we can find linear operators L and B, which depend on the solution u of the equation (2.23) and satisfy the operator equation (commonly known as the Lax operator equation)

$$iL_t = BL - LB = [B, L] \ (i = \sqrt{-1}) \tag{2.24}$$

where B is a self-adjoint operator, then the eigenfunction ψ corresponding to the eigenvalue E of the operator L can be derived from (2.24), i.e.

$$L\psi = E\psi \tag{2.25}$$

If the change of ψ with time satisfies the equation

$$i\psi_t = B\psi \tag{2.26}$$

then E does not change with time. In fact, differentiating (2.25) by t and multiplying by i yields

$$i\left[\psi\frac{dE}{dt} + E\frac{d\psi}{dt}\right] = i\left[L\psi_t + \frac{\partial L}{\partial t}\psi\right]$$

$$= iL\psi_t + [BL - LB]\psi = L(i\psi_t - B\psi) + EB\psi$$

From (2.26), $i\psi_t = B\psi \Rightarrow i\psi\dfrac{dE}{dt} = 0$.

To solve the initial value problem of the given equation (2.23), that is, given $u(x,0)$, find the solution $u(x,t)$ that satisfies (2.23), generally through the following steps:

(i) Find solution to the direct problem. That is, the scattering parameter of ψ at $|x| = \infty$ (such as the eigenvalue of the operator L, reflection, penetration coefficient, etc.) is calculated from the known value $u(x,0)$ of $t = 0$.

(ii) Find the change of scattering data with time. From (2.26), considering the asymptotic form of B at $|x| = \infty$, the variation of scattering data with time is calculated.

(iii) Find the solution to the inverse problem. From the known scattering data of L as a function of time, the solution $u(x,t)$ is constructed.

The steps of solving the general nonlinear evolution equations are simply shown in Figure 2-4.

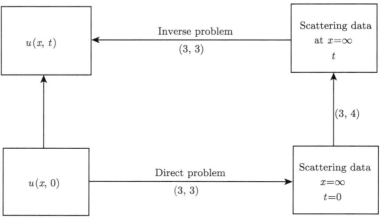

Figure 2-4

Example   Consider the solution of KdV equation

$$u_t - 6uu_x + u_{xxx} = 0 \tag{2.27}$$

with the initial condition

$$u(x,0) = u_0(x) \tag{2.28}$$

Take the family of the second-order differential operators $L(t)$:

$$L(t) \equiv -\frac{\partial^2}{\partial x^2} + u(x,t) \tag{2.29}$$

where t is the parameter, $u(x,t)$ is the solution of (2.27), and $u(\cdot,t) \in L_2$.

The eigenvalue problem of the operator L is

$$-\psi_{xx} + u(x,t)\psi = E\psi \tag{2.30}$$

As described in section 2.2, we can find the solution to its direct problem, that is, given $u_0(x)$, the scattering parameters $(k_n,\ c_n,\ n = 1,2,\cdots,N;\ a(k),\ b(k),\ 0 \leqslant k^2 < \infty)$ can be sought.

Then select the self-conjugation operator

$$B \equiv -4i\frac{\partial^3}{\partial x^3} + 3i\left(u\frac{\partial}{\partial x} + \frac{\partial}{\partial x} \cdot u\right) \tag{2.31}$$

where $u(x,t)$ satisfies the KdV equation (2.27). Then the operators L, B satisfy the operator equation (2.24). From (2.26) and the boundary condition that ψ at infinity should satisfy, $u(x,t) \to 0$ ($|x| \to \infty$), we can easily obtain the evolution of scattering data with time, that is, the result of theorem 2.2. Then from the Gelfand-Levitan integral equation, the solution $u(x,t)$ of the initial value problem of the KdV equation can be obtained.

Obviously, not all nonlinear equations (2.23) can be solved by inverse scattering method. The main difficulty is that it is not always possible to find suitable operators L and B to satisfy the operator equation (2.24) under such conditions, and sometimes for a given L, even if B is found, B may be trivial. For example, $L = \dfrac{\partial^2}{\partial x^2} + u(x,t)$, taking $B = i\dfrac{\partial}{\partial x}$, we have $L_t = u_t$, $[B,L] = i[D,u] = iu_x$, $iL_t = [B,L] \Rightarrow u_t = u_x$. This is a travelling wave equation with an expression $u = f(x-t)$ with an obvious solution. However, despite these problems, it has now been found that a wide range of nontrivial nonlinear wave equations can be solved using the inverse scattering method. The following describes the extensive generalization made by Zakharov, Shabat and AKNS.

Consider the linear problem

$$Lv = \zeta v \tag{2.32}$$

where

$$L \equiv \begin{pmatrix} i\dfrac{d}{dx} & -iq(x,t) \\ ir(x,t) & -\dfrac{d}{dx} \end{pmatrix}$$

$$v \equiv \begin{pmatrix} v_1(x,t) \\ v_2(x,t) \end{pmatrix} \tag{2.33}$$

$q(x,t)$, $r(x,t)$ are arbitrary differentiable functions, and ζ is a constant. Let B have the general form

$$B = \begin{pmatrix} a(x,t;\zeta) & b(x,t;\zeta) \\ c(x,t;\zeta) & -a(x,t;\zeta) \end{pmatrix} \tag{2.34}$$

where the coefficients (a, b, c) are also arbitrary (Initially, Zaharov, Shabat and AKNS consider the case $r = -q^*$). $i\psi_t = B\psi$ takes the form

$$i\frac{dv}{dt} = Bv \tag{2.35}$$

Now write (2.32) in scalar form

$$\begin{cases} iv_{1x} - iqv_2 = \zeta v_1 \\ irv_1 - iv_{2x} = \zeta v_2 \end{cases} \tag{2.36}$$

(2.35) is also written in scalar form

$$\begin{cases} iv_{1t} = av_1 + bv_2 \\ iv_{2t} = cv_1 - av_2 \end{cases} \tag{2.37}$$

Differentiating (2.36) with respect to t gives

$$\begin{cases} iv_{1xt} - iq_t v_2 - iqv_{2t} = \zeta v_{1t} \\ ir_t v_1 + irv_{1t} - iv_{2xt} = \zeta v_{2t} \end{cases} \tag{2.38}$$

Differentiating (2.37) with respect to x gives

$$\begin{cases} iv_{1xt} = a_x v_1 + av_{1x} + b_x v_2 + bv_{2x} \\ iv_{2xt} = c_x v_1 + cv_{1x} - a_x v_2 - av_{2x} \end{cases} \tag{2.39}$$

Eliminating v_{1xt}, v_{2xt} from (2.38), (2.39) yields

$$\begin{cases} a_x v_1 + av_{1x} + b_x v_2 + bv_{2x} = \zeta v_{1t} + iq_t v_2 + iqv_{2t} \\ c_x v_1 + cv_{1x} - a_x v_2 - av_{2x} = ir_t v_1 + irv_{1t} - \zeta v_{2t} \end{cases} \tag{2.40}$$

Then substitute (2.36), (2.37) into (2.40) and eliminate v_{1x}, v_{2x}, v_{1t}, v_{2t} to get

$$a_x v_1 + a\left(-i\zeta v_1 + qv_2\right) + b_x v_2 + b\left(i\zeta v_2 + rv_1\right)$$

$$= -\zeta i\left(av_1 + bv_2\right) + iq_t v_2 + q\left(cv_1 - av_2\right) \tag{2.41}$$

$$c_x v_1 + c\left(-i\zeta v_1 + qv_2\right) - a_x v_2 - a\left(i\zeta v_2 + rv_1\right)$$

$$= ir_t v_1 + r\left(av_1 + bv_2\right) + i\zeta\left(cv_1 - av_2\right) \tag{2.42}$$

Comparing the coefficients of v_1, v_2 for the equation (2.41) gives

$$v_1 : a_x + br = qc, \quad \therefore a_x = qc - br$$
$$v_2 : b_x + 2i\zeta b = iq_t - 2aq$$

By comparing the coefficients of v_1, v_2 in equation (2.42), we get

$$v_1 : c_x - 2i\zeta c = ir_t + 2ar$$
$$v_2 : cq - a_x = br$$

Thus, we obtain the equations that any coefficient (a, b, c, r, q) should satisfy

$$\frac{\partial a}{\partial x} = qc - rb \tag{2.43}$$

$$\frac{\partial b}{\partial x} + 2i\zeta b = i\frac{\partial q}{\partial t} - 2aq \tag{2.44}$$

$$\frac{\partial c}{\partial L} - 2i\zeta c = i\frac{\partial r}{\partial t} + 2ar \tag{2.45}$$

In the future, for the sake of writing uniformity, we order

$$\frac{dv}{dt} = Mv, \quad M = \begin{pmatrix} A & B \\ C & -A \end{pmatrix}$$

We can rewrite the equations (2.43), (2.44), (2.45) as

$$A_x = qC - rB \tag{2.46}$$

$$B_x + 2i\zeta B = q_t - 2Aq \tag{2.47}$$

$$C_x - 2i\zeta C = r_t + 2Ar \tag{2.48}$$

where $A \equiv A(x, t; \zeta)$, $B \equiv B(x, t; \zeta)$, $C \equiv C(x, t; \zeta)$, $r = r(x, t)$, $q = q(x, t)$.

Equations (2.36), (2.37) and (2.46)-(2.48) become the basis of analysis of the inverse scattering method. For the given initial values $r(x, 0)$, $q(x, 0)$, the equation set (2.36) are used to find the scattering data determined by the discrete eigenvalue (which does not change with time) and the initial time eigenfunction $v_1(x, 0; \zeta)$, $v_2(x, 0; \zeta)$ at $|x| \to \infty$. If given a set of r_t, q_t, r, q, in principle, we can find the solution A, B, C of the system of equations (2.46)-(2.48). With A, B, C, we can calculate the change of the asymptotic form ($|x| \to \infty$) of the characteristic functions v_1, v_2 with time from the system of equations (2.37), and then find the potential $r(x, t)$, $q(x, t)$ at the next time. Of course, since r, q are unknown, we cannot yet construct them so simply, but it does give us information on how to construct an exact solution to the evolution equation using the inverse scattering method.

First, we look for some very simple solutions for (2.46)-(2.48). Although it is very special, a wide range of nonlinear evolution equations can be obtained from it. If

$$A = \sum_{n=0}^{N} A^{(n)}\zeta^n, \quad B = \sum_{n=0}^{N} B^{(n)}\zeta^n, \quad C = \sum_{n=0}^{N} C^{(n)}\zeta^n \tag{2.49}$$

it is easy to obtain that $A^{(N)} = a_N$ (a_N is independent of x and can depend on t), $B^{(N)} = C^{(N)} = 0$. We first find $A^{(N-1)}$ from (2.46) and then find $B^{(N-1)}$ and $C^{(N-1)}$

from (2.47), (2.48). This process is repeated until all coefficients $A^{(n)}$, $B^{(n)}$, $C^{(n)}$ are found. In particular, the last two equations of ζ^0 are

$$\begin{cases} q_t = 2A^{(0)}q + B_x^{(0)} \\ r_t = -2A^{(0)}r + C_x^{(0)} \end{cases} \tag{2.50}$$

Here is an example to illustrate it. Taking $N = 3$, letting

$$A = A^{(0)} + A^{(1)}\zeta + A^{(2)}\zeta^2 + a_3\zeta^3 \quad (a_3 \text{ is independent of } x)$$
$$B = B^{(0)} + B^{(1)}\zeta + B^{(2)}\zeta^2$$
$$C = C^{(0)} + C^{(1)}\zeta + C^{(2)}\zeta^2$$

and substituting them into (2.46)-(2.48) to get

$$A_x^{(0)} = qC^{(0)} - rB^{(0)}$$

$$A_x^{(1)} = qC^{(1)} - rB^{(1)}$$

$$A_x^{(2)} = qC^{(2)} - rB^{(2)}$$

$$B_x^{(1)} + 2iB^{(0)} = -2A^{(1)}q$$

$$B_x^{(2)} + 2iB^{(1)} = -2A^{(2)}q$$

$$C_x^{(1)} - 2iC^{(0)} = 2A^{(1)}r$$

$$C_x^{(2)} - 2iC^{(1)} = 2A^{(2)}r$$

Thus, we can get

$$A^{(2)} = a_2, \quad B^{(2)} = ia_3q, \quad C^{(2)} = ia_3r$$

From $A^{(2)}r$, $C_x^{(2)}$, we can find $C^{(1)}$:

$$C^{(1)} = \frac{1}{2}a_3r_x + ia_2r$$

From $A^{(2)}q$, $B_x^{(2)}$, we can find $B^{(1)}$:

$$B^{(1)} = ia_2q - \frac{1}{2}a_3q_x$$

From $B^{(1)}$, $C^{(1)}$, we can find $A^{(1)}$:

$$A^{(1)} = -\frac{1}{2}a_3qr + a_1$$

Then we have $A^{(1)}q$, $B_x^{(1)} \to B^{(0)}$; $A^{(1)}r$, $C^{(1)} \to C^{(0)}$; $B^{(0)}$, $C^{(0)} \to A^{(0)}$. Thus, the expression for A, B, C is

$$
\begin{cases}
A = a_3\zeta^3 + a_2\zeta^2 + \left(\frac{1}{2}a_3qr + a_1\right)\zeta \\[2mm]
\qquad + \frac{1}{2}a_2qr - \frac{i}{4}a_3\left(qr_x - q_xr\right) + a_0 \\[2mm]
B = ia_3q\zeta^2 + \left(ia_2q - \frac{1}{2}a_3q_x\right)\zeta + ia_1q \\[2mm]
\qquad + \frac{i}{2}a_3q^2r - \frac{1}{2}a_xq_x - \frac{i}{4}a_3q_{xx} \\[2mm]
C = ia_3r\zeta^2 + \left(ia_2r + \frac{1}{2}a_3r_x\right)\zeta + ia_1r \\[2mm]
\qquad + \frac{i}{2}a_3qr^2 + \frac{1}{2}a_2r_x - \frac{i}{4}a_xr_{xx}
\end{cases}
\tag{2.51}
$$

The corresponding evolution equation (2.50) is:

$$
0 = q_t + \frac{i}{4}a_3\left(q_{xxx} - 6qrq_x\right) + \frac{1}{2}a_2(q_{xx} - 2q^2r) - ia_1q_x - 2a_0q
$$

$$
0 = r_t + \frac{i}{4}a_3\left(r_{xxx} - 6qrr_x\right) - \frac{1}{2}a_2(r_{xx} - 2qr^2) - ia_1r_x + 2a_0r
$$

Consider several special cases:

(i) $a_0 = a_1 = a_2 = 0$, $a_3 = -4i$.

(a) $r = -1$, KdV equation

$$
q_t + 6qq_x + q_{xxx} = 0
\tag{2.52}
$$

(b) $r = \mp q$, MKdV equation

$$
q_t \pm 6q^2q_x + q_{xxx} = 0
\tag{2.53}
$$

(ii) $a_0 = a_1 = a_3 = 0$, $a_2 = -2i$, $r = \mp q^*$, nonlinear Schrödinger equation

$$
q_t - iq_{xx} \mp 2iq^2q^* = 0
\tag{2.54}
$$

For the KdV equation (2.52), the scattering problem (2.36) can be reduced to the Schrödinger equation

$$
v_{2xx} + \left(\zeta^2 + q(x,t)\right)v_2 = 0
\tag{2.55}
$$

We know that for real $q(x,t)$, ζ^2 is real and its discrete eigenvalues lie on the imaginary axis, which corresponds to the stable solitons. In general, the discrete eigenvalues are complex.

Similarly, we can look for equations where A, B, C are negative powers of ζ, such as

$$A(x,t;\zeta) = \frac{a(x,t)}{\zeta}$$

$$B(x,t;\zeta) = \frac{b(x,t)}{\zeta}$$

$$C(x,t;\zeta) = \frac{c(x,t)}{\zeta}$$

Then we get

$$a_x = \frac{i}{2}(qr)_t, \quad q_{xt} = -4iaq, \quad r_{xt} = -4iar \tag{2.56}$$

Special and important cases are

(i) $a = \frac{i}{4}\cos u$, $b = c = \frac{i}{4}\sin u$, $r = -q = \frac{1}{2}u_x$, sine-Gordon equation

$$u_{xt} = \sin u \tag{2.57}$$

(ii) $a = \frac{i}{4}\cosh u$, $-b = c = \frac{i}{4}\sinh u$, $r = q = \frac{1}{2}u_x$, sinh-Gordon equation

$$u_{xt} = \sinh u \tag{2.58}$$

2.4   More General Evolutionary Equation (AKNS Equation)

In the discussion in the previous section, one is bound to ask whether the evolution equation solved by the inverse scattering method is limited to the finite expansion of A, B, C with respect to ζ? In this section, we will show that a broader class of evolutionary equations (which can be solved by the inverse scattering method) do exist.

Suppose A, B, C satisfy the boundary conditions

$$\begin{cases} A(x,t;\zeta) \to A_0(\zeta) \\ B(x,t;\zeta) \to 0 \\ C(x,t;\zeta) \to 0, \quad |x| \to \infty \end{cases} \tag{2.59}$$

The case of A, B, C taking different values when $x \to +\infty$ and $x \to -\infty$, respectively, can be considered similarly.

In order to obtain the necessary integral conditions, we solve the system of equations (2.46)-(2.48) formally, which is easily given by the particular solution of

(2.36). For this reason, we first study the fundamental solution of the eigenvalue problem (2.36).

Suppose $q(x,t)$, $r(x,t) \to 0$ ($|x| \to \infty$), for the real ζ value, define the linearly independent solution of (2.36) with the following asymptotic values

$$\begin{cases} \varphi \to \begin{pmatrix} 1 \\ 0 \end{pmatrix} e^{-i\zeta x}, & x \to -\infty \\[3mm] \bar{\varphi} \to \begin{pmatrix} 0 \\ -1 \end{pmatrix} e^{i\zeta x}, & x \to -\infty \end{cases} \tag{2.60}$$

$$\begin{cases} \psi \to \begin{pmatrix} 0 \\ 1 \end{pmatrix} e^{i\zeta x}, & x \to +\infty \\[3mm] \bar{\psi} \to \begin{pmatrix} 1 \\ 0 \end{pmatrix} e^{-i\zeta x}, & x \to +\infty \end{cases} \tag{2.61}$$

Let the scattering data $a(\zeta,t)$, $b(\zeta,t)$, $\bar{a}(\zeta,t)$, $\bar{b}(\zeta,t)$ be the coefficients of two linearly independent solutions, that is

$$\varphi = a\psi + b\psi \to \begin{pmatrix} ae^{-i\zeta x} \\ be^{i\zeta x} \end{pmatrix}, \quad x \to +\infty \tag{2.62}$$

$$\bar{\varphi} = \bar{b}\psi - \bar{a}\psi \to \begin{pmatrix} \bar{b}e^{-i\zeta x} \\ -\bar{a}e^{i\zeta x} \end{pmatrix}, \quad x \to +\infty \tag{2.63}$$

Thus, the coefficients $a(\zeta,t)$, $b(\zeta,t)$, $\bar{a}(\zeta,t)$, $\bar{b}(\zeta,t)$ can be solved by the Wronski determinants of φ, $\bar{\varphi}$, ψ, $\bar{\psi}$, that is

$$\begin{cases} a = w(\varphi,\psi)/w(\bar{\psi},\psi) = w(\varphi,\psi) \\ b = -w(\varphi,\bar{\psi}) \end{cases}$$

$$\begin{cases} \bar{a} = w(\bar{\varphi},\bar{\psi}) \\ \bar{b} = w(\bar{\varphi},\psi) \end{cases} \tag{2.64}$$

where $w(u,v) = u_1 v_2 - u_2 v_1$, $w(\bar{\psi},\psi) = 1$. Since $w(\varphi,\bar{\varphi}) = -1$, there is

$$a\bar{a} + b\bar{b} = w(\bar{\varphi},\psi)w(\bar{\varphi},\bar{\psi}) - w(\varphi,\bar{\psi})w(\bar{\varphi},\psi)$$

$$= (\varphi_1\psi_2 - \varphi_2\psi_1)(\bar{\varphi}_1\bar{\psi}_2 - \bar{\varphi}_2\bar{\psi}_1) - (\varphi_1\bar{\psi}_2 - \varphi_2\bar{\psi}_1)(\bar{\varphi}_1\psi_2 - \bar{\varphi}_2\psi_1)$$

$$= -\varphi_1\bar{\varphi}_2(\psi_1\bar{\psi}_2 - \bar{\psi}_1\psi_2) + \bar{\varphi}_1\varphi_2(\bar{\psi}_1\psi_2 - \psi_1\bar{\psi}_2)$$

$$= \bar{\varphi}_1\varphi_2 - \varphi_1\bar{\varphi}_2 = 1$$

We will prove in section 2.5 that $a(\zeta, t)$ can be analytically extended to the upper half plane, $\text{Im}\,\zeta > 0$; $\bar{a}(\zeta, t)$ can analytically extended to the lower half plane, $\text{Im}\,\zeta < 0$. And the discrete eigenvalues $\{\zeta_k\}_{k=1}^N$ of (2.36) are the zero points of $a(\zeta, t)$ in the upper half plane ($\text{Im}\,\zeta > 0$), $\varphi\left(\zeta_k, t\right) = b_k(t)\psi\left(\zeta_k, t\right)$. Similarly, $\bar{a}(\zeta, t)$ has zero points in the lower half plane, $\text{Im}\,\zeta < 0$, which are also the eigenvalues, and at zero,

$$\bar{\varphi}_k\left(\bar{\zeta}_k, t\right) = \bar{b}_k(t)\bar{\psi}_k\left(\bar{\zeta}_k, t\right)$$

Since (2.60) is selected to normalize, and we set $B, C \to 0$ ($x \to -\infty$) in (2.46)-(2.48). From (2.46), we can see $A(x, t; \zeta) \to \text{const}$ ($x \to -\infty$). Let

$$\lim_{x \to -\infty} A(x, t; \zeta) = A_-(\zeta) \tag{2.65}$$

where $A_-(\zeta)$ is an arbitrary function of ζ. Since $\varphi e^{A_- t}$, $\varphi e^{-A_- t}$ satisfy the equation (2.46), there are

$$\varphi_t = \begin{pmatrix} A - A_- & B \\ C & -A - A_- \end{pmatrix} \varphi \tag{2.66}$$

$$\bar{\varphi}_t = \begin{pmatrix} A + A_- & B \\ C & -A + A_- \end{pmatrix} \bar{\varphi} \tag{2.67}$$

Take the asymptotic value of $x \to +\infty$, and get

$$\begin{pmatrix} a_t e^{-i\zeta x} \\ b_t e^{i\zeta x} \end{pmatrix} = \begin{pmatrix} A_+ - A_- & \lim\limits_{x \to +\infty} B \\ \lim\limits_{x \to +\infty} C & -A_+ - A_- \end{pmatrix} \begin{pmatrix} a e^{-i\zeta x} \\ b e^{i\zeta x} \end{pmatrix}$$

Thus, we have

$$\begin{cases} a_t = (A_+ - A_-)\, a + B_+ b \\ b_t = C_+ a - (A_+ + A_-)\, b \end{cases} \tag{2.68}$$

In the same way,

$$\begin{cases} \bar{a}_t = -(A_+ - A_-)\, \bar{a} - C_+ \bar{b} \\ \bar{b}_t = -B_+ a + (A_+ + A_-)\, \bar{b} \end{cases} \tag{2.69}$$

where $A_+ = \lim\limits_{x \to +\infty} A$, $B_+ = \lim\limits_{x \to +\infty} B e^{2i\zeta x}$, $C_+ = \lim\limits_{x \to +\infty} C e^{-2i\zeta x}$. If the special case is considered, $A_+ = A_-$, $B_+ = C_+ = 0$, then it can be solved by (2.68) and (2.69) as

$$\begin{cases} a(\zeta, t) = a(\zeta, 0) \\ b(\zeta, t) = b(\zeta, 0) e^{-2A_-(\zeta)t} \\ \bar{a}(\zeta, t) = \bar{a}(\zeta, 0) \\ \bar{b}(\zeta, t) = \bar{b}(\zeta, 0) e^{2A_-(\zeta)t} \end{cases} \tag{2.70}$$

Now consider the general situation. Let $I(u, v)$ represent bilinear form

$$I(u, v) = \int_{-\infty}^{\infty} (-q_t u_2 v_2 + r_t u_1 v_1)\, dx \qquad (2.71)$$

Then, by the fundamental solution matrix, A_+, B_+, C_+ can be expressed as

$$\begin{cases} A_+ = -I(\psi, \bar{\psi}) + A_-(a\bar{a} - b\bar{b}) \\ B_+ = -I(\psi, \psi) + 2a\bar{b}A_- \\ C_+ = I(\psi, \bar{\psi}) + 2\bar{a}bA_- \end{cases} \qquad (2.72)$$

The inverse relation can be obtained from (2.62), (2.63):

$$\begin{cases} \psi = -a\varphi + \bar{b}\bar{\varphi} \\ \bar{\psi} = b\bar{\varphi} + \bar{a}\varphi \end{cases} \qquad (2.73)$$

After (2.73) is substituted into (2.72), A_+, B_+, C_+ are substituted into (2.68) and (2.69), the results can be obtained

$$\begin{cases} a_t = -I(\varphi, \psi) \\ b_t = I(\varphi, \psi) \end{cases} \qquad (2.74)$$

$$\begin{cases} \bar{a}_t = -I(\bar{\varphi}, \bar{\psi}) \\ \bar{b}_t = -I(\bar{\varphi}, \psi) \end{cases} \qquad (2.75)$$

Thus, the evolution of scattering data over time can be obtained. From (2.74), there is

$$\begin{aligned} \left(\frac{b}{a}\right)_t &= \frac{b_t a - a_t b}{a^2} = \frac{b}{a} \cdot \frac{1}{ab}[b_t a - a_t b] \\ &= \frac{b}{a} \cdot \frac{1}{ab}[I(\varphi, b\bar{\varphi} + \bar{a}\varphi)a - I\left(\varphi, -a\varphi + \bar{b}\bar{\varphi}\right) b] \\ &= \frac{b}{a}\frac{I(\varphi, \varphi)}{ab} \end{aligned}$$

In the same way, there is

$$\begin{cases} \left(\dfrac{\bar{b}}{\bar{a}}\right)_t = \left(\dfrac{\bar{b}}{\bar{a}}\right)\dfrac{I(\bar{\varphi}, \varphi)}{\bar{a}\bar{b}} \\[2mm] \left(\dfrac{\bar{b}}{a}\right)_t = \left(\dfrac{\bar{b}}{a}\right)\dfrac{I(\psi, \psi)}{ab} \\[2mm] \left(\dfrac{b}{\bar{a}}\right)_t = \left(\dfrac{b}{\bar{a}}\right)\dfrac{I(\psi, \psi)}{ab} \end{cases} \qquad (2.76)$$

Note that there are not many assumptions here about the functions $q(x,t)$, $r(x,t)$, just that the integral $I(u,v)$ has such a weak condition. In principle, for any q_t, r_t, we can calculate the evolution of the scattering data over time from (2.76) to ensure the conditions that $q(x,t)$ and $r(x,t)$ need at the next time.

The following seeks an analytical representation of the general evolution equation. For any complex function $\Omega(\zeta)$, $\bar{\Omega}(\zeta)$, if selected

$$I(\psi, \psi) = 2\Omega(\zeta)a\bar{b} \tag{2.77}$$

$$I(\bar{\psi}, \bar{\psi}) = -2\bar{\Omega}(\zeta)\bar{a}b \tag{2.78}$$

then we can solve the equation (2.76) linearly. (2.77) can be written in the following form:

$$\int_{-\infty}^{\infty} \left[(r_t + 2\Omega(\zeta)r) \, \psi_1^2 + (-q_t + 2\Omega(\zeta)q) \, \psi_2^2 \right] dx = 0 \tag{2.79}$$

In fact, we note that

$$I(\psi, \psi) = \int_{-\infty}^{\infty} \left(-q_t \psi_2^2 + r_t \psi_1^2 \right) dx$$

$$-\psi_1\psi_2 \Big|_{-\infty}^{\infty} = -\psi_1\psi_2 \Big|_{-\infty}$$

$$= \left(-a\varphi_1 + \bar{b}\bar{\varphi}_1 \right) \left(-a\varphi_2 + \bar{b}\bar{\varphi}_2 \right) \Big|_{-\infty}$$

$$= a\bar{b}$$

On the other hand, use (2.36) to obtain

$$-\psi_1\psi_2 \Big|_{-\infty}^{\infty} = -\int_{-\infty}^{\infty} \frac{d}{dx} \left(\psi_1\psi_2 \right) dx = -\int_{-\infty}^{\infty} \left(q\psi_2^2 + r\psi_1^2 \right) dx$$

which yields (2.79). Using the above relation and (2.36), it can be verified that vector

$$\Psi = \left(\psi_1^2, \psi_2^2 \right)^{\mathrm{T}}$$

satisfies the equation

$$L\Psi = \zeta\Psi \tag{2.80}$$

where $\left(\psi_1^2, \psi_2^2 \right)^{\mathrm{T}}$ represents the transpose of vector $\left(\psi_1^2, \psi_2^2 \right)$,

$$L = \frac{1}{2i} \begin{pmatrix} -\dfrac{\partial}{\partial x} - 2q \displaystyle\int_x^{\infty} \cdot \, r(y)dy & -2q \displaystyle\int_x^{\infty} \cdot \, q(y)dy \\ 2r \displaystyle\int_x^{\infty} \cdot \, r(y)dy & \dfrac{\partial}{\partial x} + 2r \displaystyle\int_x^{\infty} \cdot \, q(y)dy \end{pmatrix} \tag{2.81}$$

If we define $u = (r, q)^{\mathrm{T}}$ and $\sigma_3 = \begin{pmatrix} 1 & 0 \\ 0 & -1 \end{pmatrix}$, then (2.79) can be written as

$$\int_{-\infty}^{\infty} [\sigma_3 u_t + 2u\Omega(\zeta)] \Psi dx = 0 \tag{2.82}$$

If $\Omega(\zeta)$ is an integral function of ζ, then by $L\Psi = \zeta\Psi$, there is

$$\Omega(\zeta)\Psi = \Omega(L)\Psi \tag{2.83}$$

By defining the conjugation operator L^+:

$$L^+ = \frac{1}{2i} \begin{pmatrix} \dfrac{\partial}{\partial x} - 2r \displaystyle\int_{-\infty}^{x} \cdot q(y)dy & 2r \displaystyle\int_{-\infty}^{x} \cdot r(y)dy \\ -2q \displaystyle\int_{-\infty}^{x} \cdot q(y)dy & -\dfrac{\partial}{\partial x} + 2q \displaystyle\int_{-\infty}^{x} \cdot r(y)dy \end{pmatrix} \tag{2.84}$$

then (2.82) can be written as

$$\int_{-\infty}^{\infty} \left[\sigma_3 u_t + 2\Omega \left(L^+\right) u\right] \Psi dx = 0 \tag{2.85}$$

Similarly, (2.78) can be written as

$$\int_{-\infty}^{\infty} \left[\sigma_3 u_t + 2\bar{\Omega} \left(L^+\right) u\right] \bar{\Psi} dx = 0, \ \bar{\Psi} = (\bar{\psi}_1^2, \bar{\psi}_2^2)^{\mathrm{T}} \tag{2.86}$$

When taking $\Omega = \bar{\Omega}$, in order to satisfy both (2.85) and (2.86), it is required that

$$\sigma_3 u_t + 2\Omega \left(L^+\right) u = 0 \tag{2.87}$$

Without loss of generality, if $\Omega(\zeta) = A_-(\zeta)$ is selected, then

$$\sigma_3 u_t + 2A_- \left(L^+\right) u = 0 \tag{2.88}$$

(2.88) is the more general nonlinear evolution system we require, which has the linear dispersion relation $A_-(\zeta)$ and can be solved by using the inverse scattering method. For example,

$$\Omega(\zeta) = A_-(\zeta) = -2i\zeta^2$$

$$\sigma_3 u_t = \begin{pmatrix} r_t \\ -q_t \end{pmatrix}, \quad L^+ u = \frac{1}{2i} \begin{pmatrix} r_x \\ q_x \end{pmatrix}$$

$$2\left[-2i \left(L^+\right)^2 u\right] = i \begin{pmatrix} r_{xx} - 2qr^2 \\ q_{xx} - 2q^2 r \end{pmatrix} \tag{2.89}$$

Thus, we have

$$\begin{pmatrix} r_t \\ -q_t \end{pmatrix} + i \begin{pmatrix} r_{xx} - 2qr^2 \\ q_{xx} - 2q^2 r \end{pmatrix} = 0 \tag{2.90}$$

When $r = \mp q^*$, it is (2.54).

There are also

$$I(\psi, \bar{\psi}) = -2A_-(\zeta)b\bar{b} \tag{2.91}$$

$$I(\varphi, \varphi) = -2A_-(\zeta)ab \tag{2.92}$$

$$I(\bar{\varphi}, \bar{\varphi}) = 2A_-(\zeta)\bar{a}\bar{b} \tag{2.93}$$

From (2.72),

$$A_+ = A_-, \quad B_+ = C_+ = 0 \tag{2.94}$$

Therefore, the inverse scattering problem can be solved based on the scattering data. In our discussion of conservation laws, we will prove that (2.94) leads to the existence of integral densities $\{c_n\}_{n=1}^{\infty}$, which are constants of motion. Here,

$$c_1 = \int qr dx, \ c_2 = \frac{1}{2} \int (rq_x - r_x q) \, dx$$

$$c_3 = \int \left(q_x r_x + q^2 r^2 \right) dx \tag{2.95}$$

We notice that when $q \to 0$ ($|x| \to \infty$), $r = -1$. (2.36) is equivalent to the eigenvalue problem of Schrödinger equation:

$$v_{2xx} + \left(\zeta^2 + q \right) v_2 = 0 \tag{2.96}$$

In this case, the evolution equation is

$$q_t + \hat{c} \left(4L_\circ^+ \right) q_x = 0 \tag{2.97}$$

where the operator

$$L_\circ^+ = \frac{1}{4} \frac{\partial^2}{\partial x^2} - q + \frac{1}{2} q_x \int_x^{\infty} \cdot \, dy \tag{2.98}$$

$\hat{c} \left(k^2 \right) = \omega/k$, and ω is the dispersion relation of its linear equation. It is easy to verify that when $\omega = -k^3$,

$$q_t + q_{xxx} + 6qq_x = 0 \tag{2.99}$$

We can also generalize the dispersion relation $Q(\zeta)$ as a quotient of the integer functions, such as $Q(\zeta) = \Omega_1(\zeta)/\Omega_2(\zeta)$, where $\Omega_1(\zeta)$ and $\Omega_2(\zeta)$ are integer functions. Thus, analogous to (2.88),

$$\Omega_2\left(L^+\right)\sigma_3 u_t + 2\Omega_1\left(L^+\right)u = 0 \qquad (2.100)$$

If we take $\Omega = i\alpha/2\left(\zeta - \zeta_1\right)$, then (2.100) becomes

$$\frac{1}{2i}\begin{pmatrix} r_{xt} - 2r\displaystyle\int_{-\infty}^{r}(qr)_t dy - 2i\xi_1 r_t \\[2mm] q_{xt} - 2q\displaystyle\int_{-\infty}^{r}(qr)_t dy + 2i\xi_1 q_t \end{pmatrix} = -i\alpha\begin{pmatrix} r \\ q \end{pmatrix} \qquad (2.101)$$

For (2.97), while $\hat{c}\left(k^2\right) = \dfrac{1}{1+k^2}$, there is

$$q_t - q_{xxt} - 4qq_t + 2q_x\int_x^{\infty} q_t dy + q_x = 0 \qquad (2.102)$$

which reduces to the KdV equation under the long wave approximation and small amplitude assumption, and has the properties described by the PBBM equation.

Of course, it is possible to further analyze a wider class of evolution equations, for example, if we do not choose $\Omega(\zeta) = \bar{\Omega}(\zeta)$, the inverse scattering method can still be used, even for those cases where there is no constant of motion.

2.5   Solution of the Inverse Scattering Problem for AKNS Equation

In this section, we mainly discuss the solution of the inverse scattering problem. For non self-conjugate eigenvalue problem, the Marqenko integral equation can still be obtained. We first analyse the analytical properties of the scattering data.

1) Analytical properties of scattering data

For the eigenvalue problem given by (2.36) on the infinite interval $-\infty < x < \infty$, let r, q go to zero sufficiently sharply when $|x| \to \infty$. And φ, $\bar{\varphi}$, ψ, $\bar{\psi}$ are the solutions (Jost functions) of the boundary conditions (2.60), (2.61) and the system (2.36) respectively, φ, $\bar{\varphi}$ are linearly independent, as are ψ and $\bar{\psi}$. For real numbers ζ, there are

$$\varphi(\zeta, x) = a(\zeta)\bar{\psi}(\zeta, x) + b(\zeta)\psi(\zeta, x) \qquad (2.103)$$

$$\bar{\varphi}(\zeta, x) = -\bar{a}(\zeta)\psi(\zeta, x) + \bar{b}(\zeta)\bar{\psi}(\zeta, x) \qquad (2.104)$$

from which a, $\bar{a}$, b, $\bar{b}$ are determined. According to (2.36), if $u(\zeta, x)$ and $v(\zeta, x)$ are the solutions of (2.36), then

$$\frac{d}{dx}w(u, v) = 0 \qquad (2.105)$$

where

$$w(u, v) \equiv u_1(\zeta, x)v_2(\zeta, x) - u_2(\zeta, x)v_1(\zeta, x) \tag{2.106}$$

In fact,

$$[u_{1x} + i\zeta u_1 = qu_2] \cdot v_2 + u_1 [v_{2x} - i\zeta v_2 = rv_1]$$

$$- [u_{2x} - i\zeta u_2 = ru_1] \cdot v_1 - u_2 [v_{1x} + i\zeta v_1 = qv_2] = 0$$

from which (2.105) is obtained. We can also obtain the expression (2.64) for a, b, $\bar{a}$, $\bar{b}$ and $w(\bar{\varphi}, \varphi) = 1$. This leads to

$$\bar{a}(\zeta)a(\zeta) + \bar{b}(\zeta)b(\zeta) = 1 \tag{2.107}$$

Then $\psi(\zeta, x)$ and $\bar{\psi}(\zeta, x)$ can be obtained from (2.103), (2.104):

$$\psi(\zeta, x) = -a(\zeta)\bar{\varphi}(\zeta, x) + \bar{b}(\zeta)\varphi(\zeta, x) \tag{2.108}$$

$$\bar{\psi}(\zeta, x) = \bar{a}(\zeta)\varphi(\zeta, x) + b(\zeta)\bar{\varphi}(\zeta, x) \tag{2.109}$$

From $\varphi_{xx} - i\zeta\varphi_2 = r\varphi_1$ and the boundary condition $\varphi_2 \to 0$ ($x \to -\infty$), we have

$$e^{i\zeta x}\varphi(x) = \int_{-\infty}^{x} e^{2i\zeta(x-y)}r(y)e^{i\zeta y}\varphi(y)dy \tag{2.110}$$

Substituting $\varphi_2(x)$ of (2.110) into $\varphi_{1x} + i\zeta\varphi_1 = q\varphi_2$ and using $\varphi_1 \to e^{-i\zeta x}$ ($x \to -\infty$) gives

$$e^{i\zeta x}\varphi_1(x) = 1 + \int_{-\infty}^{x} M(\zeta, x, y)e^{i\zeta y}\varphi_1(y)dy \tag{2.111}$$

where

$$M(\zeta, x, y) \equiv r(y) \int_{y}^{x} e^{2i\zeta(x-y)}q(z)dz \tag{2.112}$$

Under appropriate conditions, we analytically extend φ to the upper half plane ($\eta > 0$) of the complex plane ζ ($\zeta = \xi + i\eta$). To see this, set

$$R_n(x) \equiv \int_{-\infty}^{x} |y^n| \, |r(y)|dy \tag{2.113}$$

$$Q_n(x) \equiv \int_{-\infty}^{x} |y^n| \, |q(y)|dy \tag{2.114}$$

Since r, q are assumed to go sharply enough to zero $(x \to -\infty)$, the above integrals exist when $n \geqslant 0$. For $\eta > 0$, we have

$$
\begin{aligned}
\left| e^{i\zeta x} \varphi_1(x) \right| &\leqslant 1 + \int_{-\infty}^{x} |q(z)| dz \int_{-\infty}^{z} |r(y)| \left| e^{i\zeta y} \varphi_1(y) \right| dy \\
&= 1 + \int_{-\infty}^{x} Q_0'(z) dz \int_{-\infty}^{z} R_0'(y) \left| e^{i\zeta y} \varphi_1(y) \right| dy \\
&\leqslant 1 + R_0(x) Q_0(x) + \frac{[R_0(x) Q_0(x)]^2}{(2!)^2} + \frac{[R_0(x) Q_0(x)]^2}{(3!)^2} + \cdots
\end{aligned}
$$

or

$$
\left| e^{i\zeta x} \varphi_1(x) \right| \leqslant I_0(s(x)) \tag{2.115}
$$

where $s(x) = 2 \left(R_0(x), Q_0(x) \right)^{\frac{1}{2}}$, $I_0(s)$ is a zero-order Bessel function with imaginary variables. From (2.103), (2.106), we know that $x \to +\infty$, $\varphi_1 e^{i\zeta x} \to a(\zeta)$. Thus, $a(\zeta)$ is bounded in the upper half plane of ζ $(\eta \geqslant 0)$. Let $R_0(\infty)$, $Q_0(\infty)$ be finite. The Neumann series solution of $e^{i\zeta x} \varphi_1(x)$ can be obtained from (2.111):

$$
\begin{aligned}
e^{i\zeta x} \varphi_1(x) =& 1 + \int_{-\infty}^{x} M(\zeta, x, y) dy \\
&+ \int_{-\infty}^{x} M(\zeta, x, y) dy \int_{-\infty}^{y} M(\zeta, y, z) dz + \cdots
\end{aligned} \tag{2.116}
$$

which is absolutely convergent in the upper half plane of ζ. Further, if (2.116) is differentiated by ζ, then it is easy to see that $e^{i\zeta x} \varphi_1(x)$ is analytic when $\eta > 0$. For the analyticity of $\eta = 0$, it is not enough to require that $R_0(\infty)$, $Q(\infty)$ be finite. Therefore, in (2.112), ζ is included in the exponent, and the differentiation of ζ necessarily brings the term $(z - y)$. To ensure the existence of differentiation at $\eta = 0$, we must let r, q go to zero faster than $|x|^{-2}$ $(|x| \to \pm\infty)$. $\bar{\varphi}$, ψ, $\bar{\psi}$ can be discussed in the same way as φ. This leads to the following theorem:

Theorem 2.3　*If the condition*

$$
R_0(\infty) < \infty, \quad Q_0(\infty) < \infty \tag{2.117}
$$

is satisfied, then $e^{i\zeta x} \varphi(\zeta, x)$, $e^{-i\zeta x} \psi(\zeta, x)$ are analytic functions of ζ $(\eta > 0)$, and $e^{-i\zeta x} \bar{\varphi}(\zeta, x)$, $e^{i\zeta x} \bar{\psi}(\zeta, x)$ are analytic functions of ζ $(\eta < 0)$. When $\eta = 0$, all the above four functions are bounded. For a given integer n, if

$$
R_l(\infty) < \infty, \quad Q_l(\infty) < \infty, \ l = 0, 1, 2, \cdots, n \tag{2.118}
$$

is further satisfied, then these four functions are the nth degree differentiable functions of ζ $(\eta = 0)$. If (2.118) holds for all positive integers n, then they are analytic on the ζ plane containing the real axis $(\eta = 0)$.

From (2.63), we can obtain the following corollary:

Corollary 2.1   *If the condition (2.117) is satisfied, then $a(\zeta)$ is the analytic function of ζ ($\eta > 0$), and $\bar{a}(\zeta)$ is the analytic function of ζ ($\eta < 0$). If (2.118) holds for all n, then on the ζ plane containing the real axis ($\eta = 0$), both $a(\zeta)$ and $\bar{a}(\zeta)$ are analytic.*

If r and q are given stronger conditions, the following theorem can be proved in the same way.

Theorem 2.4   *If there exist positive constants $\hat{R}$, $\hat{Q}$ and K such that*

$$|r(x)| < \hat{R}e^{-2K|x|}, \quad |q(x)| < \hat{Q}e^{-2K|x|} \tag{2.119}$$

are true for all x, then $e^{i\zeta x}\varphi(\zeta, x)$, $e^{-i\zeta x}\psi(\zeta, x)$ are analytic functions of ζ ($\eta > -K$), and $e^{-i\zeta x}\bar{\varphi}(\zeta, x)$, $e^{i\zeta x}\bar{\psi}(\zeta, x)$ are analytic functions of ζ ($\eta < +K$).

As a corollary, from (2.63), we have

Corollary 2.2   *If (2.119) is satisfied, then $a(\zeta)$ is the analytic function of ζ ($\eta > -K$), $\bar{a}(\zeta)$ is the analytic function of ζ ($\eta < +K$), $b(\zeta)$ and $\bar{b}(\zeta)$ are analytic ($-K < \eta < +K$).*

We note that if r, q are on the compact support set, then K can be chosen to be sufficiently large in (2.119), so we have the second corollary:

Corollary 2.3   *Let r, q be on the compact support set, then (2.118) holds. Therefore, $e^{i\zeta x}\varphi(\zeta, x)$, $e^{i\zeta x}\bar{\varphi}(\zeta, x)$ and $e^{i\zeta x}\bar{\psi}(\zeta, x)$ are integral functions of ζ (on the whole plane); $a(\zeta)$, $\bar{a}(\zeta)$, $b(\zeta)$ and $\bar{b}(\zeta)$ are also integral functions of ζ.*

(2.110)-(2.112) have asymptotic expansion in the upper half plane of ζ when $|\zeta| \to -\infty$,

$$\varphi_1 e^{i\zeta x} \to 1 - \frac{1}{2i\zeta}\int_{-\infty}^{x} r(y)q(y)dy + O\left(\frac{1}{\zeta^2}\right) \tag{2.120}$$

$$\varphi_2 e^{i\zeta x} \to -\frac{1}{2i\zeta}r(x) + O\left(\frac{1}{\zeta^2}\right) \tag{2.121}$$

Similarly, we have

$$\psi_1 e^{-i\zeta x} \to \frac{1}{2i\zeta}q(x) + O\left(\frac{1}{\zeta^2}\right) \tag{2.122}$$

$$\psi_2 e^{-i\zeta x} \to 1 - \frac{1}{2i\zeta}\int_{x}^{\infty} r(y)q(y)dy + O\left(\frac{1}{\zeta^2}\right) \tag{2.123}$$

In the lower half plane of ζ, when $|\zeta| \to \infty$, we have

$$\bar{\varphi}_1 e^{-i\zeta x} \to -\frac{1}{2i\zeta}q(x) + O\left(\frac{1}{\zeta^2}\right) \tag{2.124}$$

$$\bar{\varphi}_2 e^{-i\zeta x} \to -1 - \frac{1}{2i\zeta} \int_{-\infty}^{x} q(y)r(y)dy + O\left(\frac{1}{\zeta^2}\right) \tag{2.125}$$

$$\bar{\psi}_1 e^{i\zeta x} \to 1 + \frac{1}{2i\zeta} \int_{x}^{\infty} q(y)r(y)dy + O\left(\frac{1}{\zeta^2}\right) \tag{2.126}$$

$$\bar{\psi}_2 e^{i\zeta x} \to \frac{-1}{2i\zeta} r(x) + O\left(\frac{1}{\zeta^2}\right) \tag{2.127}$$

Thus, on the corresponding half plane, when $|\zeta| \to \infty$, we have

$$a(\zeta) \to 1 - \frac{1}{2i\zeta} \int_{-\infty}^{\infty} q(y)r(y)dy + O\left(\frac{1}{\zeta^2}\right) \tag{2.128}$$

$$\bar{a}(\zeta) \to 1 + \frac{1}{2i\zeta} \int_{-\infty}^{\infty} q(y)r(y)dy + O\left(\frac{1}{\zeta^2}\right) \tag{2.129}$$

When $a(\zeta)$ has zero points ζ_k $(k = 1, 2, \cdots, N)$ (N is a finite number) in the upper half plane $(\eta > 0)$, we have

$$\varphi = b_k \psi \tag{2.130}$$

where b_k is the scaling factor. When r, q are compact support functions, then $b_k \equiv b(\zeta_k)$. While $\bar{a}(\zeta)$ has zero points $\bar{\zeta}_k$ $(k = 1, 2, \cdots, \bar{N})$ in the lower half plane $(\eta < 0)$. When $\zeta = \bar{\zeta}_k$,

$$\bar{\varphi} = \bar{b}_k \psi \tag{2.131}$$

If r, q are compact support functions, then $\bar{b}_k \equiv \bar{b}(\bar{\zeta}_k)$. Here, both N and $\bar{N}$ are finite numbers. Unlike the Schrödinger equation (due to the self-conjugation of the operator L), the zeros of a, $\bar{a}$ here may be multiple. But in this case, it can be treated as the limit of simple zeros. For example, if there are double zeros at ζ_1, we can consider a to have simple zeros ζ_1, ζ_2, and let $\zeta_2 \to \zeta_1$.

When r is linear to q or q^*, the case becomes simple. First, consider the case of

$$r = \alpha q \tag{2.132}$$

where α is any non-zero, finite complex number. At this point, it is easy to know that

$$\bar{\psi}(\zeta, x) = S\psi(-\zeta, x) \tag{2.133}$$

$$\bar{\varphi}(\zeta, x) = -\frac{1}{\alpha} S\varphi(-\zeta, x) \tag{2.134}$$

where

$$S = \begin{pmatrix} 0 & 1 \\ \alpha & 0 \end{pmatrix} \tag{2.135}$$

Therefore,

$$\bar{a}(\zeta) = a(-\zeta) \tag{2.136}$$

$$\bar{b}(\zeta) = -\frac{1}{\alpha}b(-\zeta) \tag{2.137}$$

and the zeros of a, $\bar{a}$ are paired such that

$$\bar{N} = N \tag{2.138}$$

$$\bar{\zeta}_k = -\zeta_k \ (k = 1, 2, \cdots, N) \tag{2.139}$$

For the case of

$$r = \alpha q^* \tag{2.140}$$

where α is a non-zero, finite real constant, we have

$$\bar{\psi}(\zeta, x) = S\psi^*(\zeta^*, x) \tag{2.141}$$

$$\bar{\varphi}(\zeta, x) = -\frac{1}{\alpha}S\varphi^*(\zeta^*, x) \tag{2.142}$$

Thus,

$$\bar{a}(\zeta) = a^*(\zeta^*) \tag{2.143}$$

$$\bar{b}(\zeta) = -\frac{1}{\alpha}b^*(\zeta^*) \tag{2.144}$$

and the zeros of a, $\bar{a}$ are paired such that

$$\bar{N} = N \tag{2.145}$$

$$\bar{\zeta}_k = \zeta_k \ (k = 1, 2, \cdots, N) \tag{2.146}$$

$$\bar{b}_k = -\frac{1}{\alpha}b_k^* \tag{2.147}$$

Therefore, if (2.132), (2.140) are both true, and r and r^* are proportional to q, then ζ_k is either a pure imaginary number, or $-\zeta_k^*$ is also an eigenvalue.

2) Inverse scattering method

First, we obtain the integral representation of four Jost functions defined by (2.36), and then obtain the Marqenko-type inverse scattering integral equation. For simplicity, let r, q be on the compact support set, so that the solution and scattering datas of (2.36) are both integral functions of ζ. We define the contour c to be on the complex plane ζ, starting at $\zeta = -\infty + i0^+$, passing all the zeros of $a(\zeta)$ from above and ending at $\zeta = +\infty + i0^{+}$. Similarly, we define the contour $\bar{c}$, starting at $\zeta = -\infty + i0^{-}$, passing all zeros of $\bar{a}(\zeta)$ from below, and ending at $\zeta = +\infty + i0^{-}$.

Consider the integral

$$\int_{c} \frac{d\zeta'}{a(\zeta')} \frac{\varphi(\zeta', x)}{\zeta' - \zeta} e^{i\zeta x} \tag{2.148}$$

where ζ is below c, and its value is $-i\pi \begin{pmatrix} 1 \\ 0 \end{pmatrix}$ from (2.120), (2.121) and (2.128). From (2.103) $\varphi = a\bar{\psi} + b\psi$ and another integral contour including $\bar{\psi}$ in the lower half plane, together with (2.126), (2.127), we get

$$\psi(\zeta, x)e^{i\zeta x} = \begin{pmatrix} 1 \\ 0 \end{pmatrix} + \frac{1}{2\pi i} \int_{c} \frac{d\zeta'}{\zeta' - \zeta} \cdot \frac{b(\zeta')}{a(\zeta')} \psi(\zeta' x) e^{i\zeta' x} \tag{2.149}$$

where ζ is below c. Similarly, consider the integral

$$\int_{\bar{c}} \frac{d\zeta'}{\zeta' - \zeta} \frac{\bar{\varphi}(\zeta', x)}{a(\zeta')} e^{-i\zeta' x} \tag{2.150}$$

where ζ is above $\bar{c}$. In the same way, we obtain

$$\psi(\zeta, x)e^{-i\zeta x} = \begin{pmatrix} 0 \\ 1 \end{pmatrix} + \frac{1}{2\pi i} \int_{\bar{c}} \frac{d\zeta'}{\zeta' - \zeta} \cdot \frac{\bar{b}(\zeta')}{\bar{a}(\zeta')} \bar{\psi}(\zeta', x) e^{-i\zeta x} \tag{2.151}$$

Similarly, taking the above contours for $\varphi e^{i\zeta x}$, $\bar{\varphi} e^{-i\zeta x}$, respectively, yields

$$\bar{\varphi}(\zeta, x)e^{-i\zeta x} = -\begin{pmatrix} 0 \\ 1 \end{pmatrix} - \frac{1}{2\pi i} \int_{c} \frac{d\zeta'}{\zeta' - \zeta} \cdot \frac{\bar{b}(\zeta')}{a(\zeta')} \varphi(\zeta', x) e^{-i\zeta x} \tag{2.152}$$

$$\varphi(\zeta, x)e^{-i\zeta x} = \begin{pmatrix} 0 \\ 1 \end{pmatrix} - \frac{1}{2\pi i} \int_{c} \frac{d\zeta'}{\zeta' - \zeta} \cdot \frac{b(\zeta')}{\bar{a}(\zeta')} \bar{\varphi}(\zeta', x) e^{i\zeta x} \tag{2.153}$$

where ζ is located between the contours c and $\bar{c}$.

Now let ψ, $\bar{\psi}$, φ, $\bar{\varphi}$ can be expressed as integrals

$$\psi(\zeta, x) = \begin{pmatrix} 0 \\ 1 \end{pmatrix} e^{i\zeta x} + \int_{x}^{\infty} K(x, s)e^{i\zeta s} ds \tag{2.154}$$

$$\bar{\psi}(\zeta, x) = \begin{pmatrix} 1 \\ 0 \end{pmatrix} e^{-i\zeta x} + \int_{x}^{\infty} \bar{K}(x, s)e^{-i\zeta s} ds \tag{2.155}$$

$$\varphi(\zeta, x) = \begin{pmatrix} 1 \\ 0 \end{pmatrix} e^{-i\zeta x} - \int_{-\infty}^{x} L(x, s)e^{-i\zeta s} ds \tag{2.156}$$

$$\bar{\varphi}(\zeta, x) = -\begin{pmatrix} 0 \\ 1 \end{pmatrix} e^{-i\zeta x} - \int_{-\infty}^{x} \bar{L}(x, s)e^{i\zeta s} ds \tag{2.157}$$

where K, $\bar{K}$, L, $\bar{L}$ are all column vectors. Substituting these expressions into (2.149), (2.151)-(2.153) and applying the Fourier transform, we obtain the following Marqenko-type integral equation

$$\bar{K}(x,y) + \begin{pmatrix} 0 \\ 1 \end{pmatrix} F(x+y) + \int_x^\infty K(x,s) \cdot F(s+y)ds = 0 \ (y > x) \tag{2.158}$$

$$K(x,y) - \begin{pmatrix} 0 \\ 1 \end{pmatrix} \bar{F}(x+y) - \int_x^\infty \bar{K}(x,s) \cdot \bar{F}(s+y)ds = 0 \ (y > x) \tag{2.159}$$

$$\bar{L}(x,y) + \begin{pmatrix} 1 \\ 0 \end{pmatrix} G(x+y) - \int_{-\infty}^x L(x,s) \cdot G(s+y)ds = 0 \ (x < y) \tag{2.160}$$

$$L(x,y) + \begin{pmatrix} 0 \\ 1 \end{pmatrix} \bar{G}(x+y) + \int_{-\infty}^x \bar{L}(x,s) \cdot \bar{G}(s+y)ds = 0 \ (x > y) \tag{2.161}$$

where

$$F(z) = \frac{1}{2\pi} \int_c \frac{b(\zeta)}{a(\zeta)} e^{i\zeta z} d\zeta \tag{2.162}$$

$$\bar{F}(z) = \frac{1}{2\pi} \int_{\bar{c}} \frac{\bar{b}(\zeta)}{\bar{a}(\zeta)} e^{-i\zeta z} d\zeta \tag{2.163}$$

$$G(z) = \frac{1}{2\pi} \int_c \frac{\bar{b}(\zeta)}{a(\zeta)} e^{-i\zeta z} d\zeta \tag{2.164}$$

$$\bar{G}(z) = \frac{1}{2\pi} \int_{\bar{c}} \frac{b(\zeta)}{a(\zeta)} e^{i\zeta z} d\zeta \tag{2.165}$$

In the following, we prove proves the existence and uniqueness of the integral kernels $K(x,s)$, $\bar{K}(x,s)$, $L(x,s)$, $\bar{L}(x,s)$ of (2.154)-(2.157). First, start from (2.154) and require it to meet (2.36), that is, substitute (2.154) into

$$\begin{cases} \psi_{1x} + i\zeta\psi_1 = q(x)\psi_2 \\ \psi_{2x} - i\zeta\psi_2 = r(x)\psi_1 \end{cases} \tag{2.166}$$

to get

$$\int_x^\infty e^{i\zeta s} \left[(\partial x - \partial s)K_1(x,s) - q(x)K_2(x,s) \right] ds$$

$$- \left[q(x) + 2K_1(x,s) \right] e^{i\zeta x} + \lim_{s \to \infty} \left[K_1(x,s)e^{i\zeta s} \right] = 0 \tag{2.167}$$

$$\int_x^\infty e^{i\zeta s} \left[(\partial x + \partial s)K_2(x,s) - r(x)K_1(x,s) \right] ds$$

$$- \lim_{s \to \infty} \left[K_2(x,s)e^{i\zeta s} \right] = 0 \tag{2.168}$$

Thus, we have

$$(\partial x - \partial s)K_1(x, s) - q(x)K_2(x, s) = 0 \tag{2.169}$$

$$(\partial x - \partial s)K_2(x, s) - r(x)K_1(x, s) = 0 \tag{2.170}$$

and the boundary conditions are satisfied as

$$K_1(x, x) = -\frac{1}{2}q(x) \tag{2.171}$$

$$\lim_{s \to \infty} K(x, s) = 0 \tag{2.172}$$

For solutions of (2.169), (2.170) satisfying the boundary conditions (2.171),(2.172) to exist, we introduce the following coordinate transformation:

$$\mu = \frac{1}{2}(x + s), \quad \nu = \frac{1}{2}(x - s) \tag{2.173}$$

Then (2.169)-(2.172) are transformed into

$$\partial\nu K_1(\mu, \nu) - q(\mu + \nu)K_2(\mu, \nu) = 0 \tag{2.174}$$

$$\partial\mu K_2(\mu, \nu) - r(\mu + \nu)K_1(\mu, \nu) = 0 \tag{2.175}$$

$$K_1(\mu, 0) = -\frac{1}{2}q(\mu) \tag{2.176}$$

$$\lim_{\mu - \nu \to \infty} K(\mu, \nu) = 0 \tag{2.177}$$

From the characteristic theory of differential equations, we know that this solution $K(\mu, \nu)$ exists and is unique. Similarly, the existence and uniqueness of $\bar{K}$, L, $\bar{L}$ can be proved.

Finally, we prove the existence of solutions of Marqenko-type integral equations (2.158)-(2.161) under certain conditions. We are limited to the cases of

$$r(x) = -q^*(x) \tag{2.178}$$

or

$$r(x) = q^*(x) \tag{2.179}$$

and

$$Q(\infty) = \int_{-\infty}^{\infty} |q|dx < 0.523 \tag{2.180}$$

although these conditions are not sufficient and necessary.

Consider the homogeneous system of integral equations $(y > x)$ corresponding to (2.158), (2.159)

$$\begin{cases} \varphi_1(y) + \int_x^\infty \varphi_2(s)F(s+y)ds = 0 \\ \varphi_2(y) - \int_x^\infty \varphi_1(s)\bar{F}(s+y)ds = 0 \end{cases} \tag{2.181}$$

Let $\varphi(y) = \begin{pmatrix} \varphi_1 \\ \varphi_2 \end{pmatrix}$ be the solution of (2.181), which is always zero when $y < x$. By Fredholm theorem, only $\varphi(y) \equiv 0$ needs to be sufficiently proved. Multiplying (2.181) by φ_1^*, φ_2^* and integrating y, and because

$$\int_x^\infty |\varphi_i(y)|^2 \, dy = \int_{-\infty}^\infty |\varphi_i(y)|^2 \, dy, \quad i = 1, 2$$

we get

$$\int_{-\infty}^\infty \Big\{ |\varphi_1|^2 + |\varphi_2|^2 + \int_{-\infty}^\infty [\varphi_2(s)\varphi^*(y)F(s+y)$$

$$- \varphi_1(s)\varphi^*(y)\bar{F}(s+y)]ds \Big\}dy = 0 \tag{2.182}$$

Now consider two cases. First, let $r = -q^*$, then from (2.141)-(2.148) ($\alpha = -1$), we know $\bar{F}(s+y) = F^*(s+y)$, so (2.182) becomes

$$\int_{-\infty}^\infty \Big\{ |\varphi_1|^2 + |\varphi_2|^2 + 2i\,\mathrm{Im} \int_{-\infty}^\infty \varphi_1^*(y)\varphi_2(s) \cdot F(s+y)ds \Big\}dy = 0 \tag{2.183}$$

The real and imaginary parts of the above equation must be zero, so we get $\varphi(y) = 0$, that is, the solution of (2.158), (2.159) exists and is unique. Second, if $r(x) = q^*(x)$, then the problem is self-conjugated, and its spectra can only be located on the real axis, and $\bar{F}(s+y) = -F^*(s+y)$. At this point, (2.182) is

$$\int_{-\infty}^\infty \Big\{ |\varphi_1|^2 + |\varphi_2|^2 + 2\,\mathrm{Re} \int_{-\infty}^\infty \varphi_1^*(x)\varphi_2(s) \cdot F(s+y)ds \Big\}dy = 0 \tag{2.184}$$

If we require that the condition

$$|a(\zeta)| > 0 \quad (\eta \geqslant 0) \tag{2.185}$$

to be satisfied, then there are no discrete eigenvalues on the real axis, so

$$F(z) = \frac{1}{2\pi} \int_{-\infty}^\infty \frac{b(\zeta)}{a(\zeta)} e^{i\zeta z} d\zeta \tag{2.186}$$

The Fourier transform of $\varphi_i(y)$ is

$$\hat{\varphi}_i(\xi) = \int_{-\infty}^{\infty} \varphi_i(y)e^{-i\xi y}dy \tag{2.187}$$

which satisfies the Parseval relation:

$$\int_{-\infty}^{\infty} |\varphi_i|^2\, dy = \frac{1}{2\pi} \int_{-\infty}^{\infty} |\hat{\varphi}_i|^2\, d\xi \tag{2.188}$$

Substitute (2.186)-(2.188) into (2.183) and exchange the integration order to get

$$\int_{-\infty}^{\infty} \left\{ |\hat{\varphi}_1(-\xi)|^2 + |\hat{\varphi}_2^*(\xi)|^2 + 2\,\mathrm{Re}\left[\frac{b(\xi)}{a(\xi)}\hat{\varphi}_1(-\xi)\hat{\varphi}_2^*(\xi)\right] \right\} d\xi = 0 \tag{2.189}$$

If

$$\left|\frac{b(\xi)}{a(\xi)}\right| < 1 \tag{2.190}$$

then we have

$$\left| 2\,\mathrm{Re}\left[\frac{b(\xi)}{a(\xi)}\hat{\varphi}_1(-\xi)\hat{\varphi}_2(\xi)\right] \right| \leqslant 2\,|\hat{\varphi}_1(-\xi)|\,|\hat{\varphi}_2^*(\xi)| \leqslant |\hat{\varphi}_1|^2 + |\hat{\varphi}_2|^2$$

Therefore, the solution of (2.189) is $\varphi \equiv 0$, then the existence and uniqueness of the solution for (2.158), (2.159) are obtained again. From the relations $\bar{a}a + b\bar{b} = 1$ and $\bar{a} = a^*$, $\bar{b} = b^*$, for $r = q^*$, the condition (2.190) can be written as

$$|a|^2 > \frac{1}{2} \tag{2.191}$$

which is stronger than the condition (2.185). If we use the condition

$$|a(\zeta) - 1| < 1 - \frac{1}{\sqrt{2}} \tag{2.192}$$

to meet (2.191), by (2.115), there is

$$|a(\zeta) - 1| \leqslant I_0(2Q(\infty)) - 1 < 1 - \frac{1}{\sqrt{2}}$$

Therefore, the condition (2.180)

$$Q(\infty) < 0.523$$

is sufficient.

2.6   Asymptotic Solution of the Evolution Equation $(t \to \infty)$

Previously, we have illustrated the inverse scattering method and pointed out a class of evolution equations which can be solved in this way. In this section, in order to determine the asymptotic state of the solution of the evolution equation (2.88), we need to solve the system of integral equations $(y > x)$

$$\begin{cases} K(x,y;t) - \begin{pmatrix} 1 \\ 0 \end{pmatrix} \bar{F}(x+y;t) - \int_x^\infty \bar{K}(x,s;t)\bar{F}(s+y;t)ds = 0 \\[2mm] \bar{K}(x,y;t) + \begin{pmatrix} 0 \\ 1 \end{pmatrix} F(x+y;t) + \int_x^\infty K(x,s;t)F(s+y;t)ds = 0 \end{cases} \tag{2.193}$$

The proof of the asymptotic solution is similar to the KdV equation, but differs in some important respects. In the following, we discuss discrete spectrum, continuous spectrum, arbitrary spectrum (their combination) respectively, and estimate the discrete spectrum.

2.6.1   Discrete spectrum

First, consider the solvability of the system of equations (2.193). The major difference between the scattering problem (2.36) and the eigenvalue problem (2.55) is that the system of integral equations (2.193) corresponding to (2.36) does not necessarily have a solution and a unique solution, and the solution of the evolution equation may become unbounded after finite time. We explain it with an example. Let $q(x,0)$, $r(x,0)$ be smooth initial conditions, and satisfy (2.117). The spectrum consists of two kinds of eigenvalues: $\zeta \, (\text{Im}\, \zeta > 0)$ and $\bar{\zeta} \, (\text{Im}\, \zeta < 0)$, then

$$\begin{aligned} F(z,t) &= ice^{i\zeta z - 2A_0(\zeta)t} \\ \bar{F}(z,t) &= i\bar{c}e^{-i\zeta z + 2A_0(\zeta)t} \end{aligned} \tag{2.194}$$

where c, $\bar{c}$ are constants, and $A_0(\zeta)$ is the corresponding linear dispersion relation. The kernel of the integral equation (2.193) is degenerate and easy to solve. From the relations

$$\begin{cases} K_1(x,x;t) = -\dfrac{1}{2}q(x,t) \\[3mm] K_2(x,x;t) = \dfrac{1}{2}\int_x^\infty q(x,t)r(x,t)dx \\[3mm] \bar{K}_2(x,x;t) = \dfrac{1}{2}r(x,t) \end{cases} \tag{2.195}$$

we can obtain

$$\begin{cases} q(x,t) = -\dfrac{2i\bar{c}e^{2A_0(\zeta)t-2i\bar{\zeta}x}}{D(x,t)} \\[3mm] r(x,t) = \dfrac{2ice^{-2A_0(\zeta)+2i\zeta x}}{D(x,t)} \\[3mm] \displaystyle\int_x^\infty q(x,t)r(x,t)dx = \dfrac{2i\bar{c}ce^{2\left(A_0(\bar{\zeta})-A_0(\zeta)\right)t+2i(\zeta-\bar{\zeta})x}}{\left(\zeta-\bar{\zeta}\right)D(x,t)} \end{cases} \tag{2.196}$$

where

$$D(x,t) = 1 - \frac{c\bar{c}}{\left(\zeta-\bar{\zeta}\right)^2}e^{2\left[A_0(\bar{\zeta})-A_0(\zeta)\right]+2i(\zeta-\bar{\zeta})x} \tag{2.197}$$

So we can see that if $A_0(\zeta)$, $q(x,0)$, $r(x,0)$ are unrestricted and $D(x,t) = 0$ on some countable set $(x,0)$, then at these points, the homogeneous integral equation corresponding to (2.193) has infinitely many solutions. Therefore, (2.193) has no solution. By assuming that $q(x,0)$, $r(x,0)$ are smooth and approach sharply to 0 when $|x| \to \infty$, these points do not occur at $t = 0$, but after a finite time, at the special point x where $D(x,t) = 0$, $q(x,t)$, $r(x,t)$ become unbounded. Therefore, even though r, q initially satisfy (2.117), q, r will produce a "burst" at some special point x over time (according to the equation (2.88)). Such "burst" solitons do not occur in the KdV equation. Their appearance shows significant differences between the scattering problem (2.36) and the eigenvalue problem (2.55). For the eigenvalue problem (2.55), its solution is always satisfied for $t > 0$ if it satisfies the condition

$$\int_{-\infty}^{\infty} (1+|x|)|u|dx < \infty \quad (t = 0) \tag{2.198}$$

This "burst" phenomenon is complicated and needs to be verified and demonstrated in physics.

However, this does not happen when

$$r(x,t) = \alpha q^*(x,t) \ (\alpha \text{ is a real constant}) \tag{2.199}$$

Since the first conservative density $\displaystyle\int_{-\infty}^{\infty} qrdx$ is time invariant, and $\displaystyle\int_x^\infty qrdx$ is bounded, so $D(x,t) \neq 0$, and consequently (2.196) gives the global solution of the evolution equation.

We know that (2.199) includes two special cases ($\alpha = 1$, $\alpha = -1$). A unique solution exists for (2.193), but we do not have sufficient and necessary conditions for the solution of (2.193) to exist in the general case. For simplicity, henceforth, let a unique solution of (2.193) exist. The solution of (2.196) can be written as

$$q(x,t) = i\bar{c}e^{-i\varphi} \operatorname{sech}\theta \tag{2.200}$$

where

$$\varphi = i\left(A_0(\zeta) + A_0(\bar{\zeta})\right)t + (\bar{\zeta} + \zeta)x - i\gamma$$
$$\theta = \left(A_0(\zeta) - A_0(\bar{\zeta})\right)t + i(\zeta - \bar{\zeta})x + \gamma$$
$$e^{2\gamma} = -\frac{c\bar{c}}{\left(\zeta - \bar{\zeta}\right)^2}$$

It is the basic soliton solution, an undeformed, local wave with a travelling speed.

$$V = \operatorname{Re}\left\{\frac{A_0(\zeta) - A_0(\bar{\zeta})}{-i\left(\zeta - \bar{\zeta}\right)}\right\} \tag{2.201}$$

Its amplitude is proportional to $(\zeta - \bar{\zeta})$ and its wavelength is $1/(\zeta - \bar{\zeta})$. The basic characteristic of these waves is nonlinear, which does not appear in linear problems. We give two examples. In the Zaharov-Shabat Problem

$$q_t - iq_{xx} - 2iq^2 q^* = 0$$
$$q|_{t=0} = q_0(x)$$

there are $A_0(\zeta) = -2i\zeta^2$, $r = -q^*$, $\bar{c} = c^*$, $\bar{\zeta} = \zeta^* = \xi - i\eta$, and

$$q(x,t) = 2\eta e^{\left\{-4i(\xi^2 - \eta^2)t - 2i\xi x + i\varphi\right\}} \operatorname{sech}\left\{2\eta\left(x - x_0\right) + \delta\eta\xi t\right\} \tag{2.202}$$

This soliton is an envelope of a vibrating carrier whose amplitude and wavelength depend on η, and whose envelope has a constant waveform and velocity 4ξ. For another example, in the sine-Gordon equation,

$$u_{xt} = \sin u$$

The variables in physical space are

$$X = x + t, \ \ T = x - t, \ \ u = -\int_{-\infty}^{x} 2q\,dz$$

where $A_0(\zeta) = \dfrac{i}{4\zeta}$, $r = -q$, $\zeta = -\zeta = -i\eta$, $\bar{c} = -c$, the soliton solution is

$$u(X,T) = 4\tan^{-1}\left\{\exp\left[\left(c\eta + \frac{1}{4\eta}\right)(X - X_0) + \left(\eta - \frac{1}{4\eta}\right)T\right]\right\} \tag{2.203}$$

which is a kink.

For the general eigenvalue problem solved by (2.36), as long as the spectrum is purely discrete and the integral kernel is degenerate, (2.193) can be solved. Of course, the condition for the existence of the solution of (2.193) must be satisfied. Then we can get the N soliton solutions. Some special equations have been calculated by Hirota et al. When t is large, the soliton runs apart at different speeds, and their asymptotic solution is N separate waves with form (2.200). This separation process has been discussed for $r = -q^*$ by Zaharov and Shabat, who showed that the asymptotic effect of this interaction between solitons is only a phase shift.

From (2.201), we know that some eigenvalues give the same velocity to its corresponding soliton solutions, which do not separate when $t \to \infty$. This phenomenon does not occur in the KdV equation. In the Zaharov-Shabat problem, the case of $R(\zeta) = \zeta_0$ has been analysed. In the sine-Gordon equation, $|\zeta| = c_0$ has been discussed by Lamb (1971) and AKNS (1973).

$$u(X, T) = 4 \tan^{-1}\left[\frac{\eta \cos\{\xi\left(\eta\left(T - T_0\right)\right) - (4 - \nu)x\}}{\xi \cosh\{\eta\left(v\left(x - x_0\right)\right) - (4 - \nu)T\}}\right]$$

where $v = 2 + \left(1/2|\zeta|^2\right)$.

We have already seen that the quotient of any integral function is taken as $A_0(\zeta)$, and the evolution equation formed by it can be solved by the inverse scattering method. If $A_0(\zeta)$ has a pole, we can see from (2.201) that when the eigenvalue is near this pole, it has a very large velocity, such as the sine-Gordon equation, where the speed is close to the speed of light. Of course, there are other physical explanations for this high speed, but its existence makes a lot of sense.

2.6.2 Continuous spectrum

The contribution of the continuous spectrum to the asymptotic solution is now considered. We start with the simplest possible case. Let the initial values satisfy

$$R(\infty)Q(\infty) = \int_{-\infty}^{\infty} |r|dx \int_{-\infty}^{\infty} |q|dx < 0.817 \tag{2.204}$$

and

$$R(\infty)Q(\infty) < 0.383 \tag{2.205}$$

The condition (2.204) guarantees that there are no discrete eigenvalues, and (2.205) guarantees the rationality of using the inverse scattering method. The development of scattering data with time t is

$$F(x, t) = \frac{1}{2\pi} \int_{-\infty}^{\infty} \frac{b(k)}{a(k)} e^{i(kx + 2iA_0(k)t)} dk \tag{2.206}$$

$$\bar{F}(x, t) = \frac{1}{2\pi} \int_{-\infty}^{\infty} \frac{\bar{b}(k)}{\bar{a}(k)} e^{-i(kx + 2iA_0(k)t)} dk \tag{2.207}$$

When $t \to \infty$, the main contribution to the integral is near the steady point $k = K$, that is

$$\chi'(k) = \frac{x}{t} + 2iA_0(K) = 0 \quad \left(\frac{x}{t} \text{ is fixed}\right) \tag{2.208}$$

$$\chi(k) \approx \chi(K) + (k - K)^2 \chi''(K) \tag{2.209}$$

For example, $A_0(\zeta) = -2i\zeta^2$, its evolution equation is

$$\begin{cases} iq_t + q_{xx} - 2(qr)q = 0 \\ ir_t - r_{xx} + 2(qr)r = 0 \end{cases} \tag{2.210}$$

(2.208) becomes

$$x/t = -8K \tag{2.211}$$

Rotate $\frac{\pi}{4}$ for the integral path of F, rotate $-\frac{\pi}{4}$ for $\bar{F}$, so that it has the same sign as $\chi''(K)$. Thus, when $t \to \infty$, $\frac{x}{t}$ is fixed, we get

$$\begin{cases} F(x,t) = \frac{1}{4\sqrt{\pi t}} \frac{b}{a}\left(-\frac{x}{8t}\right) \cdot \exp\left[-\frac{i}{16}\left(\frac{x}{t}\right)^2 t + i\frac{\pi}{4}\right] + O\left(t^{-\frac{3}{2}}\right) \\ \bar{F}(x,t) = \frac{1}{4\sqrt{\pi t}} \frac{\bar{b}}{\bar{a}}\left(-\frac{x}{8t}\right) \cdot \exp\left[\frac{i}{16}\left(\frac{x}{t}\right)^2 t - i\frac{\pi}{4}\right] + O\left(t^{-\frac{3}{2}}\right) \end{cases} \tag{2.212}$$

The integral equation (2.193) can be combined as

$$K_1(x,y;t) - \bar{F}(x+y;t) + \iint_x^\infty K_1(x,z;t) \cdot F(z+s;t)\bar{F}(s+y;t)dzds = 0 \quad (2.213)$$

For $\bar{K}_2(x,y;t)$, there is also a similar equation. An approximate solution of (2.213) can be found with the following form:

$$K_1(X,Y;t) = \frac{1}{4\sqrt{\pi t}} f(X,Y) \cdot \exp\left[\frac{i}{16}(X+Y)^2 t - \frac{i\pi}{4}\right] + \cdots \tag{2.214}$$

where $X = \frac{x}{t}$, $Y = \frac{y}{t}$. Substitute (2.212), (2.214) into (2.213), and calculate the integral at the steady point, we get

$$f(X,Y) = \frac{\dfrac{\bar{b}}{\bar{a}}\left(-\dfrac{X+Y}{8}\right)}{1 - \alpha\dfrac{\bar{b}}{\bar{a}}\left(-\dfrac{X+Y}{8}\right)\dfrac{b}{a}\left(-\dfrac{X+Y}{8}\right)} \tag{2.215}$$

where

$$
\alpha = \begin{cases} \dfrac{1}{2}, & \text{when } X \neq Y \\[2mm] \dfrac{1}{4}, & \text{when } X = Y \end{cases}
$$

Since $q(x,t) = 2K_1(x,x;t)$, hence

$$
q(x,t) \sim \frac{1}{2\sqrt{\pi t}} \frac{\dfrac{\bar{b}}{\bar{a}}\left(-\dfrac{x}{4t}\right)}{1 - \dfrac{1}{4}\dfrac{b}{a}\left(-\dfrac{x}{4t}\right)\dfrac{\bar{b}}{\bar{a}}\left(-\dfrac{x}{4t}\right)} \cdot \exp\left[\frac{i}{4}\left(\frac{x}{t}\right)^2 t - \frac{i\pi}{4}\right] \tag{2.216}
$$

Similarly, we have

$$
r(x,t) \sim \frac{1}{2\sqrt{\pi t}} \frac{\dfrac{b}{a}\left(-\dfrac{x}{4t}\right)}{1 - \dfrac{1}{4}\dfrac{b}{a}\left(-\dfrac{x}{4t}\right)\dfrac{\bar{b}}{\bar{a}}\left(-\dfrac{x}{4t}\right)} \cdot \exp\left[-\frac{i}{4}\left(\frac{x}{t}\right)^2 t + \frac{i\pi}{4}\right] \tag{2.217}
$$

The condition (2.205) guarantees that the denominator in (2.215) is not zero. In the asymptotic approximate solutions of the KdV equation, the approximations corresponding to (2.216), (2.217) fail to hold uniformly. It is necessary for it to seek similar solutions. For example, for KdV equation

$$
u_t + 6uu_x + u_{xxx} = 0 \tag{2.218}
$$

its asymptotic approximate solution can be found as

$$
u(x,t) \sim \frac{r_0\left((i/2)\sqrt{x/3t}\right)\left(\dfrac{x}{3t}\right)^{\frac{1}{4}} e^{-2\left(\frac{x}{3t}\right)^{\frac{3}{2}}t}}{2\sqrt{3\pi t}}\left[1 + O\left(\frac{1}{t}\right)\right] \tag{2.219}
$$

where $r_0(k)$ is the initial reflection coefficient. The similar solution of (2.218) is

$$
u = \frac{1}{(3t)^{2/3}}\left[f(\eta) - \frac{1}{(3t)^{1/3}}f_1(\eta) + \frac{1}{(3t)^{2/3}}f_2(\eta) + \cdots\right] \tag{2.220}
$$

where $f(\eta)$ satisfies the nonlinear equation

$$
f''' + 6ff' - (2f + \eta f') = 0 \tag{2.221}
$$

All other $f_k(\eta)$ satisfy linear equations, $\eta = \dfrac{x}{(3t)^{1/3}} = O(1)$. If $|r_0(0)| > 1$, $f(\eta)$ has second-order poles and becomes unbounded at finite positions. If $|r_0(0)| < 1$, $f(\eta)$ is oscillating when $\eta \to -\infty$ and has the form

$$
f(\eta) = 2d(-\eta)^{1/4}\cos\theta - 2d^2(-\eta)^{-\frac{1}{2}}(1 - \cos 2\theta) + O\left((\eta)^{-\frac{5}{4}}\right) \tag{2.222}
$$

where

$$\theta = \frac{2}{3}(-\eta)^{3/2} - 3d^2 \ln(-\eta) + \theta_0 + O\left((-\eta)^{-\frac{3}{2}}\right)$$

both d and θ_0 are constants that depend on $r_0(0)$. If $|r(0)| = 1$, then $f(\eta)$ has the asymptotic expression $(\eta \to -\infty)$

$$f(\eta) = \frac{1}{2}\eta - \frac{1}{2}(-2\eta)^{-\frac{1}{2}} + \frac{1}{2}(-2\eta)^{-2} - \frac{5}{2}(-2\eta)^{-\frac{7}{2}} + O\left((-2\eta)^{-5}\right) \qquad (2.223)$$

In our problem, (2.216), (2.217) hold uniformly, but we still expect a similar solution as an asymptotic representation of the solution, which has the form

$$q(x,t) = Q_0 t^{-\frac{1}{2}} \exp\left(\frac{i}{4}\frac{x^2}{t} + 2iQ_0 R_0 \log t\right) \qquad (2.224)$$

$$r(x,t) = R_0 t^{-\frac{1}{2}} \exp\left(-\frac{i}{4}\frac{x^2}{t} - 2iQ_0 R_0 \log t\right) \qquad (2.225)$$

where Q_0, R_0 are constants. When Q_0, R_0 are real numbers, (2.224), (2.225) are consistent with (2.216), (2.217). In fact, if $|\mathrm{Im}\,(Q_0, R_0)| > \frac{1}{4}$, then r or q is boundless growth when $t \to \infty$. This behaviour reflects some instability of the evolution equation (2.210). In regions where the spatial curvature (q_{xx}, r_{xx}) is small, (2.210) can be approximated by a simple system of equations

$$\begin{cases} iq_t - 2(qr)q = 0 \\ ir_t + 2(qr)r = 0 \end{cases} \qquad (2.226)$$

When qr is a constant and $\mathrm{Im}(qr) \neq 0$, then q or r grows exponentially, but the condition (2.205) guarantees that this instability does not occur and that the solution has good behaviour.

Therefore, if the initial conditions satisfy the conditions (2.204), (2.205), then the solution of the nonlinear evolution equation (2.210) can be approximated by (2.216), (2.217). If the initial condition is "small", for example,

$$R(\infty)Q(\infty) = \int_{-\infty}^{\infty} |r|dx \int_{-\infty}^{\infty} |q|dx \ll 1$$

and the nonlinear term of the evolution equation is not important, we expect its solution to be approximated by the solution of the linearization problem.

2.6.3 Estimation of discrete spectrum

The important feature of AKNS nonlinear evolution equations is that they can be solved by inverse scattering method. Their solutions have simple asymptotic properties (when $t \to \infty$), the continuous spectrum decays algebraically, and the characteristics of the asymptotic solutions are determined by the discrete spectrum of the scattering problem at the initial time. In this section, we will review some estimates of discrete eigenvalues of (2.36), that is, estimates of the zero positions of $a(\zeta)$, $\bar{a}(\zeta)$.

We know that if $r(x)$ and $q(x)$ are correlated, then the zero of $\bar{a}(\zeta)$ can be transformed into the zero of $a(\zeta)$. If $r(x)$, $q(x)$ are independent, then the zero of $\bar{a}(\zeta)$ must be recalculated.

Now let

$$\begin{cases} \displaystyle\int_{-\infty}^{\infty} |x|^n |q(x)| dx < \infty \\ \displaystyle\int_{-\infty}^{\infty} |x|^n |r(x)| dx < \infty \end{cases} \quad \forall n \tag{2.227}$$

then $a(\zeta)$ and $\bar{a}(\zeta)$ are analytic on the full plane of ζ (including the real axis, $\mathrm{Im}\,\zeta = 0$). Introduce notations

$$R = \int_{-\infty}^{\infty} |r| dx, \quad Q = \int_{-\infty}^{\infty} |q| dx \tag{2.228}$$

We make the following analysis:

1) At $\mathrm{Im}(\zeta) \geqslant 0$, $a(\zeta)$ has a finite number of zeros. As noted earlier, the condition (2.227) guarantees that $a(\zeta)$ is analytic on $\mathrm{Im}\,\zeta \geqslant 0$, and $a(\zeta) \to 1$ ($|\zeta| \to \infty$). Thus the zeros of $a(\zeta)$ are isolated and within a bounded region, and $a(\zeta)$ has at most a finite number of zeros.

2) $a(\zeta)$ has zero at $\mathrm{Im}\,\zeta = 0$, where there is no square-integrable eigenfunction and no soliton solution.

3) Let N represent the number of zeros of $a(\zeta)$ on $\mathrm{Im}\,\zeta > 0$, including non-simple multiple zeros. Let $|\zeta_0|$ represent the radius of a circle containing all zeros of $a(\zeta)$. Taking $\xi_+ > |\zeta_0|$, $\xi_- < -|\zeta_0|$, when $|\xi_\pm| \to \infty$, we have

$$\frac{1}{2\pi} \{\arg(a(\xi_+)) - \arg(a(\xi_-))\} \to N \tag{2.229}$$

4) As mentioned earlier, if r is proportional to q or q^*, then the zero of $\bar{a}(\zeta)$ is paired with the zero of $a(\zeta)$. If $r(x)$ and $q(x)$ are both real, then the zeros of $a(\zeta)$ are themselves paired. The particular solution of the evolution equation corresponding to this pair of eigenvalues is called the "breather", 0π wave, etc., which has special characteristics different from the usual soliton.

5) If $r = +q^*(x)$, then the eigenvalue problem (2.36) is self-conjugate. Therefore, there is no eigenvalue when $\mathrm{Im}\,\zeta > 0$.

6) For arbitrary r, q, if

$$RQ = \int_{-\infty}^{\infty} |r| dx \int_{-\infty}^{\infty} |q| dx < 0.817 \tag{2.230}$$

or more precisely, if

$$I_0(2\sqrt{RQ}) < 2 \tag{2.231}$$

then $a(\zeta)$ has no zero on $\operatorname{Im} \zeta \geqslant 0$.

To prove this, we derive $\operatorname{Im} \zeta \geqslant 0$ from (2.231), having

$$|a(\zeta) - 1| < 1 \tag{2.232}$$

it is proven from it. In fact, from (2.110), (2.111), (2.115) and

$$\lim_{x \to \infty} \varphi_1(x) e^{i\zeta x} = a(\zeta)$$

we have

$$a(\zeta) - 1 = \int_{-\infty}^{\infty} r(z) \int_{z}^{\infty} q(y) e^{2i\zeta(y-z)} dy \left(\varphi_1 e^{i\zeta z} \right) dz$$

$$|a(\zeta) - 1| \leqslant \int_{-\infty}^{\infty} |r(z)| \int_{z}^{\infty} |q(y)| dy \left| \varphi_1 e^{i\zeta z} \right| dz \leqslant I_0(2\sqrt{RQ}) - 1 \tag{2.233}$$

Therefore, just

$$I_0(2\sqrt{RQ}) < 2$$

will give (2.232).

7) In order to ensure that the asymptotic method can be used, it is not only required to satisfy (2.230) or (2.231), but also to satisfy the condition

$$\left| \frac{b\bar{b}}{a\bar{a}}(\xi) \right| < 2 \ (\xi \text{ is real}) \tag{2.234}$$

thus deriving

$$RQ < 0.383$$

In fact, since $a\bar{a} + b\bar{b} = 1$, (2.234) can be written as

$$|1 - a\bar{a}(\xi)| < 2|a\bar{a}(\xi)|$$

If we set $a\bar{a}(\xi) = \alpha + i\beta$, there is

$$\left(\alpha + \frac{1}{2}\right)^2 + \beta^2 > \left(\frac{2}{3}\right)^2$$

Therefore, request $|a\bar{a}| > \frac{1}{3}$, and $|a| > \frac{1}{\sqrt{3}}$, $|\bar{a}| > \frac{1}{\sqrt{3}}$.

Finally, we require that

$$|a(\xi) - 1| < 1 - \frac{1}{\sqrt{3}}, \quad |\bar{a}(\xi) - 1| < 1 - \frac{1}{\sqrt{3}}$$

and this inequality is satisfied only if

$$I_0(2\sqrt{RQ}) < 2 - \frac{1}{\sqrt{3}}$$

that is,

$$RQ < 0.383$$

8) Based on the estimation of the maximum eigenvalue ζ_0 of the module, we require some additional smoothness for $q(x)$, $r(x)$ to obtain three upper bounds for ζ.

a) Let $q(x)$ be continuously differentiable with respect to x, and take

$$q_m' \equiv \max_x |q'(x)|, \quad A = \int_{-\infty}^{\infty} |qr|dx, \quad B = I_0(2\sqrt{RQ}) \tag{2.235}$$

If

$$|\zeta| > \frac{B}{4}\left[A + \left\{A^2 + \frac{4Rq_m'}{B}\right\}^{\frac{1}{2}}\right] = \zeta_0 \tag{2.236}$$

then $a(\zeta) \neq 0$. Therefore, all eigenvalues must lie inside the circle: $|\zeta| \leqslant \zeta_0$, where ζ_0 is determined by (2.236).

It is proved that $|a(\zeta) - 1| < 1$ can be derived from (2.236). In fact, from (2.233), there is

$$I = \int_z^{\infty} q(y)e^{2i\zeta(y-z)}dy = \int_0^{\infty} q(z+p)e^{2i\zeta p}dp$$

$$= q(z)\int_0^{\infty} e^{2i\zeta p}dp + \int_0^{\infty} q'(z+m)pe^{2i\zeta p}dp$$

where $0 \leqslant m < p$.

$$|I| \leqslant \frac{|q(z)|}{2|\zeta|} + \frac{q_m'}{4\eta^2} \tag{2.237}$$

where $\eta = \text{Im}\,\zeta$. Substitute (2.115), (2.237) into (2.233), and let

$$I_0(2\sqrt{RQ})\left\{\frac{1}{2|\zeta|}\int_{-\infty}^{\infty}|rq|dx + \frac{1}{4\eta^2}Rq_m'\right\} < 1$$

Since $|\zeta|^2 \geqslant \eta^2$, from (2.235), the above formula is true only if

$$|\zeta|^2 > B\left\{\frac{A|\zeta|}{2} + \frac{Rq_m'}{4}\right\}$$

which is (2.236).

b) If $q(x) \in c^2$, then a better bound than (2.236) can be obtained. Let $q_m'' = \max\limits_{x}|q''(x)|$ and assume ζ_1 satisfies

$$I_0(2\sqrt{RQ})\left\{\frac{1}{2|\zeta_1|}\int_{-\infty}^{\infty}|rq|dx + \frac{1}{4|\zeta_1|^2}\int_{-\infty}^{\infty}|rq'|\,dx + \frac{1}{8\eta_1^3}Rq_m''\right\} < 1 \qquad (2.238)$$

then $a(\zeta) \neq 0$ (when $|\zeta| > |\zeta_1|$).

c) We note that the quantity obtained from (2.237), (2.238) is related to the conserved quantity of the polynomial integral, so another estimate of $|\zeta_0|$ can be obtained directly by using the conservation law. If

$$\sum \frac{|c_n|}{|2\zeta_0|^n} < \infty \qquad (2.239)$$

is satisfied for some $\zeta_0 > 0$, then, $a(\zeta) \neq 0$ when $|\zeta| > \zeta_0$.

2.7 Mathematical Theory Basis of Inverse Scattering Method

In the preceding sections, we have introduced the general process of solving problems by inverse scattering method, which is only a process of finding formal solutions. Many of these steps are subject to mathematical rigour, such as: The existence of the solution of eigenvalue problem of Schrödinger equation in one-dimensional quantum mechanics; Can the potential $q(x)$ be uniquely determined from the bounded state and reflection coefficient? What conditions should the scattering matrix satisfy so that the potential $q(x) \in L_2^1$ (we introduce the notation $L_2^1: \{p(x): \int_{-\infty}^{\infty}|p(x)|(1+x^2)dx < \infty\}$)? Under what conditions does the Gelp1fand-Marqenko integral equation have a unique solution? and so on. In particular, if we are to find solutions for the theoretical study of differential equations from the inverse scattering method (which is a good new way), we must also carefully test and prove the differentiability of reflection coefficient and the solution of integral equation, the differentiability of the function constructed by the inverse scattering method, and whether it satisfies the differential equation. All these questions need to be answered in mathematical theory, which we call "mathematical theoretical basis of inverse scattering method". In this section, we mainly introduce the relevant important results, and the general methods in the proof are accompanied by a few proofs, for details, see [12].

Lemma 2.1   *For each k, $\operatorname{Im} k \geqslant 0$, integral equation*

$$m(x,k) = 1 + \int_{-\infty}^{\infty} D_k(t-x)q(t)m(t,k)dt$$

has a solution $m(x,k)$, which is the unique solution of Schrödinger equation

$$m'' + 2ikm' = q(x)m$$

with boundary condition $m(x,k) \to 1$ $(x \to +\infty)$. Here,

$$D_k(y) \equiv \int_0^y e^{2ikt}dt = \frac{1}{2ik}\left(e^{2iky} - 1\right)$$

$m(x,k)$ satisfies condition $m(x,k) = m(x,-k)$ and has estimates:

(i) $|m(x,k) - 1| \leqslant e^{\eta(x)/|k|} \cdot \dfrac{\eta(x)}{|k|} \leqslant e^{\mathrm{const}/|k|} \cdot \dfrac{\mathrm{const}}{|k|}$, $k \neq 0$.

(ii) $|m(x,k) - 1| \leqslant K \dfrac{\left(1 + \max(-x,0)\displaystyle\int_x^{\infty}(1+|t|)\,|q(t)|dt\right)}{1+|k|}$

$\qquad\qquad \leqslant K_1 \dfrac{(1 + \max(-x,0))}{1+|k|}$ $\qquad\qquad\qquad\qquad$ (2.240)

(iii) $|m'(x,k)| = \left|\dfrac{dm(x,k)}{dx}\right| \leqslant K_2 \dfrac{\displaystyle\int_x^{\infty}(1+|t|)|q(t)|dt}{1+|k|}$

$\qquad\qquad\qquad\qquad \leqslant \dfrac{K_3}{1+|k|}$, $\quad -\infty < x < \infty$

(iv) $|m'(x,k)| \leqslant K_4 \dfrac{\displaystyle\int_x^{\infty}|q(t)|dt}{1+|k|}$, $\quad 0 \leqslant x < \infty$,

where $\eta(x) = \displaystyle\int_x^{\infty}|q(t)|dt$, the constants K and k_j depend only on modules

$$\int_{-\infty}^{\infty}\left(1+|x|^j\right)|q(x)|dx$$

$j = 0$, 1, or 2. For each x, $m(x,k)$ is analytic in $\operatorname{Im} k > 0$ and is continuous on $\operatorname{Im} k \geqslant 0$. In particular, by (ii), $m(x,k) - 1 \in H^{2+}$, where H^{2+} denotes the Hardy space of $h(k)$, $H^{2+} = \{h(k) \in L^2(-\infty, \infty), \operatorname{supp} \hat{h} \in (-\infty, 0)\}$,

$$\hat{h} = \frac{1}{\pi} \int_{-\infty}^{\infty} e^{2iky} h(k) dk$$

Finally, $\dot{m}(x,k) = \dfrac{d}{dk} m(x,k)$ exists $k \neq 0$ for all $\operatorname{Im} k \geqslant 0$, $k\dot{m}(x,k)$ is continuous for all $\operatorname{Im} k \geqslant 0$, such as $q(x) \in L_2^1$, then $\dot{m}(x,k)$ also exists and is continuous at $k = 0$. We have estimates:

(v) $|\dot{m}(x,k)| \leqslant \operatorname{const}\left(1 + x^2\right)$, $\forall \operatorname{Im} k \geqslant 0$, $q \in L_2^1$.

Proof   The iteration of Volterra integral equation is always convergent, and we have

$$m(x,k) = 1 + \sum_{n=1}^{\infty} g_n(x,k)$$

where

$$g_n(x,k) = \int_{x \leqslant x_1 \leqslant \cdots \leqslant x_n} D_k\left(x_1 - x_2\right) \cdots D_k\left(x_n - x_{n-1}\right)$$
$$\cdot q(x_1) \cdots q(x_n) dx_1 \cdots dx_n$$

$$|g_n(x,k)| \leqslant \int_{x \leqslant x_1 \leqslant \cdots \leqslant x_n} \frac{1}{|k|^n} |q(x_1)| \cdots |q(x_n)| dx_1 \cdots dx_n$$

$$= \frac{1}{|k|^n} \frac{\left(\int_x^{\infty} |g(t)| dt\right)^n}{n!}$$

Here, we used the estimate $|D_k(y)| \leqslant \dfrac{1}{|k|}$; $\operatorname{Im} k \geqslant 0$, which gives (i).

In addition,

$$|g_n(x,k)| \leqslant \int_{x \leqslant x_1 < \cdots \leqslant x_n} \left(x_1 - x\right)\left(x_2 - x\right) \cdots \left(x_n - x_{n-1}\right)$$
$$\cdot |q(x_1)| \cdots |q(x_n)| dx_1 \cdots dx_n$$

$$\leqslant \int_{x \leqslant x_1 < \cdots \leqslant x_n} \left(x_1 - x\right)\left(x_2 - x\right) \cdots \left(x_n - x\right)$$
$$\cdot |q(x_1)| \cdots |q(x_n)| dx_1 \cdots dx_n$$

$$= \frac{\left(\int_x^\infty (t-x)|q(t)|dt \right)^n}{n!}$$

where we used the estimate $|D_k(y)| \leqslant y$, $k \geqslant 0$, $y \geqslant 0$, which gives

$$|m(x,k) - 1| \leqslant e^{\gamma(x)}\gamma(x)$$

Here,

$$\gamma(x) = \int_x^\infty (t-x)|q(t)|dt$$

We know that

$$|m(x,k)| \leqslant 1 + \int_x^\infty (t-x)|q(t)||m(t,k)|dt$$

$$= 1 + \int_x^\infty t|q(t)||m(t,k)|dt + \int_x^\infty (-x)|q(t)||m(t,k)|dt$$

$$\leqslant 1 + \int_0^\infty t|q(t)||m(t,k)|dt + \int_x^\infty (-x)|q(t)||m(t,k)|dt$$

It is noticed that the second inequality holds for both positive and negative x, and

$$1 + \int_0^\infty t|q(t)||m(t,k)|dt$$

$$\leqslant 1 + \left(\left(1 + e^{\gamma(0)}\gamma(0) \right) \int_0^\infty t|q(t)|dt \right) = K < \infty$$

Setting $M(x,k) = m(x,k)/K(1+|x|)$, $p(x) = (1+|x|)|q(x)| \in L^1$, we have

$$|M(x,k)| \leqslant 1 + \int_x^\infty p(t)|M(t,k)|dt$$

By solving iteratively, we can obtain

$$|M(x,k)| \leqslant \exp\left\{ \int_x^\infty (1+|t|)|q(t)|dt \right\} \leqslant K_1 < \infty$$

that is,

$$|m(x,k)| \leqslant K_2(1+|x|)$$

As we did above,

$$|m-1| \leqslant \int_0^\infty t|q(t)||m(t,k)|dt + \int_x^\infty (-x)|q(t)||m(t,k)|dt$$

$$\leqslant e^{\gamma(0)}\gamma(0) \int_0^\infty t|q(t)|dt + (-x)K_2 \int_x^\infty (1+|t|)|q(t)|dt$$

Then for $x \leqslant 0$, there is

$$|m - 1| \leqslant K_3(1 + |x|) \int_x^\infty (1 + |t|)|q(t)|dt$$

for $x \geqslant 0$,

$$|m - 1| \leqslant e^{\gamma(0)}\gamma(x) \leqslant e^{\gamma(0)} \int_x^\infty t|q(t)|dt$$

Combining (i), we get (ii). Then (iii) and (iv) are estimated in terms of

$$m'(x, k) = -\int_x^\infty e^{2ik(t-x)}q(t)m(t, k)dt$$

as (ii). Through direct calculation, it can be proved that m is the unique solution of Schrödinger equation with $m \to 1$ ($x \to +\infty$). And it can be seen from the uniform convergence of the expansion series that it is analytic at $\text{Im } k > 0$ and is continuous on $\text{Im } k \geqslant 0$.

Next, we consider the estimate of $\dot{m}(x, k)$ and have

$$\dot{m}(x, k) = \int_x^\infty D_k(t - x)q(t)\dot{m}(t, k)dt + \int_x^\infty \dot{D}_k(t - x)q(t)m(t, k)dt \qquad (2.241)$$

For $q \in L_1^1$, using the inequality

$$|k\dot{D}_k(t - x)| = \int_0^{t-x} u\left[\frac{\partial}{\partial u}e^{2iku}\right]du \leqslant 2|t - x|$$

we get

$$\left|\int_x^\infty k\dot{D}_k(t - x)q(t)m(t, k)dt\right|$$

$$\leqslant K(1 + \max(-x, 0))\int_x^\infty (t - x)|q(t)|dt \leqslant K(x) < \infty$$

We see that $\dot{m}(x, k)$ exists ($k \neq 0$, $\text{Im } k \geqslant 0$) and that $k\dot{m}(x, k)$ is continuous (even $k \to 0$). In fact, there is $\lim_{k \to 0} k\dot{m}(x, k) = 0$. For $q \in L_2^1$, by using

$$|\dot{D}_k(t - x)| \leqslant \left|\int_0^{t-x} 2iue^{2iku}du\right| \leqslant (t - x)^2$$

then we have

$$\left|\int_x^\infty \dot{D}_k(t - x)q(t)m(t, k)dt\right| \leqslant \int_x^\infty (t - x)^2|q(t)||m(t, k)|dt$$

Let $x < 0$, then

$$\int_x^\infty t^2 |q(t)||m(t,k)|dt = \int_0^\infty t^2 |q(t)||m(t,k)|dt + \int_x^0 t^2 |q(t)||m(t,k)|dt$$

$$\leqslant \int_0^\infty t^2 |q(t)||m(t,k)|dt + x^2 \int_x^0 |q(t)||m(t,k)|dt$$

$$\leqslant \text{const} \left(1 + x^2\right)$$

The last inequality above is derived from $|m(t,k)| \leqslant K(1 + \max(-t,0))$. If $x \geqslant 0$, then

$$\int_x^\infty t^2 |q(t)||m(t,k)|dt \leqslant K \int_0^\infty t^2 |q(t)|dt$$

Therefore, for all x, we have

$$\int_x^\infty t^2 |q(t)||m(t,k)|dt \leqslant K(1 - x \max(-x,0))$$

Now let $x \geqslant 0$, then

$$\int_x^\infty (t-x)^2 |q(t)||m(t,k)|dt \leqslant \int_x^\infty t^2 |q(t)||m(t,k)|dt \leqslant K$$

If $x \leqslant 0$, then

$$\int_x^\infty (t-x)^2 |q(t)||m(t,k)|dt$$

$$\leqslant 2 \int_0^\infty t^2 |q(t)||m(t,k)|dt + 2x^2 \int_x^\infty |q(t)||m(t,k)|dt$$

$$\leqslant 2K \left(1 + x^2\right) + 2x^2 K \int_{-\infty}^\infty |q(t)|(1 + |t|)dt$$

$$\leqslant K_1 \left(t + x^2\right)$$

Therefore,

$$\int_x^\infty (t-x)^2 |q(t)||m(t,k)|dt \leqslant K_2(1 - x) \max(-x,0)$$

which gives

$$\dot{m}(x,k)| \leqslant K_2(1 - x \max(-x,0)) + \int_x^\infty (t-x)|q(t)||\dot{m}(t,k)|dt$$

Iterating it, we have

$$|\dot{m}(x,k)| \leqslant K_2(1 - x\max(-x,0))e^{\gamma(x)}$$

This bound guarantees the uniform convergence (for k) of the successive approximation sequence of $\dot{m}(x,k)$ in (2.241), so $\dot{m}(x,k)$ exists and is continuous on $\operatorname{Im} k \geqslant 0$, including $k = 0$.

Finally, for any x,

$$|\dot{m}(x,k)| \leqslant K_2\left(1 + x^2\right) + \int_x^\infty t|q(t)||\dot{m}(t,k)|dt + (-x)\int_x^\infty |q(t)||\dot{m}(t,k)|dt$$

$$\leqslant K_2\left(1 + x^2\right) + \int_0^\infty t|q(t)||\dot{m}(t,k)|dt + |x|\int_x^\infty |q(t)||\dot{m}(t,k)|dt$$

$$\leqslant K_2\left(1 + x^2\right) + K_2e^{\gamma(0)}\left(1 + x^2\right)\int_0^\infty t|q(t)|dt + |x|\int_0^\infty |q(t)||\dot{m}(t,k)|dt$$

i.e.,

$$h(x,k) \leqslant 1 + \int_x^\infty \left(1 + t^2\right)|q(t)||h(t,k)|dt$$

where $h = |\dot{m}(x,k)|/K_3\left(1 + x^2\right)$. Iterating it, we have

$$|h(x,k)| \leqslant \exp\left\{\int_x^\infty \left(1 + t^2\right)|q(t)|dt\right\}$$

i.e.,

$$|\dot{m}(x,k)| \leqslant K_4\left(1 + x^2\right)$$

Thus, conclusion (v) and the lemma are proved.   □

The following proves the properties of the zeros of $m(x,k)$ on $\operatorname{Im} k \geqslant 0$.

Lemma 2.2   *For any x, $m(x,k)$ has a finite number of zeros on $\operatorname{Im} k \geqslant 0$. These zeros are simple and lie on the imaginary k axis. If $k = i\beta$ $(\beta > 0)$ is the zero of $m(x,k)$, then $k^2 = -\beta^2$ is the non-degenerate eigenvalue for operator $H \equiv -\dfrac{d^2}{dy^2} + q(y)$ in L^2 $(x < y < \infty)$ with Dirichlet boundary condition on $y = x$. For any x, there is no zero for real k (except at $k = 0$). If $m(x,0) = 0$, we say that the Dirichlet operator $-\dfrac{d^2}{dy^2} + q(y)$ has a "virtual level" on L^2 $(x < y < \infty)$; $k^2 = 0$ is not the eigenvalue of this operator.*

From Lemma 2.1, $m - 1 \in H^{2+}$. Thus, for translation transformation

$$m(x, k) = 1 + \int_0^\infty B(x, y)e^{2iky}dy$$

for each x, $B(x, y) \in L^2(0 < y < \infty)$, B has the following set of properties.

Lemma 2.3   *The integral equation*

$$B(x, y) = \int_{x+y}^\infty q(t)dt + \int_0^y dz \int_{x+y-z}^\infty dt\, q(t)B(t, z), \quad y \geqslant 0$$

has a real unique solution $B(x, y)$, *which satisfies*

$$|B(x, y)| \leqslant e^{\gamma(x)}\eta(x + y)$$

In particular, $B(x, y) \in L^1 \cap L^\infty$ $(0 < y < \infty)$, *and*

$$\|B(x, \cdot)\|_\infty \leqslant e^{\gamma(x)}\eta(x), \quad \|B(x, \cdot)\|_1 \leqslant e^{\gamma(x)}\gamma(x)$$

$B(x, y)$ *is absolutely continuous with respect to* x, y *and has*

$$\left|\frac{\partial}{\partial x}B(x, y) + q(x + y)\right| \leqslant e^{\gamma(x)}\eta(x + y)\eta(x)$$

$$\left|\frac{\partial}{\partial y}B(x, y) + q(x + y)\right| \leqslant 2e^{\gamma(x)}\eta(x + y)\eta(x)$$

$B(x, y)$ *is the solution of the wave equation*

$$\frac{\partial^2}{\partial x \partial y}B(x, y) - \frac{\partial^2}{\partial x^2}B(x, y) + q(x)B(x, y) = 0, \quad y \geqslant 0$$

with boundary condition

$$-\frac{\partial B(x, 0^+)}{\partial x} = -\frac{\partial B(x, 0^+)}{\partial y} = q(x)$$

Finally, $m(x, k) = 1 + \int_0^\infty B(x, y)e^{2iky}dy$ *is the Jost function in Lemma 2.1.*

Proof   Similar to the iterative solving of *Agranoviq, Marqenko*[16].

$$B(x, y) = \sum_{n=0}^\infty K_n(x, y), \quad K_0(x, y) = \int_{x+y}^\infty q(t)dt$$

$$K_{n+1}(x, y) = \int_0^y dz \int_{x+y-z}^\infty q(t)K_n(t, z)dt, \quad n = 0, 1, \cdots$$

We prove that

$$|K_n(x,y)| \leqslant \frac{\gamma^n(x)}{n!}\eta(x+y), \quad n \geqslant 0 \tag{2.242}$$

If (2.242) is true for n (obviously true for $n = 0$), then we have

$$|K_{n+1}(x,y)| \leqslant \int_0^y dz \int_{x+y-z}^\infty |q(t)|\eta(t+z)\frac{\gamma^n(t)}{n!}dt$$

$$\leqslant \eta(x+y) \int_0^y dz \int_{x+y-z}^\infty |q(t)|\frac{\gamma^n(t)}{n!}dt$$

$$= \eta(x+y) \left(\int_x^{x+y} |q(t)|\frac{\gamma^n(t)}{n!} \left(\int_{x+y-t}^y dz \right) dt \right.$$

$$\left. + \int_{x+y}^\infty |q(t)|\frac{\gamma^n(t)}{n!} \left(\int_0^y dz \right) dt \right)$$

$$= \eta(x+y) \left(\int_x^{x+y} |q(t)|\frac{\gamma^n(t)}{n!}(t-x)dt + \int_{x+y}^\infty |q(t)|\frac{\gamma^n(t)}{n!}ydt \right)$$

$$\leqslant \eta(x+y) \int_x^\infty |q(t)|(t-x)\frac{\gamma^n(t)}{n!}dt$$

$$\leqslant \eta(x+y) \int_x^\infty |q(t)|(t-x)\frac{\left(\int_0^\infty (u-x)|q(u)|du \right)^n}{n!}dt$$

$$= \eta(x+y)\frac{(\gamma(x))^{n+1}}{(n+1)!}$$

This completes the proof of induction. Also, we have $|B(x,y)| \leqslant e^{\gamma(x)} \cdot \eta(x+y)$. In particular, $\|B(x,\cdot)\|_\infty \leqslant e^{\gamma(x)}\eta(x)$, and

$$\|B(x,\cdot)\|_1 \leqslant e^{\gamma(x)} \int_0^\infty \eta(x+y)dy = c^{\gamma(x)}\gamma(x)$$

Obviously, B is an absolutely continuous function of x and y, and it is easy to directly verify that B is the solution of the wave equation. And there are

$$\left| \frac{\partial}{\partial x}B(x,y) + q(x+y) \right| = \left| -\int_0^y q(x+y-z)B(x+y-z,z)dz \right|$$

$$\leqslant \int_0^y |q(x+y-z)|e^{\gamma(x+y-z)}\eta(x+y)dz$$

$$\leqslant e^{\gamma(x)}\eta(x+y)\eta(x)$$

The calculation for $\dfrac{\partial B(x,y)}{\partial y}$ is similar. Finally, if we define

$$m(x,k) = 1 + \int_0^\infty B(x,y)e^{2iky}dy$$

then the above estimate shows that $m'(x,k)$ exists, and

$$
\begin{aligned}
m'(x,k) &= \int_0^\infty \left[\frac{\partial}{\partial x}B(x,y)\right]e^{2iky}dy \\
&= \int_0^\infty \left[\frac{\partial}{\partial x}B(x,y) - \frac{\partial}{\partial y}B(x,y)\right]e^{2iky}dy + \int_0^\infty \left[\frac{\partial}{\partial y}B(x,y)\right]e^{2iky}dy \\
&= -\int_0^\infty \left[\int_x^\infty q(t)B(t,y)dt\right]e^{2iky}dy - B(x,0) - 2ik\int_0^\infty B(x,y)e^{2iky}dy
\end{aligned}
$$

It follows that $m''(x,k)$ exists almost everywhere and has

$$m'' + 2ikm' = qm$$

while $|m(x,k) - 1| \leqslant \|B(x,\cdot)\|_1 \leqslant e^{\gamma(x)}\gamma(x) \to 0 \ (x \to +\infty)$. Therefcre, m is the (unique) Jost function of Lemma 2.1. $\square$

Now let $m_1(x,k)$, $m_2(x,k)$ be the Jost function in Lemma 2.1. Taking $f_1(x,k) \equiv e^{ikx}m_1(x,k)$, $f_2(x,k) \equiv e^{-ikx}m_2(x,k)$, then $f_1(x,k)$, $f_2(x,k)$ are the solutions of Schrödinger equation

$$-f_j'' + qf_j = k^2 f_j, \quad j = 1,2$$

with $f_1 \sim e^{ikx}$ $(x \to +\infty)$, $f_2 \sim e^{-ikx}$ $(x \to -\infty)$. At this point, $f_1(x,k)$ and $f_1(x,-k)$ are two linearly independent solutions of the equation $(k \neq 0)$. In fact, Wronski determinant

$$
\begin{aligned}
&[f_1(x,k), f_1(x,-k)] \\
&\equiv f_1'(x,k)f_1(x,-k) - f_1(x,k)f_1'(x,-k) \\
&= \text{const} = \lim_{x\to+\infty}\left(e^{ikx}(ik)e^{-ikx} - e^{ikx}(-ik)e^{ikx} + o(1)\right) \\
&= 2ik \neq 0
\end{aligned}
$$

Similarly, $[f_2(x,k), f_2(x,-k)] = -2ik \neq 0$.

Thus, there exist (unique) functions $T_1(k)$, $T_2(k)$, $R_1(k)$, $R_2(k)$ (they are called penetration coefficients and reflection coefficients, respectively) such that

$$f_2(x,k) = \frac{R_1(k)}{T_1(k)}f_1(x,k) + \frac{1}{T_1(k)}f_1(x,-k)$$

$$f_1(x,k) = \frac{R_2(k)}{T_2(k)} f_2(x,k) + \frac{1}{T_2(k)} f_2(x,-k)$$

where $k \neq 0$. For m_1, m_2, there are relations

$$T_1(k)m_2(x,k) = R_1(k)e^{2ikx}m_1(x,k) + m_1(x,-k)$$

$$T_2(k)m_1(x,k) = R_2(k)e^{-2ikx}m_2(x,k) + m_2(x,-k)$$

We define the scattering matrix as

$$S(k) = \begin{pmatrix} T_1(k) & R_2(k) \\ R_1(k) & T_2(k) \end{pmatrix}, \quad k \neq 0$$

Then

$$\frac{1}{T_1(k)} = \frac{1}{2ik}[f_1(x,k), f_2(x,k)] = \frac{1}{T_2(k)}$$

$$\frac{R_1(k)}{T_1(k)} = \frac{1}{2ik}[f_2(x,k), f_1(x,k)]$$

$$\frac{R_2(k)}{T_2(k)} = \frac{1}{2ik}[f_2(x,-k), f_1(x,k)]$$

From the above we have

$$T_1(k) = T_2(k) = T(k)$$

$$R_1(k)T_2(-k) + R_2(-k)T_1(k) = 0$$

and

$$\overline{T(k)} = T(-k), \quad \overline{R_1(k)} = R_1(-k), \quad \overline{R_2(k)} = R_2(-k)$$

Substituting one of these algebraic relations into the other yields

$$|T(k)|^2 + |R_1(k)|^2 = 1 = |T(k)|^2 + |R_2(k)|^2$$

Therefore, for real $k \neq 0$, $S(k)$ is a U-matrix.

By using

$$m_1(x,k) = 1 + \int_{-\infty}^{\infty} \left(\frac{e^{2ik(t-x)} - 1}{2ik} \right) q(t)m_1(t,k)dt$$

$$= e^{-2ikx} \left(\frac{1}{2ik} \int_{-\infty}^{\infty} q(t)m_1(t,k)dt \right)$$

$$+ 1 - \frac{1}{2ik} \int_{-\infty}^{\infty} q(t)m_1(t,k)dt + o(1)$$

and

$$m_1(x,k) = \frac{R_2(k)}{T(k)} e^{-2ikx} m_2(x,k) + \frac{1}{T(k)} m_2(x,-k)$$

$$= e^{-2ikx} \frac{R_2(k)}{T(k)} + \frac{1}{T(k)} + o(1)$$

then we get the integral expressions of the scattering coefficients

$$\frac{R_2(k)}{T(k)} = \frac{1}{2ik} \int_{-\infty}^{\infty} e^{2ikx} q(t) m_1(t,k) dt$$

$$\frac{1}{T(k)} = 1 - \frac{1}{2ik} \int_{-\infty}^{\infty} q(t) m_1(t,k) dt$$

The main properties of scattering matrix S can be summarized as the following theorems.

Theorem 2.5 *Let $q(x)$ be the real potential, $q(x) \in L_1^1$, then*

$$S(k) = \begin{pmatrix} T_1(k) & R_2(k) \\ R_1(k) & T_2(k) \end{pmatrix}$$

is continuous for all $k \neq 0$ (if $q \in L_2^1$, then $S(k)$ is also continuous on $k = 0$) and has the following properties:

I *(Symmetry)* $T_1(k) = T_2(k) \equiv T(k)$.
II *(U-property)* $T(k)\overline{R_2(k)} + R_1(k)\overline{T(k)} = 0$,

$$|T(k)|^2 + |R_1(k)|^2 = 1 = |T(k)|^2 + |R_2(k)|^2$$

Thus,

$$|T(k)|, |R_j(k)| \leqslant 1, \ \ j = 1,2$$

III *(Analyticity)* $T(k)$ *is meromorphic at* $\operatorname{Im} k > 0$ *and has a finite number of simple zeros* $i\beta_1, \cdots, i\beta_n$, $\beta_j > 0$. *On the imaginary axis, its residue is*

$$i \left(\int_{-\infty}^{\infty} f_1(x, i\beta_j) f_2(x, i\beta_j) dx \right)^{-1}, \ \ j = 1,2,\cdots,n$$

$-\beta_1^2, \cdots, -\beta_n^2$ *are simple eigenvalues of the operator H, and $T(k)$ is continuous on* $\operatorname{Im} k \geqslant 0$, $k \neq 0, i\beta_1, \cdots, i\beta_n$ *(if $q \in L_2^1$, then $T(k)$ is continuous on* $\operatorname{Im} k \geqslant 0$, $k \neq i\beta_1, \cdots, i\beta_n$).

IV *(Asymptotics)*

(i) $T(k) = 1 + o\left(\dfrac{1}{k}\right)$, *when* $|k| \to \infty$, $\operatorname{Im} k \geqslant 0$.

(ii) $R_j(k) = o\left(\dfrac{1}{k}\right)$, $j = 1, 2$, when $|k| \to \infty$, k is real. Further, if $q(x)$ has Nth derivative and belongs to $L^1(-\infty < x < \infty)$, then $R_j(k) = o\left(1/k^{N+1}\right)$ when $|k| \to \infty$, k is real.

(iii) If H has no eigenvalue, then
$$T(k) - 1 \in H^{2^+}, \quad |T(k)| \leqslant 1, \text{ almost everywhere for } \operatorname{Im} k \geqslant 0$$

V (The estimate at $k = 0$) $|T(k)| > 0$ for all $\operatorname{Im} k \geqslant 0$, $k \neq 0$, $|k| \leqslant \text{const} |T(k)|$ when $k \to 0$. If $q \in L_2^1$, there are two possibilities:

(i) $0 < \text{const} \leqslant |T(k)|$, so $|R_j(k)| \leqslant \text{const} < 1$, $j = 1, 2$, or

(ii) $T(k) = ak + o(k)$, $a \neq 0$, when $k \to 0$, $\operatorname{Im} k \geqslant 0$.

$1 + R_j(k) = a_j k + o(k)$, $j = 1, 2$, when $k \to 0$, k is real.

VI (Realness)
$$\overline{T(k)} = T(-k), \quad \overline{R_j(k)} = R_j(-k), \quad j = 1, 2.$$

Theorem 2.6

$$\frac{R(k)}{T(k)} = \frac{1}{2ik} \int_{-\infty}^{\infty} e^{-2ikt} \pi_1(t) dt$$

$$\frac{1}{T} = 1 - \frac{1}{2ik} \int_{-\infty}^{\infty} q(t) dt - \frac{1}{2ik} \int_{0}^{\infty} \pi_2(t) e^{2ikt} dt$$

where

$$|\pi_1(y)| \leqslant |q(y)| + kL(y) \in L^1(-\infty < y < \infty)$$

$$|\pi_2(y)| \leqslant K \left(\int_{y/2}^{\infty} |q(t)| dt + \int_{-\infty}^{y/2} |q(t)| dt \right) \in L^1(0 < y < \infty)$$

$$L(y) = \int_{y}^{\infty} |q(t)| dt, \; y \geqslant 0$$

$$L(y) = \int_{-\infty}^{y} |q(t)| dt, \; y < 0$$

For the asymptotics of $T(k)$, $m_1(x, k)$, $m_2(x, k)$, there is the following result.

Theorem 2.7 (i) If $q \in L_1^1$, then

$$m_1(x, k) = 1 + \frac{1}{2ik} \int_{x}^{\infty} \left(e^{2ik(t-x)} - 1 \right) q(t) dt$$

$$+ \frac{1}{2(2ik)^2} \left(\int_{x}^{\infty} q(t) dt \right)^2 + o\left(\frac{1}{k^2}\right)$$

$$m_2(x, k) = 1 + \frac{1}{2ik} \int_{-\infty}^{x} \left(e^{2ik(x-t)} - 1 \right) q(t) dt$$

$$+ \frac{1}{2(2ik)^2} \left(\int_x^\infty q(t)dt \right)^2 + o\left(\frac{1}{k^2} \right)$$

$$T(k) = 1 + \frac{1}{2ik} \int_{-\infty}^\infty q(t)dt + \frac{1}{2(2ik)^2} \left(\int_{-\infty}^\infty q(t)dt \right)^2 + o\left(\frac{1}{k^2} \right)$$

(ii) *If $q \in L_1^1$, $q' \in L^1$, then*

$$m_1(x,k) = 1 - \frac{1}{2ik} \int_x^\infty q(t)dt + \frac{1}{2(2ik)^2}$$

$$\cdot \left(\int_x^\infty q(t)dt \right)^2 - \frac{q(x)}{(2ik)^2} + o\left(\frac{1}{k^2} \right)$$

$$m_2(x,k) = 1 - \frac{1}{2ik} \int_{-\infty}^x q(t)dt + \frac{1}{2(2ik)^2}$$

$$\cdot \left(\int_{-\infty}^x q(t)dt \right)^2 - \frac{q(x)}{(2ik)^2} + o\left(\frac{1}{k^2} \right)$$

The proof is straightforward.

For the potential $q(x)$, we have the following important expressions and estimates of its relation to the scattering data.

Theorem2.8   *Let $q \in L_1^1$ have the bounded state $-\beta_n^2 < \cdots < -\beta_1^2$, gauge constants c_j, $j = 1, 2, \cdots, n$, and reflection coefficient R, then*

(i) $$q(x) = \lim_{a \to \infty} \frac{2i}{\pi} \int_{-a}^a kR(k)e^{2ikx}m^2(x,k)dk$$

$$+ \sum_{j=1}^n (2c_j \exp\{-2\beta_j x\})' \, m^2(x, i\beta_j)$$

$$= \lim_{b \to +\infty} \frac{1}{b} \int_0^b da \left(\frac{2i}{\pi} \int_{-a}^a kR(k)e^{2ikx}m^2(x,k)dk \right)$$

$$+ \sum_{j=1}^n (2c_j \exp\{-i\beta_j x\})' \, m^2(x, i\beta_j)$$

where the convergence is convergence almost everywhere in the sense of Césaro mean.

(ii)

$$q(x) = F'(x) + 2 \int_0^\infty F'(x+t)B(x,t)dt$$

$$+ \int_0^\infty F'(x+t)\,(B_x * B_x)\,(t)dt$$

$$+ \sum_{j=1}^{n} (2c_j \exp\{-2\beta_j x\})' \cdot m_1^2(x, i\beta_j)$$

$$= \Omega'(x) + 2 \int_0^\infty \Omega'(x+t)B(x,t)dt$$

$$+ \int_0^\infty \Omega'(x+t)(B_x * B_x)(t)dt$$

$$\Omega(t) = F(t) + \sum_{j=1}^{n} 2c_j \exp\{-2\beta_j t\}$$

(iii) *If* $kR(k) \in L^1$, *then*

$$q(x) = \frac{2i}{\pi} \int_{-\infty}^\infty kR(k)e^{2ikx}m^2(x,k)dk$$

$$+ \sum_{j=1}^{n} (2c_j \exp\{-2\beta_j x\})' \cdot m^2(x, i\beta_j)$$

Theorem 2.9   *Let* $q(x) \in L_1^1$, *and*

$$\Omega(y) = F(y) + \sum_{j=1}^{n} 2c_j \exp\{-2\beta_j y\}$$

then $\Omega(y)$, $F(y)$ *are absolutely continuous functions and*

$$|q(x) - \Omega'(x)| \leqslant K_1(x) \left(\int_x^\infty |q(t)|dt \right)^2$$

Here, $K_1(x)$ *is a non-increasing function and*

$$\int_a^\infty |F(t)|(1+|t|)dt \leqslant K_2(a) < \infty$$

$$\int_a^\infty |F(t)|dt \leqslant K_3(a) < \infty$$

Theorem 2.10   *The necessary and sufficient condition for matrix*

$$\begin{pmatrix} T_1(k) & R_2(k) \\ R_1(k) & T_2(k) \end{pmatrix}, \quad -\infty < k < \infty$$

to be a scattering matrix of the real potential $q(x) \in L_2^1$ *(without bounded state) is that* T_1, T_2, R_1, R_2 *satisfy the following conditions:*

(i) (*Realness*) $T_1(k) = T_2(k) = T(k)$.

(ii) (*U-property*)

$$|T(k)|^2 + |R_1(k)|^2 = |T(k)|^2 + |R_2(k)|^2 = 1$$

$$R_1(k)\overline{T(k)} + \overline{R_2(k)}T(k) = 0$$

(iii) (*Analyticity*) $T(k)$ *is analytic in the upper half plane and continuous to the axis.*

(iv) (*Asymptotics*) $T(k) = 1 + o(1/|k|)$, $\operatorname{Im} k \geqslant 0$,

$$R_i(k) = o\left(\frac{1}{k}\right), \ k \ is \ real, \ i = 1,2$$

(v) (*The estimate at* $k = 0$) $|T(k)| > 0$, $\operatorname{Im} k \geqslant 0$, $k \neq 0$, *and*

(1) $0 < c < |T(k)|$, *for all* $\operatorname{Im} k \geqslant 0$, *or*

(2) $T(k) = T(0)k + o(k)$, $T(0) \neq 0$, $\operatorname{Im} k \geqslant 0$,

$$1 + R_1(k) = \rho_i k + o(k), \ i = 1,2, \ k \ is \ real$$

(vi) (*Realness*) $T_j(k) = \overline{T_j(-k)}$, $R_j(k) = \overline{R_j(-k)}$, $j = 1,2$.

(vii) $F_j(y) = \dfrac{1}{\pi} \displaystyle\int_{-\infty}^{\infty} R_j(k)e^{2iky}dk$, $j = 1,2$ *is absolutely continuous, and*

$$\int_a^{\infty} |F_1'(t)| \left(1 + t^2\right) dt < \infty$$

$$\int_{-\infty}^a |F_2'(t)| \left(1 + t^2\right) dt \leqslant c(a) < \infty, \ -\infty < a < \infty$$

For KdV equation, we construct its solution by inverse scattering method, which leads to the decay of the initial function when $|x| \to \infty$ and the decay rate and smoothness of its corresponding solution.

For the initial value problem of KdV equation

$$u_t - 6uu_x + u_{xxx} = 0$$

$$u|_{t=0} = U(x)$$

We assume that the initial function $U(x)$ satisfies

(i) $U(x) \in c^s(R)$, $s > 3$.

(ii) $U^{(s+1)}(x)$ is piecewise continuous.

(iii) To a certain $N > 0$, $U^{(j)}(x) = o\left(|x|^{-N}\right)$, $\forall \ j \leqslant s + 1$.

We solve for $B_\pm(x, y, t)$ by Marqenko equation

$$B_\pm(x, y, t) \pm \int_0^{\pm\infty} \Omega_\pm(x + y + z, t)B_\pm(x, z, t)dz + \Omega_\pm(x + y, t) = 0 \qquad (2.243)$$

where

$$\Omega_\pm(x,t) = F_\pm(x,t) + 2\sum_{j=0} c_j(t)e^{\pm 2\eta_j x}$$

$$F_\pm(x,t) = \frac{1}{\pi}\int_{-\infty}^{\infty} r_\pm(\xi,t) = e^{\pm 2i\xi t} d\xi$$

Thus, let

$$u(x,t) = -B_+^{(1,0,0)}(x,0,t) = B_-^{(1,0,0)}(x,0,t)$$

it is the solution to the initial value problem of KdV equation with

$$B^{(j,k,l)}(x,y,t) = \partial_x^j \partial_y^k \partial_t^l B = \left(\frac{\partial}{\partial x}\right)^j \left(\frac{\partial}{\partial y}\right)^k \left(\frac{\partial}{\partial t}\right)^l B$$

We resort to the following results of scholars *Faddeev*[8] and *Agranoviq, Marqenko*[16] to establish the existence and smoothness of the solution to the initial value problem.

Theorem 2.11 (Agranoviq Marqenko) *For fixed x, t, let the condition*

$$\pm\int_x^{+\infty} (1+|s|)\left|\Omega_\pm^{(1,0)}(s,t)\right| ds < \infty, \quad \forall x \in R \qquad (2.244)$$

be satisfied, then the corresponding equation (2.243) has solution $B_\pm(x,y,t)$, and

$$\pm\int_x^{\infty} (1+|x|)\left|B_\pm^{(1,0,0)}(x,0,t)\right| dx < \infty, \quad \forall x \in R$$

Theorem 2.12 (Faddeev) *For fixed t, we assume that*

$$\pm\int_x^{+\infty} (1+s^2)\,|\Omega_\pm(s,t)|\, ds < \infty, \quad \forall x \in R \qquad (2.245)$$

be satisfied and conditions

$$\text{If } \lim_{\xi\to 0} \xi a_+(\xi) \neq 0, \text{ then } r_+(\xi) = -1 + A\xi + o(\xi) \qquad (2.246)$$

$$t(\xi) = \alpha\xi + o(\xi), \quad \alpha \neq 0, \quad \xi \to 0$$

be satisfied. Thus, the solutions of (2.243) satisfy

$$-B_+^{(1,0,0)}(x,0,t) = B_-^{(1,0,0)}(x,0,t)$$

Further, if the function u is defined as

$$u(x,t) = -B_+^{(1,0,0)}(x,0,t) = B_-^{(1,0,0)}(x,0,t)$$

then the Schrödinger equation

$$L_u \psi \equiv \psi_{xx} + u(x,t)\psi = \zeta^2 \psi$$

with potential $u(x,t)$ has scattering data (2.70).

Using the integral expressions of the reflection coefficients $r_\pm(\xi)$ with respect to the initial function $U(x)$, we can test that all the conditions in Theorem 2.11 and Theorem 2.12 are true, from which we obtain the solution $u(x,t)$ of the initial value problem of the KdV equation. And it satisfies

$$\int_{-\infty}^{\infty} \left(1 + |x|^2\right) |u(x,t)| dx < \infty, \quad \forall\, t \in R$$

We have the following theorem

Theorem 2.13   (a) *If $j + 3l \leqslant 2[N] - 6 - \mu$, then the solution $u(x,t)$ to the initial value problem exists, and $u^{(j,l)}(x,t)$ exists, for $t \neq 0$.*

(b) *$u^{(j,0)}(x,t) \to U^{(j)}(x)$ $(t \to 0)$, $j = 0,1,2$.*

(c) *For $t > 0$,*

$$u^{(j,0)}(x,t) = \begin{cases} O\left(|x|^{[\frac{j}{2}]+3-[N]+\frac{\mu}{2}}\right), & x \to +\infty,\ j \leqslant 2[N] - 6 - \mu \\ O\left(|x|^{-\frac{1}{2}(4-j)-\delta}\right), & x \to -\infty,\ j \leqslant 2 \end{cases}$$

(d) *For $t < 0$,*

$$u^{(j,0)}(x,t) = \begin{cases} O\left(|x|^{-\frac{1}{2}(4-j)-\delta}\right), & x \to +\infty,\ j \leqslant 2 \\ O\left(|x|^{[\frac{j}{2}]+3-[N]+\frac{\mu}{2}}\right), & x \to -\infty,\ j \leqslant 2[N] - 6 - \mu \end{cases}$$

where $\delta = \dfrac{1}{16}$.

$$\mu = \begin{cases} 0, & \text{Jost functions } f_\pm(x,0) \text{ are linearly dependent} \\ 2, & \text{Jost functions } f_\pm(x,0) \text{ are linearly independent} \end{cases}$$

and $N > 6 + \dfrac{\mu}{2}$.

As can be seen from Theorem 2.13, if $t < 0$, then u decays quickly when $x \to +\infty$, but slowly when $x \to -\infty$.

2.8   High-Order and Multidimensional Scattering Inversion Problems

In section 2.3, we consider only the second-order scattering inversion problem. This can be written in the following matrix form:

$$V = \begin{pmatrix} V_1 \\ V_2 \end{pmatrix}$$

$$V_x = i\zeta \begin{pmatrix} -1 & 0 \\ 0 & 1 \end{pmatrix} V + \begin{pmatrix} 0 & q \\ r & 0 \end{pmatrix} V \tag{2.247}$$

$$V_t = QV, \quad Q = \begin{pmatrix} A & B \\ C & -A \end{pmatrix} \tag{2.248}$$

Now consider the high-order scattering inversion problem:

$$V = \begin{pmatrix} V_1 \\ \vdots \\ V_n \end{pmatrix}$$

$$\begin{cases} V_x = i\zeta DV + NV \\ V_t = QV \end{cases} \tag{2.249}$$

where $D = (d_i \delta_{ij})$, $N_{ii} = 0$, d_i are constants. By using $V_{xt} = V_{tx}$ and $\zeta_t = 0$, we have

$$Q_x = N_t + i\zeta(DQ - QD) + (NQ - QN) \tag{2.250}$$

or

$$Q_x = N_t + i\zeta[D, Q] + [N, Q] \tag{2.251}$$

We seek Q so that it satisfies (2.251), in which case the two equations of (2.249) are compatible, and thus we can obtain the nonlinear evolution equation. Expand Q into the following form

$$Q = Q^{(1)}\zeta + Q^{(0)} \tag{2.252}$$

Substituting the above formula into (2.251) yields

$$Q_x^{(1)}\zeta + Q_x^{(0)} = N_t + \left[N, Q^{(0)}\right] + i\zeta \left\{ \left[D, Q^{(0)}\right] \right.$$
$$\left. + \left[N, Q^{(0)}\right] \right\} + i\zeta^2 \left[D, Q^{(1)}\right]$$

Comparing the coefficient of ζ^2 yields

$$i\left[D, Q^{(1)}\right] = 0$$

or

$$\sum_k \left(D_{ik}Q_{kj}^{(1)} - Q_{ik}^{(1)}D_{kj}\right) = 0$$

From $D_{ik} = \delta_{ik}d_i$, we have

$$(d_i - d_j)\, Q_{ij}^{(1)} = 0$$

Thus,

$$Q_{ij}^{(1)} = q_i\delta_{ij} \tag{2.253}$$

Taking $q_i = \text{const}$ and comparing the coefficients of ζ, we get

$$Q_x^{(1)} = i\left[D, Q^{(0)}\right] + \left[N, Q^{(1)}\right]$$

Therefore,

$$Q_{ijx}^{(1)} = i\sum_k \left(D_{ik}Q_{kj}^{(0)} - Q_{ik}^{(0)}D_{kj}\right) + \sum_k \left(N_{ik}Q_{kj}^{(1)} - Q_{ik}^{(1)}N_{kj}\right)$$

$$i\left(d_i - d_j\right) Q_{ij}^{(0)} + (q_j - q_i)\, N_{ij} = 0$$

or

$$Q_{ij}^{(0)} = \frac{q_i - q_j}{i\,(d_i - d_j)}N_{ij} \quad (i \neq j) \tag{2.254}$$

When $i = j$, take $Q_{ij}^{(0)} = 0$.

Definition. Letting $a_{ij} = \dfrac{1}{i}\dfrac{q_i - q_j}{d_i - d_j} = a_{ji}$, we have

$$Q_{ij}^{(0)} = a_{ij}N_{ij} \quad (i \neq j) \tag{2.255}$$

Comparing the coefficients of ζ^0, from $Q_x^{(0)} = N_t + \left[N, Q^{(0)}\right]$, we obtain

$$a_{ij}N_{ijx} = N_{ijt} + \sum_k \left(N_{ik}a_{kj}N_{kj} - a_{ik}N_{ik}N_{kj}\right)$$

Therefore, we get $N(N-1)$ evolution equations

$$N_{ijt} = a_{ij}N_{ijx} + \sum_k (a_{ik} - a_{kj}) N_{ik}N_{kj} \tag{2.256}$$

or

$$N_{ijt} = a_{ij}N_{ijx} + \sum_{k \neq i,j} (a_{ik} - a_{kj}) N_{ik}N_{kj}$$

Let $N_{ij} = \sigma_{ij}N_{ji}^*$ $(i > j)$. If $\sigma_{ik}\sigma_{kj} = -\sigma_{ij}$ $(i > k > j)$ and a_{ij} is a real number, in this case, (2.256) is compatible with (2.256)*, so the number of equations can be reduced. In fact, multiplying

$$N_{ijt}^* = a_{ji}N_{jix}^* + \sum_{k \neq j,i} (a_{ij} - a_{ki}) N_{jk}^*N_{ki}^*$$

by σ_{ij} and using $a_{ij} = a_{ji}$, we get

$$N_{ijt} = a_{ij}N_{ijx} + \sum_{k \neq j,i} (a_{jk} - a_{ki}) \sigma_{ik}N_{jk}^*N_{ki}^*$$

Using $\sigma_{ij} = -\sigma_{ik}\sigma_{kj}$, we obtain

$$N_{ijt} = a_{ij}N_{ijx} + \sum_{k \neq i,j} (a_{ik} - a_{kj}) N_{ik}N_{kj}$$

which can also be rewritten as

$$N_{ijt} = a_{ij}N_{ijx} + \sum_{k>j>i} (a_{ik} - a_{kj}) N_{ik}\sigma_{kj}N_{jk}^*$$

$$+ \sum_{j>k>i} N_{ij}N_{jk} (a_{ik} - a_{kj})$$

$$+ \sum_{j>i>k} \sigma_{ik}N_{ki}^*N_{kj} (a_{ik} - a_{kj})$$

Example 1 $n = 3$

$$N = \begin{pmatrix} 0 & N_{12} & N_{13} \\ \sigma_{21}N_{12}^* & 0 & N_{23} \\ \sigma_{31}N_{13}^* & \sigma_{32}N_{23}^* & 0 \end{pmatrix} = \begin{pmatrix} 0 & A_1 & A_2 \\ \sigma_{21}A_1^* & 0 & A_3 \\ \sigma_{31}A_2^* & \sigma_{32}A_3^* & 0 \end{pmatrix}$$

Letting $a_{12} = V_1$, $a_{13} = V_2$, $a_{23} = V_3$, from (2.256), we have

$$\begin{cases} A_{1t} = V_1 A_{1x} + \sigma_{32} (V_2 - V_3) A_2 A_3^* \\ A_{2t} = V_2 A_{2x} + (V_1 - V_3) A_1 A_3 \\ A_{3t} = V_3 A_{3x} + \sigma_{21} (V_1 - V_2) A_1^* A_2 \end{cases} \tag{2.257}$$

where $-\sigma_{31} = \sigma_{21}\sigma_{32}$. (2.257) is a three-wave equation. For convenience, sometimes we set $A_1 = ia_1u_1$, $A_2 = -ia_2u_2$, $A_3 = ia_3u_3$, and

$$a_1^2 = \frac{q^2}{(V_1 - V_3)(V_1 - V_2)}, \quad a_2^2 = \frac{q^2}{(V_2 - V_3)(V_1 - V_2)}$$

$$a_3^2 = \frac{a_1q}{(V_2 - V_3)a_2}$$

then (2.256) becomes

$$\begin{cases} u_{1t} = V_1u_{1x} + \sigma_{32}iqu_2u_3^* \\ u_{2t} = V_2u_{2x} - iqu_1u_3 \\ u_{3t} = V_3u_{3x} + \sigma_{21}iqu_1^*u_2 \end{cases} \tag{2.258}$$

If we expand Q to the ζ^2, then we get

$$Q = Q^{(2)}\zeta^2 + Q^{(1)}\zeta + Q^{(0)}$$

Substituting the above formula into $Q_x = N_t + i\zeta[D, Q] + [N, Q]$ yields

$$\beta_{ij}N_{ijxx} + \varepsilon_{ij}N_{ijx} - \sum_{k \neq i,j} \gamma_{ijk}(N_{ik}N_{kj})_x$$

$$= N_{ijt} + \sum_{k \neq i,j} (\varepsilon_{kj} - \varepsilon_{ik}) \cdot N_{ik}N_{kj}$$

$$+ N_{ij}\left\{2\beta_{ij}N_{ik}N_{kj} + \sum_{k \neq i,j} (\beta_{kj} + \gamma_{ikj})N_{jk}N_{kj}\right.$$

$$\left. - (\beta_{ki} + \gamma_{kji}) \cdot N_{ik}N_{ki}\right\} + \sum_{k \neq i,j} (\beta_{kj}N_{ik}N_{kjx}$$

$$- \beta_{ik}N_{ki}N_{ikx}) + \sum_{k \neq i,j} \sum_{m \neq i,j} (\gamma_{ikm}N_{kj}N_{im}N_{mk}$$

$$- \gamma_{kjm}N_{ik}N_{km}N_{mj}) \tag{2.259}$$

where

$$a_{ij} = \frac{q_i^{(2)} - q_j^{(2)}}{i(d_i - d_j)} = a_{ji}$$

$$\beta_{ij} = \frac{d_{ij}}{i(d_i - d_j)} = -\beta_{ji}$$

$$\gamma_{ijk} = \frac{a_{kj} - a_{ik}}{i(d_i - d_j)} = \gamma_{jik} = \gamma_{kij}$$

$$\varepsilon_{ij} = \frac{q_i^{(1)} - q_j^{(1)}}{i\,(d_i - d_j)} = \varepsilon_{ji}$$

In fact, comparing the coefficients of ζ^3, ζ^2, ζ^1, ζ^0 yields

$$i\left[D, Q^{(2)}\right] = 0, \quad Q_x^{(1)} = \left[iD, Q^{(0)}\right] + \left[N, Q^{(1)}\right]$$

$$Q_x^{(2)} = i\left[D, Q^{(1)}\right] + \left[N, Q^{(2)}\right]$$

$$Q_x^{(0)} = N_t + \left[N, Q^{(0)}\right], \quad Q_{ik}^{(2)} = q_i^{(2)}\delta_{ik}, \quad D_{ik} = d_i\delta_{ik}$$

From $Q_x^{(2)} = i\left[D, Q^{(1)}\right] + \left[N, Q^{(2)}\right]$, i.e.,

$$i\sum_k \left(D_{ik}Q_{kj}^{(1)} - Q_{ik}^{(1)}D_{kj}\right) + \sum_k \left(N_{ik}Q_{kj}^{(2)} - Q_{ik}^{(2)}N_{kj}\right) = 0$$

we have

$$Q_{ij}^{(1)} = \frac{q_i^{(2)} - q_j^{(2)}}{i\,(d_i - d_j)}N_{ij} = a_{ij}N_{ij}$$

From

$$Q_x^{(1)} = i\left[D, Q^{(0)}\right] + \left[N, Q^{(1)}\right]$$

$$a_{ij}N_{ijx} = i\,(d_i - d_j)\,Q_{ij}^{(0)} + \sum_{k \neq i,j} (a_{kj} - a_{ik})\,N_{ik}N_{kj} + \left(Q_{ji}^{(1)} - Q_{ii}^{(1)}\right)N_{ij}$$

there are

$$Q_{ij}^{(0)} = \frac{a_{ij}}{i\,(d_i - d_j)}N_{ijx} + \sum_{k \neq i,j} \frac{(a_{ik} - a_{kj})}{i\,(d_i - d_j)}N_{ik}N_{kj} - \frac{Q_{ij}^{(1)} - Q_{ii}^{(1)}}{i\,(d_i - d_j)}$$

$$N_{ij} = \beta_{ij}N_{ijx} - \sum_{k \neq i,j} \gamma_{ijk}N_{ik}N_{kj} + \varepsilon_{ij}N_{ij}$$

Substituting $Q_x^{(0)} = N_t + \left[N, Q^{(0)}\right]$ into the above equations gives (2.259).

Example 2 $n = 3$, taking

$$N = \begin{pmatrix} 0 & A_1 & A_2 \\ \sigma_{21}A_1^* & 0 & 0 \\ \sigma_{31}A_2^* & 0 & 0 \end{pmatrix}$$

and $\varepsilon_{ij} = 0$, then we have

$$
\begin{cases}
iA_{1t} = A_{1xx} + 2A_1 \left(\sigma_{21} |A_1|^2 + \sigma_{31} |A_2|^2 \right) \\
iA_{2t} = A_{2xx} + 2A_2 \left(\sigma_{21} |A_1|^2 + \sigma_{31} |A_2|^2 \right)
\end{cases}
\tag{2.260}
$$

For the Boussinesq equation

$$
W_{tt} - W_{xx} - 6 \left(W^2 \right)_{xx} + W_{xxxx} = 0
\tag{2.261}
$$

it corresponds to the case of taking

$$
N = \begin{pmatrix}
0 & 0 & 1 \\
N_{21} & 0 & (1 + W_3) N_{31} \\
N_{31} & 1 & 0
\end{pmatrix}
\tag{2.262}
$$

where $W_3 = e^{-2\pi i/3}$. In this case, its eigenvalue problem is

$$
\Psi_{xxx} + (\lambda + Q_1) \Psi + Q_2 \Psi_x = 0
\tag{2.263}
$$

with $Q_1 = N_{31x} + N_{21}$, $Q_2 = (2 + W_3) N_{31}$.

Example 3   If we take

$$
N = \begin{pmatrix}
0 & A & iB \\
0 & 0 & A^* \\
-i & 0 & 0
\end{pmatrix}
\tag{2.264}
$$

then we can obtain the system of equations

$$
\begin{cases}
iA_t + \lambda A_{xx} = AB \\
B_t = -\alpha \left(|A|^2 \right)_x
\end{cases}
\tag{2.265}
$$

This problem arises from the interaction of long gravity waves and short capillary waves in shallow water.

The scattering inversion problem in higher dimensions is considered below. Here, we only consider the case involving the parameter y. In this case, the eigenvalue problem is

$$
\frac{\partial v}{\partial x}(x, y, t) = i\zeta d(y)v(x, y, t) + \int_{-\infty}^{\infty} N(x, y, z; t)v(x, z, t)dz
\tag{2.266}
$$

$$
\frac{\partial v}{\partial t}(x, y, t) = \int_{-\infty}^{\infty} Q(x, y, z; t)v(x, y, z; t)dz
\tag{2.267}
$$

From $v_{xt} = v_{tx}$ and $\zeta_t = 0$, we get

$$Q_x(x, y, z; t) = N_t(x, y, z; t) + i(d(y) - d(z))Q(x, y, z; t)$$

$$+ \int_{-\infty}^{\infty} [Q(x, z', z; t) N(x, y, z'; t)$$

$$- N(x, z', z; t) Q(x, y, z'; t)]dz' \qquad (2.268)$$

Expanding according to ζ, $Q = Q^{(1)} + \zeta Q^{(0)}$, the integral differential equation can be written as

$$N_t(x, y, z; t) = \alpha(y, z)N_x(x, y, z; t) + \int_{-\infty}^{\infty} [\alpha(y, z') - \alpha(y, z)]$$

$$\cdot N(x, y, z'; t) N(x, z', z; t) dz' \qquad (2.269)$$

where $\alpha(y, z) = [c(z) - c(y)]/i[d(z) - d(y)] = \alpha(z, y)$. When σ satisfies $\sigma(y, z') \sigma(z', z) = -\sigma(y, z)$, $y > z' > z$, then the symmetry condition $N(x, y, z; t) = \sigma(y, z)N^*(x, z, y; t)$ $(y > z)$ is satisfied. We now consider the two-dimensional case

$$\begin{cases} V_x = i\zeta DV + NV + BV_y \\ V_t = QV + CV_y \end{cases} \qquad (2.270)$$

Let $\zeta_t = 0$, and B, D, C are all constants. We have

$$V_{xt} = i\zeta D [QV + CV_y] + N_t V + N (QV + CV_y)$$

$$+ B (Q_y V + QV_y + CV_{yy})$$

$$V_{tx} = Q_x V + Q [i\zeta DV + NV + BV_y]$$

$$+ C [i\zeta DV_y + N_y V + NV_y + BV_{yy}]$$

Using $V_{xt} = V_{tx}$ and equalizing the coefficients of V, V_y, V_{yy}, we get

$$V_{yy}: \quad [C, B] = 0 \qquad (2.271)$$

$$V_y: \quad i\zeta[C, D] + [Q, B] + [C, N] = 0 \qquad (2.272)$$

$$V: \quad i\zeta[Q, D] + [Q, N] + Q_x + CN_y - BQ_y = N_t \qquad (2.273)$$

Next, we consider the simplest case

$$C = c_i \delta_{ij}, \quad B = b_i \delta_{ij}, \quad D = d_i \delta_{ij}, \quad N_{ii} = 0$$

where a_i, b_i, d_i are constants. Then, from (2.271), we have

$$\sum_k (C_{ik} B_{kj} - B_{ik} C_{kj}) = 0 = c_i b_i - c_i b_i$$

From (2.272), there is

$$i\zeta[C,D] + \sum_k (Q_{ik}B_{kj} - B_{ik}Q_{kj}) \sum_k (c_{ik}N_{kj} - N_{ik}c_{kj}) = 0$$

Therefore,

$$Q_{ij} = \frac{c_i - c_j}{b_i - b_j} N_{ij} \ (i \neq j), \quad Q_{ii} = q_i \ (\text{take } q_i \text{ as constant})$$

Defining

$$a_{ij} = \frac{c_i - c_j}{b_i - b_j} = a_{ij}$$

then we have

$$Q_{ij} = a_{ij}N_{ij} \ (i \neq j)$$

From (2.273), there is

$$i\zeta \sum_k (Q_{ik}D_{kj} - D_{ik}Q_{kj}) + \sum (Q_{ik}N_{kj} - N_{ik}Q_{kj})$$

$$+ Q_{ij,x} + \sum (C_{ik}N_{kj,y} - B_{ik}Q_{kj,y}) = N_{ij,t} \tag{2.274}$$

For $i \neq j$, we have

$$i\zeta \left[a_{ij}N_{ij} \left(d_j - d_i \right) \right] + (q_i - q_j) N_{ij} + \sum_{k \neq i,j} (a_{ik} - a_{kj}) N_{ik}N_{kj}$$

$$+ a_{ij}N_{ij,x} + c_i N_{ij,x} - b_i a_{ij}N_{ij,y} = N_{ij,t} \tag{2.275}$$

When $i = j$, the equation is naturally satisfied. We notice that (2.275) contains ζ, so we choose $q_i = q_i(\zeta)$. Since q_i is undetermined, specifically take $q_i - q_j = i\zeta a_{ij} (d_i - d_j)$. From (2.275), we can obtain

$$N_{ij,t} = a_{ij}N_{ij,x} + \beta_{ij}N_{ij,y} + \sum_{k \neq i,j} (a_{ik} - a_{kj}) N_{ik}N_{kj} \tag{2.276}$$

where

$$\beta_{ij} = c_i - b_i a_{ij} = \frac{b_j c_j - c_i b_j}{b_i - b_j} \ (\text{group velocity in the } y \text{ direction})$$

$$a_{ij} = \frac{c_i - c_j}{b_i - b_j} \ (\text{group velocity in the } x \text{ direction})$$

If $\sigma_{ij} = -\sigma_{ik}\sigma_{kj}$ $(i > k > j)$, then $N_{ij} = \sigma_{ij}N_{ji}^*$ is compatible. Here, a_{ij}, β_{ij} are real.

Example 4 Three-wave equation, $n = 3$. Taking $N_{12} = A_1$, $N_{13} = A_2$, $N_{23} = A_3$, we have

$$
\begin{cases}
A_{1t} = a_{12}A_{1x} + \beta_{12}A_{1y} + a_{32}\left(a_{13} - a_{23}\right)A_2A_3^* \\
A_{2t} = a_{13}A_{2x} + \beta_{13}A_{2y} + \left(a_{12} - a_{23}\right)A_1A_3 \\
A_{3t} = a_{23}A_{3x} + \beta_{23}A_{3y} + \sigma_{21}\left(a_{12} - a_{23}\right)A_1^*A_3
\end{cases}
\tag{2.277}
$$

In (2.270), if the time dependence of V is selected as

$$
V_t = QV + c_1V_y + c_2V_{yy}
\tag{2.278}
$$

and let B, c_1, c_2 all be diagonal constant matrices, then for

$$
N = \begin{pmatrix} 0 & A \\ \pm A^* & 0 \end{pmatrix}
\tag{2.279}
$$

there is a system of evolution equations

$$
\begin{cases}
iA_t + A_{xx} + A_{yy} + \left(Q_1 - Q_2\right)A = 0 \\
Q_{1x} + k_1Q_{1y} = \mp\left[(AA^*)_x - k_1(AA^*)_y\right] \\
Q_{2x} + k_2Q_{2y} = \pm\left[(AA^*)_x - k_2(AA^*)_y\right]
\end{cases}
\tag{2.280}
$$

where

$$
k_1 = \frac{ib_1}{\sqrt{b_1b_2}}, \quad k_2 = \frac{ib_2}{\sqrt{b_1b_2}}
$$

If it is independent of y, then (2.280) is

$$
iA_t + A_{xx} \mp 2A^2A^* = 0
$$

(Similarly, if it is independent of x, there is $iA_t + A_{yy} \pm 2A^2A^* = 0$), that is, the ordinary nonlinear Schrödinger equation is obtained. In any case, (2.280) is a high-dimensional nonlinear Schrödinger equation

$$
iA_t + \nabla_x^2 A + kA^2A^* = 0
$$

For another form of the evolution equation

$$
\begin{cases}
A_t = D_1A + WA \\
D_0W = \pm 2D_1|A|^2
\end{cases}
\tag{2.281}
$$

where

$$D_1 = \frac{(e_1 - e_2)\,\partial_x^2 + 2\,(b_1 e_2 - e_1 b_2)\,\partial_{xy} + \left(e_1 b_2^2 - b_1^2 e_2\right)\partial_y^2}{(b_2 - b_1)^2}$$

$$D_0 = -\partial_x^2 + (b_1 + b_2)\,\partial_{xy} - b_1 b_2 \partial_y^2$$

if we take $W = C|A|^2 + \hat{W}$, $C = -2\,(e_1 - e_2)\,/\,(b_1 - b_2)$, $e_1 = -e_2$, $b_1 = -b_2$, $e_1 = 2i$, $\hat{W} = -2iQ$, we have

$$\begin{cases} iA_t = \left(\dfrac{1}{b_1^2}A_{xx} + A_{yy}\right) + 2QA \mp \dfrac{2}{b_1^2}|A|^2 A^* \\[4mm] Q_{xx} - b_1^2 Q_{yy} = \mp 2\left(|A|^2\right)_{yy} \end{cases} \tag{2.282}$$

(2.282) has physical meaning for b_1 to be real or purely imaginary.

For the two-dimensional KdV equation

$$u_{xt} + 6\,(uu_x)_x + u_{xxxx} + 3b^2 u_{yy} = 0 \tag{2.283}$$

its corresponding eigenvalue problem is

$$v_{xx} + (\lambda + u)v + bv_{yy} = 0 \tag{2.284}$$

which can be obtained by taking

$$B = \begin{pmatrix} 0 & 0 \\ -b & 0 \end{pmatrix}, \quad N = \begin{pmatrix} 0 & 1 \\ -u & 0 \end{pmatrix}, \quad D = \begin{pmatrix} 1 & 0 \\ 0 & -1 \end{pmatrix}$$

from our equation.

Li Yi-Shen et al. have carried out a series of research work on the scattering inversion method that the eigenvalue λ is related to time t, which can be seen in [11],[17],[18].

References

[1] Gardner C S, Greene J M, Kruskal M D, et al. Method for solving the Korteweg-deVries equation[J]. Physical review letters, 1967, 19(19): 1095.

[2] Lax P D. Integrals of nonlinear equations of evolution and solitary waves[J]. Communications on Pure and Applied Mathematics, 1968, 21(5): 467-490.

[3] Shabat A, Zakharov V. Exact theory of two-dimensional self-focusing and one-dimensional self-modulation of waves in nonlinear media[J]. Sov. Phys. JETP, 1972, 34(1): 62.

[4] Ablowitz M J, Kaup D J. AC Newell and H. Segur[J]. Stud. Appl. Math, 1974, 53: 249.

[5] Miura R M. The Korteweg–deVries equation: a survey of results[J]. SIAM review, 1976, 18(3): 412-459.

[6] Ablowitz M J. Lectures on the inverse scattering transform[J]. Studies in Applied Mathematics, 1978, 58(1): 17-94.

[7] Calogero F, Degasperis A. Coupled nonlinear evolution equations solvable via the inverse spectral transform, and solitons that come back the boomeron[J]. Lett. Nuovo Cim.;(Italy), 1976, 16(14).

[8] Faddeev L D. Properties of the S-matrix of the one-dimensional Schrodinger equation [J]. Trudy Matematicheskogo Instituta imeni VA Steklova, 1964, 73: 314-336.

[9] Tanaka S. Korteweg-de Vries equation: construction of solutions in terms of scattering data[J]. Osaka, 11 (1974), 49-59.

[10] Zakharov V E, Faddeev L D. Korteweg–de Vries equation: A completely integrable Hamiltonian system[J]. Funktsional'nyi Analiz i ego Prilozheniya, 1971, 5(4): 18-27.

[11] Y. S. Li, Proceedings of the 1980 Beijing Symposium on Differential Geometry and Differential Equations[M], Science Press, 1982, 3: 1297.

[12] Deift P, Trubowitz E. Inverse scattering on the line[J]. Communications on Pure and Applied Mathematics, 1979, 32(2): 121-251.

[13] Dodd R K, Morris H C. Bäcklund transformations[C]//Geometrical Approaches to Differential Equations: Proceedings of the Fourth Scheveningen Conference on Differential Equations, The Netherlands August 26–31, 1979. Berlin, Heidelberg: Springer Berlin Heidelberg, 2006: 63-94.

[14] Ma Y C. The complete solution of the long-wave–short-wave resonance equations[J]. Studies in Applied Mathematics, 1978, 59(3): 201-221.

[15] Kaup D J. The three-wave interaction—A nondispersive phenomenon[J]. Studies in applied mathematics, 1976, 55(1): 9-44.

[16] Agranovich Z S, Marchenko V A. The inverse problem of scattering theory[M]. Courier Dover Publications, 2020.

[17] 李翊神. 一类发展方程和谱的变形 [J]. 中国科学 (A 辑数学物理学天文学技术科学), 1982, 25(5): 385-390.

[18] 李翊神, 庄大蔚. 与位势依赖于能量的特征值问题相联系的非线性发展方程 [J]. 数学学报, 1982, 25(4): 464-474.

Chapter 3
Interaction of Solitons and Its Asymptotic Properties

3.1 Interaction of Solitons and Asymptotic Properties of $t \to \infty$

The soliton does not change its original amplitude and waveform after nonlinear interaction, which was first discovered by Kruskal and Zabusky[1] in numerical calculation. Later, P. D. Lax[2] gave rigorous analytical proof in theory. Lax also analyzed the detailed process of the nonlinear interaction of two solitons in detail and pointed out:

(i) If the wave speed $c_1 \gg c_2$, the first wave is higher than (and therefore faster than) the second wave. If the first wave is initially positioned to the left of the second wave, the first wave will eventually catch up with the second wave. During the interaction, the larger wave first absorbs the smaller (second) wave and then spits it out again, resulting in a single peak (maximum amplitude).

(ii) If the wave speed $c_1 \approx c_2$, the large wave catches up with the small wave. The large wave peak decreases and the small wave peak rises when they interact with each other. There are two peaks, and then the process is exchanged. Lax also analyzed the behaviour of the solution of the KdV equation at $t \to \infty$. He noted: If $u(x, t)$ is the solution to the KdV equation

$$u_t + uu_x + u_{xxx} = 0 \tag{3.1}$$

which is defined for all x, t and disappears at $x = \pm\infty$, then there are discrete positive numbers $c_1, c_2, ..., c_N$ (they are called the characteristic velocities of u) and a phase shift $\theta_j^{\pm}$ such that

$$\lim_{t \to \pm\infty} u(x + ct, t) = \begin{cases} s(\xi - \theta_j^{\pm}, c_j), & c = c_j \\ 0, & c \neq c_j \end{cases} \tag{3.2}$$

where s represents the soliton solution of (3.1), $\xi = x - c_j t$.

The interaction diagram of $c_1 \gg c_2$ is shown in Figure 3-1.

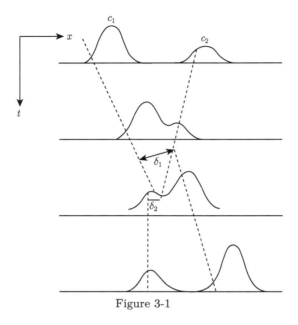

Figure 3-1

The following are the results obtained by solving N solitons using the inverse scattering method, and the Lax theory is proved by algebraic analysis: If as $t \to \infty$, the solution of the KdV equation decomposes into N solitons corresponding to the eigenvalues $k_1, k_2, \cdots, k_N$, with only their phases shifted.

From (2.19) of Chapter 2, we have

$$K(x, y, t) = - \sum_{m=1}^{N} c_m \psi_m e^{-k_m y}$$

$$u = - 2\frac{d}{dx} K(x, x; t) = 2\frac{d}{dx} \sum_{m=1}^{N} c_m \psi_m e^{-k_m x}$$

$$\equiv 2\frac{d}{dx} \sum_{m=1}^{N} f_m(x) \equiv 2 \sum_{m=1}^{N} f'_m(x)$$

where

$$f_m(x) = c_m \psi_m e^{-k_m x}$$

In order to compute $\sum_{m} f'_m$ $(t \to \pm\infty)$, we need to determine f_m, so rewrite

$$\psi_m(x) + \sum_{n=1}^{N} c_m c_n \frac{e^{-(k_m+k_n)x}}{k_m + k_n} \psi_n = c_m e^{-k_m x} \qquad (3.3)$$

as

$$c_m^{-2} e^{2k_m x} f_m(x) + \sum_{n=1}^{N} \frac{f_n(x)}{k_m + k_n} = 1 \quad (m = 1, 2, \cdots, N) \tag{3.4}$$

By taking the derivative of x in the above equation, we obtain

$$c_m^{-2} e^{2k_m x} f_m'(x) + \sum_{n=1}^{N} \frac{f_n'(x)}{k_m + k_n} = -2k_m c_m^{-2} e^{2k_m x} f_m \tag{3.5}$$

To consider the asymptotic state of $|t| \to \infty$, take the moving coordinate system

$$\xi \equiv x - 4k_p^2 t, \quad p = 1, 2, \cdots, N \tag{3.6}$$

where $\lambda_p = -k_p^2$ is the eigenvalue of the pth soliton, $4k_p^2$ is its velocity, and $2k_p^2$ is its amplitude. Thus there is

$$c_m^{-2} e^{2k_m x} = c_m(0)^{-2} \exp\left\{-8k_m \left(k_m^2 - k_p^2\right) t + 2k_m \xi\right\}$$

$$\equiv c_m(\xi) \exp\left\{-8k_m \left(k_m^2 - k_p^2\right) t\right\}$$

where $c_m(\xi) = c_m(0)^{-2} e^{2k_m \xi}$. So (3.4), (3.5) can be rewritten as

$$c_m(\xi) e^{-8k_m \left(k_m^2 - k_p^2\right) t} f_m + \sum_{n=1}^{N} \frac{f_n}{k_m + k_n} = 1 \tag{3.7}$$

$$c_m(\xi) e^{-8k_m \left(k_m^2 - k_p^2\right) t} f_m' + \sum_{n=1}^{N} \frac{f_n'}{k_m + k_n} = -2k_m c_m(\xi) e^{-8k_m \left(k_m^2 - k_p^2\right) t} f_m \tag{3.8}$$

Let $k_1 > k_2 \cdots > k_N > 0$, consider
1) The asymptotic state of $t \to \infty$
Taking the limit of (3.7), we get

$$\begin{cases} \displaystyle\sum_{n=1}^{N} \frac{f_n}{k_m + k_n} = 1, & m = 1, 2, \cdots p - 1 \\ \displaystyle c_p f_p + \sum_{n=1}^{N} \frac{f_n}{k_p + k_n} = 1, & m = p \\ f_m = 0, & m = p + 1, \cdots N \end{cases}$$

The above formula can be simply written as

$$\sum_{n=1}^{p} \frac{f_n}{k_m + k_n} = 1 - c_p \delta_{mp} f_p \quad (m = 1, 2, \cdots, p) \tag{3.9}$$

Similarly, we have

$$\sum_{n=1}^{p} \frac{f'_n}{k_m + k_n} = -c_p \delta_{mp} \left(2k_p f_p + f'_p\right) \ (m = 1, 2, \cdots, p) \tag{3.10}$$

$$f'_m = -2k_m f_m = 0 \ (m = p+1, \cdots, N) \tag{3.11}$$

It is easy to know that the matrix $K_p = \left(\dfrac{1}{k_m + k_n}\right)$ $(m = 1, 2, \cdots, p)$ has a positive determinant. In fact,

$$0 < \det C = \det \left(c_m c_n \frac{e^{-(k_m+k_n)x}}{k_m + k_n}\right)$$

$$= \det \left(\frac{1}{k_m + k_n}\right) \prod_{m=1}^{N} c_m^2 e^{-2\sum_{m=1}^{N} k_m x}$$

So $\det \left(\dfrac{1}{k_m + k_n}\right) > 0$. By Cramer's rule, f_m, f'_m can be solved from (3.9) and (3.10)

$$f_m \det K_p = \sum_{n=1}^{p} K_{mn} - c_p K_{pm} f_p \ (m = 1, 2, \cdots, p) \tag{3.12}$$

$$f'_m \det K_p = -c_p K_{pm} \left(2k_p f_p - f'_p\right) \ (m = 1, 2, \cdots, p) \tag{3.13}$$

where K_{mn} is the algebraic cofactor of $\left(\dfrac{1}{k_m + k_n}\right)$ element. We use L_p to represent a matrix in which the last row of K_p is all 1. Setting $m = p$, f_p and f'_p can be given

$$f_p = \frac{\det L_p}{\det K_p + c_p \det K_{p-1}}$$

$$f'_p = -\frac{2c_p k_p f_p \det K_{p-1}}{\det K_p + c_p \det K_{p-1}}$$

Sum (3.13) and substitute the above expression for f_p, f'_p

$$\sum_{m=1}^{p} f'_m \det K_p = -\sum_{m=1}^{p} c_p K_{pm} \left(2k_p f_p - f'_p\right)$$

$$= -\sum_{m=1}^{p} c_p K_{pm} 2k_p f_p \left(1 - \frac{c_p \det K_{p-1}}{\det K_p + c_p \det K_{p-1}}\right)$$

$$= -\sum_{m=1}^{p} c_p K_{pm} f_p 2k_p \frac{\det K_p}{\det K_p + c_p \det K_{p-1}}$$

$$= -c_p \det L_p f_p 2k_p \frac{\det K_p}{\det K_p + c_p \det K_{p-1}}$$

$$= -2k_p c_p \left(\det L_p\right)^2 \frac{\det K_p}{\left(\det K_p + c_p \det K_{p-1}\right)^2}$$

So there is

$$\lim_{\substack{t\to\infty \\ \xi \text{ is fixed}}} \sum_{m=1}^{p} f'_m = -\frac{2k_p c_p}{\left[\dfrac{\det K_p}{\det L_p} + c_p \dfrac{\det K_{p-1}}{\det L_p}\right]^2}$$

Subtract the last row from each row in K_p to get

$$\det K_p = \frac{\prod\limits_{m=1}^{p-1} (k_p - k_m)}{\prod\limits_{m=1}^{p} (k_p + k_m)} \det L_p$$

Similarly, subtracting the last column from each row in L_p yields

$$\det L_p = \frac{\prod\limits_{m=1}^{p-1} (k_p - k_m)}{\prod\limits_{m=1}^{p-1} (k_p + k_m)} \det K_{p-1}$$

$$\frac{\det K_p}{\det L_p} = \frac{\prod\limits_{m=1}^{p-1} (k_p - k_m)}{\prod\limits_{m=1}^{p} (k_p + k_m)}$$

$$\frac{\det K_{p-1}}{\det L_p} = \frac{\prod\limits_{m=1}^{p-1} (k_p + k_m)}{\prod\limits_{m=1}^{p-1} (k_p - k_m)}$$

Notice that $c_p = c_p(0)^{-2} e^{2k_p \xi}$, ξ_p is defined as

$$e^{2k_p \xi_p} \equiv \frac{c_p^2(0)}{2k_p} \prod_{m=1}^{p-1} \left(\frac{k_p - k_m}{k_p + k_m}\right)^2$$

Thus, we have

$$\lim_{\substack{t \to \infty \\ \xi \text{ is fixed}}} u = \lim_{\substack{t \to \infty \\ \xi \text{ is fixed}}} 2\sum_{m=1}^{p} f'_m = \dfrac{4k_p c_p}{\left[\dfrac{\prod\limits_{m=1}^{p-1}(k_p-k_m)}{\prod\limits_{m=1}^{p}(k_p+k_m)} + c_p \dfrac{\prod\limits_{m=1}^{p-1}(k_p+k_m)}{\prod\limits_{m=1}^{p-1}(k_p-k_m)} \right]^2}$$

$$= -\dfrac{4k_p c_p}{\prod\limits_{m=1}^{p-1}\left(\dfrac{k_p-k_m}{k_p+k_m}\right)^2 \left[\dfrac{1}{4k_p^2} + \dfrac{2c_p}{2k_p}\prod\limits_{m=1}^{p-1}\left(\dfrac{k_p+k_m}{k_p-k_m}\right)^2 + c_p^2\prod\limits_{m=1}^{p-1}\left(\dfrac{k_p+k_m}{k_p-k_m}\right)^4 \right]}$$

$$= -16k_p^3 c_p \prod_{m=1}^{p-1}\left(\dfrac{k_p+k_m}{k_p-k_m}\right)^2 \left[1 + 2k_p c_p \prod_{m=1}^{p-1}\left(\dfrac{k_p+k_m}{k_p-k_m}\right)^2 \right]^{-2}$$

$$= -8k_p^2 \left[2k_p c_p \prod_{m=1}^{p-1}\left(\dfrac{k_p+k_m}{k_p-k_m}\right)^2 \right] \cdot \left[1 + 2k_p c_p \prod_{m=1}^{p-1}\left(\dfrac{k_p+k_m}{k_p-k_m}\right)^2 \right]^{-2}$$

$$= -8k_p^2 e^{2k_p(\xi-\xi_p)} \left[1 + e^{2k_p(\xi-\xi_p)} \right]^{-2}$$

$$= -2k_p^2 \operatorname{sech}^2 \left[k_p \left(\xi - \xi_p \right) \right]$$

$$= -2k_p^2 \operatorname{sech}^2 \left[k_p \left(x - 4k_p^2 t - \xi_p \right) \right]$$

That is to say, around $x = 4k_p^2 t$, it is a soliton with amplitude $2k_p^2$ and velocity $4k_p^2$.

2) The asymptotic state of $t \to -\infty$

By (3.7), (3.8), we have

$$\sum_{m=p}^{N} \dfrac{f_n}{k_m + k_n} = 1 - c_p \delta_{mp} f_p \ (m = p, \cdots, N)$$

$$\sum_{m=p}^{N} \dfrac{f'_n}{k_m + k_n} = -c_p \delta_{mp} \left(2k_p f_p + f'_p \right) \ (m = p, \cdots, N)$$

$$f'_m = -2k_m f_m = 0, \ m = 1, 2, \cdots, p-1$$

As discussed in $t \to \infty$, define $\bar{\xi}_p$ as

$$\bar{\xi}_p e^{2k_p \bar{\xi}_p} \equiv \dfrac{c_p^2(0)}{2k_p} \prod_{m=p+1}^{N} \left(\dfrac{k_p - k_m}{k_p + k_m} \right)^2$$

Then we get

$$\lim_{t \to -\infty} u(x, t) = -2k_p^2 \operatorname{sech}^2 \left[k_p \left(x - 4k_p^2 t - \bar{\xi}_p \right) \right]$$

The phase difference is

$$\xi_p - \bar{\xi}_p = \frac{1}{k_p} \left[\sum_{m=1}^{p-1} \log \left(\frac{k_m - k_p}{k_m + k_p} \right) - \sum_{m=p+1}^{N} \log \left(\frac{k_p - k_m}{k_p + k_m} \right) \right]$$

We can also analyse the interactions of finite solitons in other way. Let KdV equation be

$$u_t + \delta u u_x + u_{xxx} = 0 \tag{3.14}$$

Take $u = p_x$, then $(p_t)_x + \delta \left(\frac{1}{2} p_x^2 \right)_x + (p_{xxx})_x = 0$. Integrate it to get

$$p_t + \frac{1}{2} \delta p_x^2 + p_{xxx} = 0 \tag{3.15}$$

Make a transformation similar to Hopf-Cole, $\delta p = 12(\log F)_x$, substitute it into the above formula, eliminate some terms and merge it into

$$F (F_t + F_{xxx})_x - F_x (F_t + F_{xxx}) + 3 \left(F_{xx}^2 - F_x F_{xxx} \right) = 0 \tag{3.16}$$

Note that (3.16) contains the operator $L = \dfrac{\partial}{\partial t} + \dfrac{\partial^3}{\partial x^3}$ and $F = 1 + e^{-\alpha(x-s)+\alpha^3 t}$ (α, s are real) is the particular solution of $F_t + F_{xxx} = 0$. If (3.16) is linear, we can expect the summation of α, s to produce its solution. Since (3.16) has an interaction term due to its nonlinearity, we can approximate the expansion in terms of the interaction term in the usual way:

$$F = 1 + F^{(1)} + F^{(2)} + \cdots$$

Substituting it to (3.16) yields a series of equations:

$$\{F_t^{(1)} + F_{xxx}^{(1)}\}_x = 0$$
$$\{F_t^{(2)} + F_{xxx}^{(2)}\}_x = -3\{F_{xx}^{(1)} - F_x^{(1)} F_{xxx}^{(1)}\}$$
$$\cdots\cdots$$

If only two terms of $F^{(1)}$ is taken: $F^{(1)} = f_1 + f_2$, $f_j = e^{-\alpha_j(x-s_j)+\alpha_j^3 t}$ ($j = 1, 2$). Obviously, $F^{(1)}$ selected in this way satisfies the first equation, and substituting it into the equation of $F^{(2)}$ yields

$$\{F_t^{(2)} + F_{xxx}^{(2)}\}_x = 3\alpha_1\alpha_2 (\alpha_2 - \alpha_1)^2 f_1 \cdot f_2$$

It can be solved as

$$F^{(2)} = \frac{(\alpha_2 - \alpha_1)^2}{(\alpha_2 + \alpha_1)^2} f_1 \cdot f_2$$

Surprisingly, $F^{(3)} = F^{(4)} = \cdots = 0$, so we get the exact solution of (3.16)

$$F = 1 + f_1 + f_2 + \frac{(\alpha_2 - \alpha_1)^2}{(\alpha_2 + \alpha_1)^2} f_1 \cdot f_2 \qquad (3.17)$$

We notice that the interaction terms contain only the term $f_1 f_2$, not f_1^2, f_2^2 in this expression. This result can be generalized to N terms f_i. Let $F^{(1)} = \sum_{j=1}^{N} f_j$, then $F^{(2)}$ contains the terms $f_j f_k$, $(j \neq k)$, but no terms f_j^2. $F^{(3)}$ contains $f_j f_k f_l$ $(j \neq k \neq l)$, but no f_j^3, f_j^2, f_k^2, and so on. $F^{(N)} \propto f_1, f_2, \cdots, f_N$, then

$$F = 1 + \sum_j f_j + \sum_{j \neq k} a_{ij} f_j f_k + \sum_{j \neq k \neq l} a_{jkl} f_j f_k f_l + \cdots + \alpha_{1,2,\cdots,N} f_1 f_2 \cdots f_N$$

It can be proved that $F = \det |F_{mn}|$, where $F_{mn} = \delta_{mn} + \dfrac{2\alpha_m}{\alpha_m + \alpha_n} f_m$. Matrix (F_{mn}) and $C = \left(\delta_{mn} + \dfrac{e^{-(k_m + k_n)x}}{k_m + k_n} \right)$ obtained by the above inverse scattering method are consistent. Now considering $N = 2$, from $\delta u = \delta p_x = 12 (\log F)_{xx}$ and the expression (3.17) of F, we can obtain the expression of solution to KdV equation (3.14)

$$\begin{aligned}
\frac{\delta u}{12} &= \{ \alpha_1^2 f_1 + \alpha_2^2 f_2 + 2 (\alpha_2 - \alpha_1)^2 f_1 f_2 + [(\alpha_2 - \alpha_1) / \\
&\quad \cdot (\alpha_2 + \alpha_1)]^2 (\alpha_2^2 f_1^2 f_2 + \alpha_1^2 f_1 f_2^2) \} / \{ [1 + f_1 + f_2 \\
&\quad + ((\alpha_2 - \alpha_1) / (\alpha_2 + \alpha_1))^2 f_1 f_2]^2 \} \\
f_j &= \exp \left[-\alpha_j (x - s_j) + \alpha_j^3 t \right] \qquad (3.18)
\end{aligned}$$

We consider a soliton solution, $\delta u = 3\alpha^2 \operatorname{sech}^2 \dfrac{\theta - \theta_0}{2}$, where $\theta = \alpha x - \alpha^3 t$, $\theta_0 = s\alpha$, which can be expressed as $f = e^{-\alpha(x-s) + \alpha^3 t}$

$$\frac{\delta u}{12} = \frac{\alpha^2 f}{(1 + f)^2}$$

When $f = 1$, δu takes the maximum value, its maximum amplitude is $\delta u = 3\alpha^2$. The position of maximum value is $-\alpha(x - s) + \alpha^3 t = 0$, that is, $x = s + \alpha^2 t$, and the wave velocity $c = \alpha^2$. In order to investigate their interaction and the asymptotic state of $t \to \pm\infty$, one should make full use of the expression (3.18). We will discuss several cases below.

1. On the (x, t) region, $f_1 \approx 1$, f_2 is large or small.

(1) $f_1 \approx 1$, $f_2 \ll 1$. From (3.18), we know that

$$\frac{\delta u}{12} \approx \frac{\alpha_1^2 f_1}{(1 + f_1)^2} \text{ is the soliton } \alpha_1 \text{ wave}$$

(2) $f_1 \approx 1$, $f_2 \gg 1$. From (3.18), we know that

$$\frac{\delta u}{12} \approx \frac{[(\alpha_2 - \alpha_1)/(\alpha_2 + \alpha_1)]^2 \alpha_1^2 f_1 f_2^2}{\left(f_2 + [(\alpha_2 - \alpha_1)/(\alpha_2 + \alpha_1)]^2 f_1 f_2\right)^2} = \frac{\alpha_1^2 \tilde{f}_1}{(1 + \tilde{f}_1)^2}$$

$$\tilde{f}_1 = \left(\frac{\alpha_2 - \alpha_1}{\alpha_2 + \alpha_1}\right)^2 f_1$$

This is still the soliton α_1 wave, and only the phase s_1 is changed to

$$\tilde{s}_1 = s_1 - \frac{1}{\alpha_1} \log \left(\frac{\alpha_2 + \alpha_1}{\alpha_2 - \alpha_1}\right)^2$$

2. On the (x, t) region, $f_2 \approx 1$, f_1 is large or small. For an analysis similar to 1, this is the soliton α_2 wave.

3. f_1, f_2 are both small or large, in which case $\delta u \approx 0$.

4. $f_1 \approx 1$, $f_2 \approx 1$, which represents the interaction zone.

Now, we set $\alpha_2 > \alpha_1 > 0$, that is, the case where the α_2 wave chases the α_1 wave. When $t \to -\infty$,

α_1 wave: $f_1 \approx 1$, $x = s_1 + \alpha_1^2 t$,

$$f_2 = e^{-\alpha_2(x - s_2) + \alpha_2^3 t} = e^{-\alpha_2[s - s_2] - \alpha_2[\alpha_1^2 - \alpha_2^2]t} \ll 1 \ (t \to -\infty). \text{ Following}$$

the previous discussion, it is shown that it is the soliton α_1 wave at $x = s_1 + \alpha_1^2 t$.

α_2 wave: $f_2 \approx 1$, $x = s_2 - \frac{1}{\alpha_2} \log \left(\frac{\alpha_2 + \alpha_1}{\alpha_2 - \alpha_1}\right)^2 + \alpha_2^2 t$. In this case, $f_2 \gg 1 \ (t \to -\infty)$, which indicates it is the soliton α_2 wave at

$$x = s_2 - \frac{1}{\alpha_2} \log \left(\frac{\alpha_2 + \alpha_1}{\alpha_2 - \alpha_1}\right)^2 + \alpha_2^2 t$$

Elsewhere, $\delta u \approx 0$ ($f_1 f_2$ is large or small).

When $t \to \infty$, there is

α_1 wave: $x = s_1 - \dfrac{1}{\alpha_1} \log\left(\dfrac{\alpha_2 + \alpha_1}{\alpha_2 - \alpha_1}\right)^2 + \alpha_1^3 t$, $f_1 \approx 1$, $f_2 \gg 1$.

α_2 wave: $x = s_2 + \alpha_2^2 t$, $f_2 \approx 1$, $f_1 \ll 1$.

Elsewhere: $\delta u \approx 0$.

The above results show that the soliton does not change the original parameter α_1, α_2, and the fast wave α_2 is in front. The collision interaction process only enables

α_2 wave shifts forward $\dfrac{1}{\alpha_2} \log\left(\dfrac{\alpha_2 + \alpha_1}{\alpha_2 - \alpha_1}\right)^2$,

α_1 wave shifts backward $\dfrac{1}{\alpha_1} \log\left(\dfrac{\alpha_2 + \alpha_1}{\alpha_2 - \alpha_1}\right)^2$.

When $f_1 \approx 1$, $f_2 \approx 1$, the time and place of interaction is approximately

$$x = s_1 + \alpha_1^2 t = s_2 + \alpha_2^2 t, \quad t = \frac{-(s_2 - s_1)}{\alpha_2^2 - \alpha_1^2}$$

$$x = \frac{\alpha_2^2 s_1 - \alpha_1^2 s_2}{\alpha_2^2 - \alpha_1^2}$$

3.2 Behaviour State of the Solution to KdV Equation Under Weak Dispersion and WKB Method

We know that the Burgers equation $u_t + uu_x = \varepsilon u_{xx}$ ($\varepsilon > 0$), when $\varepsilon \to 0$, $u_\varepsilon \to u$, where u is the generalized solution of the equation

$$u_t + uu_x = 0$$

For the KdV equation

$$u_t + uu_x = \varepsilon u_{xxx}$$

when $\varepsilon \to 0$, $u_\varepsilon(x, t) \to u(x, t)$? Is $u(x, t)$ a generalized solution of $u + uu_x = 0$? This problem is of great concern to us. For the KdV equation, in general, the answer is no, that is, the solution of the KdV equation does not tend to any discontinuous solution with shock waves when $\varepsilon \to 0$. Let's make a rough analysis of the problem and then explain it.

For the solution u_ε of the KdV equation

$$u_t + uu_x = \varepsilon u_{xxx} \tag{3.19}$$

if u_ε and some of its derivatives tend to 0 ($|x| \to \infty$), then

$$\int_{-\infty}^{\infty} u_\varepsilon(x, t)dx = \int_{-\infty}^{\infty} u(x, 0)dx = M_0$$

$$\int_{-\infty}^{\infty} \frac{1}{2} u_\varepsilon^2(x,t)dx = \int_{-\infty}^{\infty} \frac{1}{2} u^2(x,0)dx = E$$

Now for $u_t + uu_x = 0$ on $(-\infty, \infty)$, integrate with x, differentiate with t, and exchange the order of integration. Notice that there is a jump on the discontinuity line $x = x(t)$, we have

$$\frac{dM}{dt} + D[u] = \frac{1}{2}\left[u^2\right]$$

where $D = \dfrac{dx(t)}{dt}$, $[f] = f_+ - f_- = f(x(t)+0) - f(x(t)-0)$, $x(t)$ is the discontinuity line, $M = \displaystyle\int_{-\infty}^{\infty} u(x,t)dx$. Since momentum is conserved, $\dfrac{dM}{dt} = 0$, we get the shock wave relation

$$D = \frac{dx}{dt} = \frac{1}{2}\frac{\left[u^2\right]}{[u]}$$

which is the Hugoniot condition. Multiply u by $u_t + uu_x = 0$ and integrate over x to get

$$\frac{dE}{dt} + D\left[\frac{1}{2}u^2\right] = \frac{1}{3}\left[u^3\right], \quad E = \int_{-\infty}^{\infty} \frac{1}{2}u^2(x,t)dx$$

Substituting $D = \dfrac{1}{2}\dfrac{\left[u^2\right]}{[u]}$ into the above formula, we get $\dfrac{dE}{dt} = \dfrac{1}{12}[u]^3$. From the entropy condition $u_- > u_+$, there is $\dfrac{dE}{dt} < 0$. But for solution of KdV equation (3.19), there is always $\dfrac{dE}{dt} = 0$ $(E = E_0)$. It can therefore be asserted that when $\varepsilon \to 0$, the solution u_ε of (3.19) does not approach the discontinuous solution containing shock wave of the limit equation $u_t + uu_x = 0$. Then we use the method of Gelefind[4] to see what the limit of the solution of (3.19) is. Let the solution of (3.19) have a smooth transition region with thickness $\Delta(\varepsilon)$, which is connected to two different states, when $\varepsilon \to 0$, $\Delta(\varepsilon) \to 0$. Using the moving coordinate system, let $\xi = \dfrac{x - x(t)}{\Delta}$, $t' = t$, $x(t)$ be the unknown shock trajectory. $u(x,t) = u(\xi, t')$, multiply Δ by (3.19) to get

$$\varepsilon\Delta^{-2}u_{\xi\xi\xi} - (u - D)u_\xi = \Delta \cdot u_{t'} \tag{3.20}$$

Let $u_{t'}$ be bounded, so that when $\Delta \to 0$, $\Delta \cdot u_{t'} \to 0$. If $\Delta = o\left(\varepsilon^{\frac{1}{2}}\right) = \varepsilon^{1/2}$, then (3.20) tends to the differential equation

$$u_{\xi\xi\xi} - (u - D)u_\xi = 0$$

when $\varepsilon \to 0$ and $u \to u_1$, $\xi \to +\infty$; $u \to u_0$, $\xi \to -\infty$. In this case, the solution $u(\xi, t')$ consists of soliton vibration solutions, as shown in Figure 3-2.

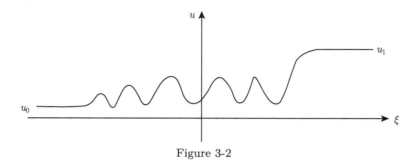

Figure 3-2

We now analyse this problem from qualitative theory, and there are the following theorems.

Theorem 3.1 *The solution of the quasi-linear hyperbolic equation*

$$u_t + (f(u))_x = 0 \tag{3.21}$$

with shock waves can never be taken as the limit of the travelling wave solution of the dispersive equation (KdV equation)

$$u_t + f(u)_x = \varepsilon^2 u_{xxx} \tag{3.22}$$

Proof Here, the shock wave solution is

$$\bar{u}(x, t) = \begin{cases} u_0, & x - Dt < 0 \\ u_1, & x - Dt > 0 \end{cases}$$

and satisfies $D(u_1 - u_0) = f(u_1) - f(u_0)$, $f'(u_1) < D < f'(u_0)$. Now consider the travelling wave solution of (3.22), $u(x, t) = u(\xi)$, where

$$\xi = \frac{x - Dt}{\varepsilon}, \quad u_t = \left(\frac{-D}{\varepsilon}\right)\frac{du}{d\xi}, \quad u_x = \frac{1}{\varepsilon}\frac{du}{d\xi}$$

Then we have

$$u''' = [-Du + f(u)]'$$

Integrating once gives

$$u'' = -Du + f(u) + C$$

$u(\xi) \to u_0$, $\xi \to -\infty$; $u(\xi) \to u$, $\xi \to +\infty$. We ask whether continuous solutions exist for the boundary value problem of this ordinary differential equation satisfying $f(u_1) - f(u_0) = D(u_1 - u_0)$. From

$$\begin{cases} u' = v \\ v' = -Du + f(u) \end{cases}$$

the above second-order ordinary differential equation problem can be reduced to the boundary value problem of the ordinary differential system. Letting

$$P(u) = \frac{1}{2}Du^2 - F(u), \quad F'(u) = f(u)$$

$$H(u,v) = \frac{1}{2}v^2 + P(u)$$

it can be reduced to a system of canonical equations:

$$\begin{cases} u' = v = H_v \\ v' = -Du + f(u) = -H_u \end{cases}$$

It can be considered that $v \to 0$ ($|\xi| \to \infty$), so the two critical points are $(u_0, 0)$, $(u_1, 0)$. $H(u(\xi), v(\xi))$ is constant along the orbital, so if two critical points $(u_0, 0)$, $(u_1, 0)$ are connected by a trajectory, they must be on the same "energy surface", that is, deducing $P(u_0) = P(u_1)$ from $H(u_0, 0) = H(u_1, 0)$, we can show that this equation does not hold, at least along the weak shock wave curve. In fact, the weak shock wave quantity associated with the left status can be expressed as a single parameter σ. $u = u(\sigma)$, $s(\sigma)(u(\sigma) - u_0) = f(u(\sigma)) - f(u_0)$, $s(0) = f'(u_0)$, p is also a function of σ,

$$p(u(\sigma)) = \frac{1}{2}s(\sigma)u^2(\sigma) - F(u(\sigma))$$

Differentiating with σ yields

$$\dot{p}(u(\sigma)) = \frac{1}{2}\dot{s}(\sigma)u^2 + s(\sigma)u\dot{u} - f(u(\sigma))\dot{u}$$

From R-H condition, $s_u = f$ (always choose $u_0 = f(u_0) = 0$),

$$\dot{p}(u(\sigma)) = \frac{1}{2}\dot{s}(\sigma)u^2 \neq 0$$

Let $f''(u) \neq 0$, $\dot{s}(\sigma) \neq 0$, hence $P(u_0) \neq P(u_1)$. Therefore, the above integral curve connecting the two critical points does not exist.

From the graph in Figure 3-3,

$$H(u,v) = \frac{1}{2}v^2 + P(u)$$

One can draw the equipotential line of H on the u, v plane. $(u_1, 0)$ is the center point of the linearised matrix $\begin{pmatrix} H_{uu} & H_{uv} \\ H_{uv} & H_{vv} \end{pmatrix}$, so that no trajectory enters it.

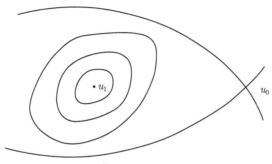

Figure 3-3

For one-dimensional isothermal aerodynamic equations

$$\begin{cases} u_t - v_x = 0 \\ v_t + P(u)_x = 0 \end{cases} \tag{3.23}$$

where $P'(u) < 0$, $P''(u) > 0$, there is a similar relationship as follows.

Theorem 3.2 *The shock wave solution of the equations* (3.23) *cannot be taken as the limit of the traveling wave solution of the corresponding dispersion equations.*

It seems that for the hyperbolic equation with the higher-order derivative term of the viscosity, the convergence of the viscosity to zero must be of even order, at least for the second and fourth orders.

We use WKB method to construct the approximate solution u_ε

$$u_t + u u_x = \varepsilon^2 u_{xxx} \tag{3.24}$$

where $\varepsilon^2 \ll 1$. As we know, WKB method is an important approximation method in mathematical physics, which has been used to solve the linear ordinary differential equation

$$\varepsilon^2 \psi_{xx} + V(x)\psi = 0, \quad \varepsilon^2 \ll 1 \tag{3.25}$$

where $V(x)$ is a function that varies slowly over length $o(\delta)$. Let the form of the solution be

$$\psi(x;\varepsilon) \sim \psi(\theta, x; \varepsilon) \equiv W(x)e^{i\theta}, \ \theta \equiv \frac{B(x,\varepsilon)}{\varepsilon} \qquad (3.26)$$

Two unknown functions W and B are introduced instead of an unknown function ψ, with two independent variables, θ, x instead of an variable x. We can choose in such a way that the solution θ is periodic. For nonlinear partial differential equations, an exponential solution (3.26) cannot be expected. Let $u(x,t;\varepsilon)$ can be expanded into the following series

$$u(x,t;\varepsilon) \sim U(\theta, x, t; \varepsilon) = U^{(0)}(\theta, x, t) + \varepsilon U^{(1)}(\theta, x, t) + \cdots \qquad (3.27)$$

where $\theta = \theta(x,t;\varepsilon)$. One can choose θ to have a period of 1, i.e.,

$$U(\theta, x, t; \varepsilon) = U(\theta + 1, x, t; \varepsilon)$$

If $\theta(x,t;\varepsilon)$ is determined, the approximate solution is obtained as $u(x,t;\varepsilon) \simeq U(\theta(x,t;\varepsilon), x, t; \varepsilon)$. For the KdV equation (3.24), let

$$\theta = \frac{B(x,t;\varepsilon)}{\varepsilon}$$

where $B = o(1)$. Set $L \equiv B_t$, $K \equiv B_x$, $l = \dfrac{B_t}{B_x} = \dfrac{L}{K}$ is independent of θ.

$$\frac{\partial}{\partial t} \rightarrow \frac{L}{\varepsilon}\frac{\partial}{\partial \theta} + \frac{\partial}{\partial t}, \quad \frac{\partial}{\partial x} \rightarrow \frac{K}{\varepsilon}\frac{\partial}{\partial \theta} + \frac{\partial}{\partial x}$$

$u = U(\theta, x, t; \varepsilon)$. Multiply (3.24) by ε/K to get

$$lU_0 + UU_\theta + K^2 U_{\theta\theta\theta} + \varepsilon\left[\frac{1}{K}(U_t + UU_x) + 3(KU)_{x\theta\theta}\right]$$

$$+ \varepsilon^2\left\{\frac{1}{K}[K_{xx} - U + 3(KU_x)_x]\right\}_\theta + \varepsilon^3\frac{1}{K}U_{xxx} = 0 \qquad (3.28)$$

Substituting (3.27) into (3.28), we obtain a series of differential equations for $U^{(i)}$ ($i = 0, 1, \cdots$) with respect to θ, the first of which is

$$l^{(0)}U_\theta^{(0)} + U^{(0)}U_\theta^{(0)} + K^{(0)2}U_{\theta\theta\theta}^{(0)} = 0 \qquad (3.29)$$

where $l^{(0)}$, $K^{(0)}$ are the first terms of the expansion series for l, K. Integrate twice to obtain

$$\frac{1}{2}l^{(0)}\left[U^{(0)}\right] + \frac{1}{6}\left[U^{(0)}\right]^3 + \frac{1}{2}\left[K^{(0)}\right]\left[U_\theta^{(0)}\right]^2 = mU^{(0)} + n \qquad (3.30)$$

where m and n are integral constants that depend on x and t. Fixed x, t, (3.30) has the solution

$$U^{(0)}(\theta, x, t) = -l^{(0)} - \left(l^{(0)} + 2m\right)$$
$$\cdot \left\{a + (b-a)\,\mathrm{Cn}^2\left[2\mathscr{K}(k)\,(\theta - \theta_0)\,;k\right]\right\} \tag{3.31}$$

where $k^2 = \dfrac{b-a}{c-a}$, Cn is Jacobian elliptic function, $\mathscr{K}$ is the first complete elliptic integral, a, b, c are the roots of $U_\theta^{(0)} = 0$ in (3.30).

When $i \geqslant 1$, there is an equation for $U^{(i)}$:

$$\mathscr{L}U^{(i)} = N_i\left(U^{(i-1)}, \cdots, l^{(i)}, \cdots, K^{(i)}, \cdots\right)$$

where

$$\mathscr{L} \equiv l^{(0)}\frac{\partial}{\partial\theta} + \frac{\partial}{\partial\theta}U^{(0)} + \left[K^{(0)}\right]^2\frac{\partial^3}{\partial\theta^3}$$

The non-homogeneous term N_i contains only lower-order solutions, and in principle can be found forever. The important fact is that when $\varepsilon \to 0$, the limiting solution that should be obtained from (3.24) is (3.29), not $u_t + uu_x = 0$.

3.3   Stability Problem of Soliton

Here, we consider the linear stability problem of solitons, that is, the Lyaplov stability problem of small perturbations of the KdV equation with respect to the steady wave solution. Let the KdV equation be

$$u_t + uu_x + \mu u_{xxx} = 0 \tag{3.32}$$

which has a solitary wave solution (stationary wave)

$$u_0(x) = -u_\infty\left(1 - 3\,\mathrm{sech}^2\sqrt{\frac{u_\infty}{4\mu}}\,x\right), \quad u_\infty > 0 \tag{3.33}$$

Suppose there is a small perturbation $v(x,t)$ with respect to this stationary wave solution, i.e.,

$$u = u_0(x) + v(x,t), \quad |v| \ll |u_0| \tag{3.34}$$

Substituting it into (3.32) gives the linear equation for v

$$v_t + u_0 v_x + u_{0x}v + \mu v_{xxx} = 0 \tag{3.35}$$

Suppose $v(x, t) = f(x)g(t)$, $g(t) \propto e^{\sigma t}$, $\sigma = \text{const}$, then the equation

$$\frac{d^3 f}{dy^3} - 4\left(1 - 3\,\text{sech}^2 y\right) \frac{df}{dy} - \left(24\,\text{sech}^2 y \tanh y + \alpha\right) f = 0 \qquad (3.36)$$

satisfied by the perturbed function f can be obtained. Here, $y \equiv \sqrt{\dfrac{u_\infty}{4\mu}} x$, $\alpha \equiv$
$(-8\sigma/u_\infty)\sqrt{\mu/u_\infty}$, the boundary condition is

$$f \to 0, \quad |y| \to \infty \qquad (3.37)$$

There are three independent solutions for this equation

$$f_k = \lambda_k \left(\lambda_k - 2\right)^2 e^{\lambda_k y} + 4 \frac{d^2}{dy^2}\left\{ e^{(\lambda_k - 1)y} \,\text{sech}\, y \right\}$$

$$= e^{\lambda_k y} \left[\lambda_k \left(\lambda_k - 2\right)^2 + 4e^{-y}\,\text{sech}\,\left\{ \lambda_k \left(\lambda_k - 2\right) \right.\right.$$

$$\left.\left. -2\left(\lambda_k - 1\right)\tanh y + 2\tanh^2 y \right\} \right], \quad k = 1, 2, 3 \qquad (3.38)$$

where λ_k is the root of the cubic equation

$$\lambda^3 - 4\lambda - \alpha = 0 \qquad (3.39)$$

In particular, if $\alpha = 0$, this is the stable case ($\sigma = 0$). In the case of $\lambda_k = 0$, 2 or -2, $f_k \propto \text{sech}^2 y \tanh y$ $(k = 1, 2, 3)$. So there are three independent solutions

$$\begin{cases} f_1 = \text{sech}^2 y \tanh y \equiv f_0 \\[2mm] f_2 = 3y f_0 + \tanh^2 y - 2\,\text{sech}^2 y \\[2mm] f_3 = 15 y f_0 + 2\sinh^2 y + 7\tanh^2 y - 8\,\text{sech}^2 y \end{cases} \qquad (3.40)$$

Note that (3.39) has the double root $\lambda = \pm 2/\sqrt{13}$ when $\alpha = \mp 16/3\sqrt{3}$, then two of (3.40) are equal, for example, $f_1 \equiv f_2$. Therefore, three new independent solutions must be selected as

$$\begin{cases} f_1 = \lambda_0 \left(\lambda_0 - 2\right)^2 e^{\lambda_0 y} + 4 \frac{d^2}{dy^2}\left\{ e^{(\lambda_0 - 1)y}\,\text{sech}y \right\} \\[3mm] f_2 = y f_1 + \left(3\lambda_0^2 - 8\lambda_0 + 4\right) e^{\lambda_0 y} + 8 \frac{d}{dy}\left\{ e^{(\lambda_0 - 1)y}\,\text{sech}y \right\} \\[3mm] f_3 = \lambda_3 \left(\lambda_3 - 2\right)^2 e^{\lambda_0 y} + 4 \frac{d^2}{dy^2}\left\{ e^{(\lambda_3 - 1)y}\,\text{sech}y \right\} \end{cases} \qquad (3.41)$$

Here, λ_0 is $2/\sqrt{3}$ or $-2/\sqrt{3}$, λ_3 is $-4/\sqrt{3}$ or $4/\sqrt{3}$, which depending on whether λ_0 is $2/\sqrt{3}$ or $-2/\sqrt{3}$. Obviously, (3.39) has no triple root. It is easy to see that any linear combination of solutions from (3.38), (3.40) or (3.41) cannot satisfy the boundary condition (3.37) unless f_0 takes the form (3.40) and is therefore a stable solution. It follows that the soliton solution of the KdV equation is stable in the Lyaplov sense with respect to small perturbations. For the nonlinear stability of the soliton wave and the stability of the "Cnoidal wave", see [4], [5].

3.4   Wave Equation under Water Wave and Weak Nonlinear Effect

The KdV equation was first obtained by Korteweg-de Vries in 1895 in water waves on the assumption of long wave approximations, small but finite amplitudes. Let's first derive this equation in water waves. Consider an non-viscous, incompressible fluid (water) in the constant gravity field. The space coordinate system is (x_1, x_2, y), the component of velocity u is (u_1, u_2, v), and the acceleration of gravity is in the negative direction of y. So we have the equation

$$\nabla \cdot \mathbf{u} = 0 \tag{3.42}$$

$$\frac{\partial \mathbf{u}}{\partial t} + (\mathbf{u} \cdot \nabla)\mathbf{u} = -\frac{1}{\rho}\nabla P - g\boldsymbol{j} \tag{3.43}$$

Now consider irrotational motion, i.e. $\mathrm{rot}\,\mathbf{u} = 0$, so there is a velocity potential $\mathbf{u} = \nabla\varphi$. From

$$\nabla\left(\frac{1}{2}\mathbf{u}^2\right) = (\mathbf{u} \cdot \nabla)\mathbf{u} - \mathrm{rot}\,\mathbf{u}_x \cdot \mathbf{u} = (\mathbf{u} \cdot \nabla)\mathbf{u} \tag{3.44}$$

and the integral of (3.43), we get

$$\frac{p - p_0}{\rho_0} = B(t) - \varphi_t - \frac{1}{2}(\nabla\varphi)^2 - gy$$

where $B(t)$ is an arbitrary function and p_0 is an arbitrary constant. Letting

$$\varphi' = \varphi - \int B(t)dt$$

we have

$$\mathbf{u} = \nabla\varphi, \quad \frac{p - p_0}{\rho_0} = -\varphi'_t - \frac{1}{2}\left(\nabla\varphi'\right)^2 - gy \tag{3.45}$$

Henceforth, φ' is still denoted as φ. From (3.42), there is

$$\nabla \cdot \mathbf{u} = 0 \Longrightarrow \nabla^2\varphi = 0 \tag{3.46}$$

Let the surface equation of water be

$$f(x_1, x_2, y, t) = 0 \tag{3.47}$$

On this surface, a fluid particle cannot pass through it, so the velocity of the fluid orthogonal to this surface must be equal to the normal velocity of the surface. The normal velocity of (3.47) is

$$\frac{-f_t}{\sqrt{f_{x_1}^2 + f_{x_2}^2 + f_y^2}}$$

and the normal velocity of the fluid is

$$\frac{u_1 f_{x_1} + u_2 f_{x_2} + v f_y}{\sqrt{f_{x_1}^2 + f_{x_2}^2 + f_y^2}}$$

The condition for the two to be equal is

$$f_t + u_1 f_{x_1} + u_2 f_{x_2} + v f_y = 0 \tag{3.48}$$

In particular, when $y = \eta(x_1, x_2, t)$, $f(x_1, x_2, y, t) \equiv \eta(x_1, x_2, t) - y$, (3.48) gives

$$\eta_t - u_1 \eta_{x_1} + u_2 \eta_{x_2} = v \tag{3.49}$$

In addition, on the free plane, $p = p_0$ (ignoring the movement of air), hence

$$\begin{cases} \eta_t + \varphi_{x_1} \eta_{x_1} + \varphi_{x_2} \eta_{x_2} = \varphi_y \\ \varphi_t + \dfrac{1}{2}\left(\varphi_{x_1}^2 + \varphi_{x_2}^2 + \varphi_y^2\right) + g\eta = 0 \end{cases} \qquad y = \eta(x_1, x_2, t) \tag{3.50}$$

where $u_1 = \varphi_{x_1}$, $u_2 = \varphi_{x_2}$, $v = \varphi_y$. By solid boundary conditions, the normal velocity of the fluid must be 0, $\mathbf{n} \cdot \nabla \varphi = 0$. Especially at the bottom $y = -h_0(x_1, x_2)$, we have $\varphi_y + \varphi_{x_1} h_{0x_1} + \varphi_{x_2} h_{0x_2} = 0$. For a horizontal base, $\varphi_y = 0, y = -h_0$. So our whole problem is proposed as follows: Seeking the velocity potential φ and surface η satisfies

$$\nabla^2 \varphi = 0 \tag{3.51}$$

$$\begin{cases} \eta_t + \varphi_{x_1} \eta_{x_1} + \varphi_{x_2} \eta_{x_2} = \varphi_y \\ \varphi_t + \dfrac{1}{2}\left(\varphi_{x_1}^2 + \varphi_{x_2}^2 + \varphi_y^2\right) - g\eta = 0 \end{cases} \qquad y = \eta(x_1, x_2, t) \tag{3.52}$$

$$\varphi_y = 0, \quad y = -h \tag{3.53}$$

For simplicity, we consider the one-dimensional case below, that is, $\eta = \eta(x, t)$. y is measured from the horizontal base, at this point, $\varphi_y = 0$, $y = 0$, two parameters are introduced:

$$\alpha = \frac{a}{h_0}, \quad \beta = \frac{h_0^2}{l^2}$$

where a is the amplitude of the wave, l is the wavelength, $y = h_0 + \eta$. Let $x = lx'$, $y = h_0 y'$, $t = \dfrac{lt'}{c_0}$, $\eta = a\eta'$, $\varphi = \dfrac{gla\varphi'}{c_0}$, $c_0^2 = gh_0$, and then ignore the notation", ".
From (3.46), (3.52), (3.53), we have

$$\beta\varphi_{xx} + \varphi_{yy} = 0, \quad 0 < y < 1 + \alpha\eta \tag{3.54}$$

$$\varphi_y = 0, \quad y = 0 \tag{3.55}$$

$$\left.\begin{array}{l} \eta_t + \alpha\varphi_x\eta_x - \dfrac{1}{\beta}\varphi_y = 0 \\[4mm] \eta + \varphi_t + \dfrac{1}{2}\alpha\varphi_x^2 + \dfrac{1}{2}\dfrac{\alpha}{\beta}\varphi_y^2 = 0 \end{array}\right\} \quad y = 1 + \alpha\eta \tag{3.56}$$

The formal solution of (3.54), (3.55) is

$$\varphi = \sum_0^\infty (-1)^m \frac{y^{2m}}{(2m)!} \frac{\partial^{2m} f}{\partial x^{2m}} \beta^{2m} \tag{3.57}$$

where $f = f_0(x, t)$. Substituting (3.57) into the first equation of (3.56) yields

$$\eta_t + \alpha\left[f_x - (1 + \alpha\eta)\eta_x f'_{xx}\beta - \frac{(1 + \alpha\eta)^2}{2} f_{xxx}\beta + \cdots\right]\eta_x$$
$$+(1 + \alpha\eta)f'' - \frac{1}{3!}(1 + \alpha\eta)^3 \cdot f_{xxxx}\beta + o\left(\beta^2\right) = 0$$

i.e.,

$$\eta_t + \{(1 + \alpha\eta)f_x\}_x - \left\{\frac{1}{6}(1 + \alpha\eta)^3 f_{xxxx}\right.$$
$$\left. + \frac{1}{2}\alpha(1 + \alpha\eta)^2 f_{xxx}\eta_x\right\}\beta + o\left(\beta^2\right) = 0 \tag{3.58}$$

Similarly, substituting it into the second equation in (3.56) yields

$$\eta + f_t + \frac{1}{2}\alpha f_x^2 - \frac{1}{2}(1 + \alpha\eta)^2 \left\{f_{xxt} + \alpha f_x f_{xxx} - \alpha f_{xx}^2\right\}\beta + o\left(\beta^2\right) = 0 \tag{3.59}$$

In (3.58), (3.59), if we ignore the first-order terms of β and differentiate (3.59) respect to x, we have

$$\begin{cases} \eta_t + \{(1 + \alpha\eta)w\}_x = 0 \\ w_t + \alpha w w_x + \eta_x = 0 \end{cases} \quad w = f_x \qquad (3.60)$$

If the first-order term β is kept, then there is

$$\begin{cases} \eta_t + \{(1 + \alpha\eta)w\}_x - \dfrac{1}{6}\beta w_{xxx} + O\left(\alpha\beta, \beta^2\right) = 0 \\[2mm] w_t + \alpha w w_x + \eta_x - \dfrac{1}{2}\beta w_{xxt} + O\left(\alpha\beta, \beta^2\right) = 0 \end{cases} \qquad (3.61)$$

If the first-order terms of α, β are ignored in (3.61), then when $w = \eta$, there is the same equation $\eta_t + \eta_x = 0$, so w can be expanded by α, β,

$$w = \eta + \alpha A + \beta B + O\left(\alpha^2 + \beta^2\right)$$

where A, B are functions of η and the derivative of η with respect to x. From (3.61), we can obtain

$$\eta_t + \eta_x + \alpha\left(A_x + 2\eta\eta_x\right) + \beta\left(B_x - \frac{1}{6}\eta_{xxx}\right) + O\left(\alpha^2 + \beta^2\right) = 0$$

$$\eta_t + \eta_x + \alpha\left(A_t + \eta\eta_x\right) + \beta\left(B_t - \frac{1}{2}\eta_{xxt}\right) + O\left(\alpha^2 + \beta^2\right) = 0$$

Since $\eta_t = -\eta_x + O(\alpha, \beta)$, the derivative with respect to t can be converted into the derivative with respect to x in the first-order terms. In particular, when $A = -\dfrac{1}{4}\eta^2$, $B = \dfrac{1}{3}\eta_{xx}$, the above two equations are consistent. There is

$$\eta_t + \eta_x + \frac{3}{2}\alpha\eta\eta_x + \frac{1}{6}\beta\eta_{xxx} + O\left(\alpha^2 + \beta^2\right) = 0 \qquad (3.62)$$

At this point,

$$w = \eta - \frac{1}{4}\alpha\eta^2 + \frac{1}{3}\beta\eta_{xx} + O\left(\alpha^2 + \beta^2\right)$$

If the second-order terms are ignored in equation (3.62), the typical KdV equation is obtained as

$$\eta_t + \eta_x + \frac{3}{2}\alpha\eta\eta_x + \frac{1}{6}\beta\eta_{xxx} = 0 \qquad (3.63)$$

If we set $\eta_{xxx} = -\eta_{xxt}$ in (3.63), we have

$$\eta_t + \eta_x + \frac{3}{2}\alpha\eta\eta_x - \frac{1}{6}\beta\eta_{xxt} = 0 \tag{3.64}$$

which is the Benjamain equation.

In the following, we derive a rather broad class of wave equations under weakly nonlinear interactions, reducing it to the KdV equation or Burgers equation. A system of equations is set as

$$\begin{cases} n_t + (nu)_x = 0 & (3.65) \\ (nu)_t + \left(nu^2 + P\right)_x = 0 & (3.66) \\ P = P\left(f, n, u, f_i, n_i, u_i, f_{ij}, n_{ij}, u_{ij}, \cdots\right) & (3.67) \\ F\left(f, n, u, f_i, n_i, u_i, f_{ij}, n_{ij}, u_{ij}, \cdots\right) = 0 & (3.68) \end{cases}$$

Here, n, u and f are state variables, n represents the number density of the mass point, u represents the velocity of the mass point, i and j represent the derivatives of space variables x and time variables t respectively, P generally represents the functions of state variables (n, u, f) and their derivatives. If f represents a parametric function, then P is a function of n, u and their derivatives. (3.65) is the conservation of mass, (3.66) is the conservation of momentum, as shown in the following examples.

(i) Gas dynamics: f is pressure p,

$$P = \frac{1}{m}\left(p - \mu u_x\right), \quad F = P - A\rho^r, \quad mn = \rho \tag{3.69}$$

ρ is the density, μ is the viscosity coefficient.

(ii) Shallow water wave: n is water depth h, now there are only two state variables h, u,

$$P = \frac{1}{2}gh^2 - \frac{1}{3}h^3\left(u_{xt} + uu_{xx} + u_x^2\right) \tag{3.70}$$

(iii) Magnetic fluid wave of cold plasma: f is magnetic intensity B, $P = \frac{1}{2}B^2$,

$$F \equiv B - n - (B_x/n)_x = 0 \tag{3.71}$$

(iv) Ionic sound waves of cold plasma: f is the electrostatic potential, ψ is the wave function,

$$P = e^\psi - \frac{1}{2}\psi_x^2, \quad F \equiv n - e^\psi + \psi_{xx} = 0 \tag{3.72}$$

In the state of local thermodynamic equilibrium, if all derivatives in P, F disappear, then

$$P = P(f, n), \quad F(f, n) = 0 \tag{3.73}$$

In this case, from (3.66), we have

$$nu_t + nuu_x + P_x = 0, \quad P_x = \frac{\partial P}{\partial f}\frac{\partial f}{\partial x} + \frac{\partial P}{\partial n}\frac{\partial n}{\partial x}$$

Then using

$$\frac{\partial F}{\partial f} \cdot \frac{\partial f}{\partial x} + \frac{\partial F}{\partial n}\frac{\partial n}{\partial x} = 0$$

to eliminate $\dfrac{\partial f}{\partial x}$, we get

$$u_t + uu_x + \left(a^2/n\right)n_x = 0, \quad a^2 \equiv [P_n - (F_n/F_f)\, P_f]$$

If $a^2 > 0$, then

$$\begin{cases} n_t + (nu)_x = 0 \\ u_t + uu_x + \left(a^2/n\right)n_x = 0 \end{cases} \tag{3.74}$$

is a system of hyperbolic equations with the characteristic line $\dfrac{dx}{dt} = u \pm a$, and a is the wave velocity. From small perturbations about uniform states, the wave equation is obtained as

$$u_{tt} - a_0^2 u_{xx} = 0$$

where a_0 is the uniform wave velocity. In the following derivation of the KdV equation and Burgers equation, it is necessary to consider the nonlinear term of the small perturbation, that is, the effect of the P, F derivatives. We make the following transformations:

$$\begin{cases} \xi = \varepsilon^\alpha \left(x - a_0 t\right) \\ \tau = \varepsilon^{\alpha+1} t \end{cases} \tag{3.75}$$

where ε represents the amplitude of the initial disturbance. Set $\varepsilon < 1$, index $\alpha > 0$ to be determined, a_0 represents a certain wave speed and is regarded as a constant. Substitute (3.75) into (3.65), (3.66), and get

$$\varepsilon n_\tau + (u - a_0)\, n_\xi + nu_\xi = 0 \tag{3.76}$$

$$\varepsilon u_\tau + (u - a_0) u_\xi + n^{-1} P_\xi = 0 \tag{3.77}$$

The state variables (n, f, u) are asymptotically expanded by ε. Make the expansions around the equilibrium $A = (n, f, u) = (n_0, f_0, 0)$:

$$\begin{cases} n = n_0 + \varepsilon n^{(1)} + \varepsilon^2 n^{(2)} + \cdots \\ f = f_0 + \varepsilon f^{(1)} + \varepsilon^2 f^{(2)} + \cdots \\ u = 0 + \varepsilon u^{(1)} + \varepsilon^2 u^{(2)} + \cdots \end{cases}$$

P, F are also expanded as

$$P = P_0 + P_{f_0} (f - f_0) + P_{n_0} (n - n_0) + P_{u_0} (u - u_0) + O(\varepsilon^2)$$

$$F = F_0 + F_{f_0} (f - f_0) + F_{n_0} (n - n_0) + F_{u_0} (u - u_0) + O(\varepsilon^2)$$

Due to the invariance of the equation with respect to Galileo, $P_{u_0} = F_{u_0} = 0$. Because $P^{(1)} = P_{f_0} f^{(1)} + P_{n_0} n^{(1)}$ and $F_{f_0} \dfrac{\partial f^{(1)}}{\partial \xi} + F_{n_0} \dfrac{\partial n^{(1)}}{\partial \xi} = 0$, so

$$\frac{\partial P^{(1)}}{\partial \xi} = P_{f_0} \frac{\partial f^{(1)}}{\partial \xi} + P_{n_0} \frac{\partial n^{(1)}}{\partial \xi} = \left[P_{n_0} - \left(\frac{F_{n_0}}{F_{f_0}} \right) P_{f_0} \right] \frac{\partial n^{(1)}}{\partial \xi} = a_0^2 \frac{\partial n^{(1)}}{\partial \xi}$$

If the above expansion takes into account second-order terms, we get

$$P_\xi^{(2)} \approx a_0^2 n_\xi^{(2)} + A_n^{(1)} n_\xi^{(1)} + \varepsilon^{\alpha-1} B n_{\xi\xi}^{(1)} + \varepsilon^{2\alpha-1} c n_{\xi\xi\xi}^{(1)}$$

The constants a_0^2, A, B, C have the following table for the above example:

	a_0	A	B	C
Aerodynamics	$2KT/m$	0	$-va_0$	0
Water wave	gh_0	0	0	$\frac{1}{3} gh_0^3$
Magnetic fluid wave	B_0	1	0	1
Ionic sound wave	1	0	0	1

Comparing the first-order terms of ε from (3.76), (3.77), there is

$$a_0 n_\xi^{(1)} = n_0 u_\xi^{(1)}, \quad a_0 u_\xi^{(1)} = \left(\frac{a^2}{n_0} \right) n_\xi^{(1)}$$

Integrating it and using the boundary conditions, we have $a_0 n^{(1)} = n_0 u^{(1)}$. Substituting it into (3.76), (3.77), and taking the second-order approximation, we get

$$n_\tau^{(1)} + u^{(1)} n_\xi^{(1)} + n_0 u_\xi^{(2)} + n^{(1)} u_\xi^{(1)} - a_0 n_\xi^{(2)} = 0$$

i.e.,

$$n_\tau^{(1)} + 2\frac{a_0}{n_0}n^{(1)}n_\xi^{(1)} - a_0 n_\xi^{(2)} + n_0 u_\xi^{(2)} = 0$$

and

$$\frac{a_0}{n_0}n_\tau^{(1)} + \frac{A}{n_0}n^{(1)}n_\xi^{(1)} + \varepsilon^{\alpha-1}\frac{B}{n_0}n_{\xi\xi}^{(1)} + \varepsilon^{2\alpha-1}\frac{C_0}{n_0}n_{\xi\xi\xi}^{(1)}$$

$$+ \frac{a_0^2}{n}n_\xi^{(2)} + u^{(1)}u_\xi^{(1)} - a_0 u_\xi^{(2)} = 0$$

From

$$-a_0 u_\xi^{(2)} + \frac{a_0^2}{n_0}n_\xi^{(2)} = \frac{a_0}{n_0}\left(n_\tau^{(1)} + 2\frac{a_0}{n_0}n^{(1)}n_\xi^{(1)}\right)$$

eliminating $n_\xi^{(2)}$, $u_\xi^{(2)}$ gives the equation for $n^{(1)}$,

$$n_\tau^{(1)} + \left(\frac{A}{2a_0} + \frac{3}{2}\frac{a_0}{n_0}\right)n^{(1)}n_\xi^{(1)} + \varepsilon^{\alpha-1}\frac{B}{2a_0}n_{\xi\xi}^{(1)} + \varepsilon^{2\alpha-1}\frac{C}{2a_0}n_{\xi\xi\xi}^{(1)} = 0 \qquad (3.78)$$

In (3.78), if $B \neq 0$ (dissipation, $B < 0$), $\alpha = 1$, $C = 0$, then Burgers equation is obtained; If $B = 0$ (dispersion), $\alpha = \frac{1}{2}$, then the KdV equation is obtained as

$$\eta_\tau^{(1)} + \left(\frac{A}{2a_0} + \frac{3}{2}\frac{a_0}{n_0}\right)n^{(1)}n_\xi^{(1)} + \frac{C}{2a_0}n_{\xi\xi\xi}^{(1)} = 0$$

References

[1] Zabusky N J, Kruskal M D. Interaction of "solitons" in a collisionless plasma and the recurrence of initial states[J]. Physical review letters, 1965, 15(6): 240.

[2] Lax P D. Integrals of nonlinear equations of evolution and solitary waves[J]. Communications on Pure and Applied Mathematics, 1968, 21(5): 467-490.

[3] Berezin Y A, Karpman V I. Nonlinear evolution of disturbances in plasmas and other dispersive media[J]. Soviet Physics JETP, 1967, 24(5): 1049-1056.

[4] Гельфанд И М. Некоторые задачи теории квазилинейных уравнений[J]. Успехи мат. наук, 1959, 14(2): 86.

[5] Cohn D B, MacKenzie K R. Density-step-excited ion acoustic solitons[J]. Physical Review Letters, 1973, 30(7): 258.

[6] Lax P D. Hyperbolic systems of conservation laws II[J]. Communications on Pure and Applied Mathematics, 1957, 10(4): 537-566.

[7] 张学铭. 关于最用量原理与拉格朗日场论 [J]. 应用数学和计算数学, 1979,(3): 50.

[8] 张学铭. 在非古典变分场中关于 SOLITON 的某些性质 [J]. 数学物理学报, 1984, 4(1): 1-7.

[9] 管克英. 有外场的非线性 Schrödinger 方程孤子解、定态解及释能法 [J]. 科学通报, 1981, 26(21): 1283-1286.

[10] 徐邦清. 非线性 Schrödinger 方程孤立子解的一般稳定性理论 [J]. 数学物理学报, 1982, 2(4): 421-434.

Chapter 4
Hirota Method

4.1 Introduction

Besides the inverse scattering method and the prolongation structure method, there is another important and direct method for solving the N solitons of certain nonlinear evolutionary equations, which is the Hirota method using function transformations to find specific solutions. This method has evolved from seeking N soliton solutions for the KdV equation, MKdV equation, sine-Gordon equation, Toda lattice equation and Boussinesq equations to a widely applicable approach for finding soliton solutions across a broad range of nonlinear evolution equations. Furthermore, this method can be employed to derive the Bäcklund transformations for these specific nonlinear evolution equations.

Let's take the KdV equation as an example to briefly introduce this method. Let the KdV equation have the form

$$u_t + 6u_t u_x + u_{xxx} = 0 \tag{4.1}$$

and satisfy the boundary conditions $u = 0$, $|x| \to \infty$. We use perturbation method to solve (4.1). Let $u = \omega_x$, integrating (4.1) with respect to x yields

$$w_t + 3w_x^2 + w_{xxx} = 0 \tag{4.2}$$

here, selecting the integral constant as 0. Expanding ω in terms of the small parameter ϵ, it is the following power series

$$w = \epsilon w_1 + \epsilon^2 w_2 + \cdots \tag{4.3}$$

Substituting (4.3) into (4.2) and combining the same power terms of ϵ gives

$$\left(\frac{\partial}{\partial t} + \frac{\partial^3}{\partial x^3} \right) w_1 = 0$$

$$\left(\frac{\partial}{\partial t} + \frac{\partial^3}{\partial x^3} \right) w_2 = -3 \left(w_1 \right)_x^2$$

$$\left(\frac{\partial}{\partial t} + \frac{\partial^3}{\partial x^3}\right) w_3 = -6\left(w_1\right)_x \left(w_2\right)_x$$

$$\cdots \cdots$$

Solving these equations one by one can obtain the formal solution of this perturbation series. But this series may converge slowly or even diverge, so a method similar to Padá approximation needs to be considered.

Substituting $w = G/F$ into (4.2), we have

$$\left(G_t F - GF_t\right)/F^2 + 3\left(G_x F - GF_x\right)^2/F^4$$

$$+ \left(G_{xxx}F - 3G_{xx}F_x + 3G_x F_{xx} - GF_{xxx}\right)/F^2$$

$$+ 6\left(FG_x F_x^2 + FGF_x F_{xx} - GF_x^3\right)/F^4 = 0 \qquad (4.4)$$

At first glance, equation (4.4) appears more complex than the original equation and contains two unknown functions F and G in one equation. However, we observe that equation (4.4) can be rewritten as

$$\left[G_t F - GF_t + 3\lambda\left(G_x F - GF_x\right) + G_{xxx}F - 3G_{xx}F_x + 3G_x F_{xx} - GF_{xxx}\right]/F^2$$

$$+ 3\left(G_x F - GF_x\right)\left[G_x F - GF_x - 2\left(FF_{xx} - F_x^2\right) - 2\lambda F^2\right]/F^4 = 0 \qquad (4.5)$$

By introducing any parameter λ, two equations can be obtained

$$G_t F - GF_t + 3\lambda\left(G_x F - GF_x\right) + G_{xxx}F - 3G_{xx}F_x + 3G_x F_{xx} - GF_{xxx} = 0 \quad (4.6)$$

$$2\left(FF_{xx} - F_x^2\right) + 2\lambda F^2 - \left(G_x F - GF_x\right) = 0 \qquad (4.7)$$

They can also be represented separately as

$$\left[\frac{\partial}{\partial t} - \frac{\partial}{\partial t'} + 3\lambda\left(\frac{\partial}{\partial x} - \frac{\partial}{\partial x'}\right) + \left(\frac{\partial}{\partial x} - \frac{\partial}{\partial x'}\right)^3\right] G(x,t)F\left(x',t'\right)\big|_{x=x',t=t'} = 0 \quad (4.8)$$

and

$$\left[\left(\frac{\partial}{\partial x} - \frac{\partial}{\partial x'}\right)^2 + \lambda\right] F(x,t)F\left(x',t'\right)\bigg|_{x=x',t=t'}$$

$$- \left(\frac{\partial}{\partial x} - \frac{\partial}{\partial x'}\right) G(x,t)F\left(x',t'\right)\bigg|_{x=x',t=t'} = 0 \qquad (4.9)$$

For convenience, the operators D_t, D_x and their various products are introduced as follows:

$$D_t^n D_x^m f \cdot g = \left(\frac{\partial}{\partial t} - \frac{\partial}{\partial t'}\right)^n \left(\frac{\partial}{\partial x} - \frac{\partial}{\partial x'}\right)^m f(x,t)g\left(x',t'\right)\big|_{x=x',t=t'} \qquad (4.10)$$

Thus, using these symbols, equations (4.8) and (4.9) can be abbreviated as

$$\left(D_t + 3\lambda D_x + D_x^3\right) G \cdot F = 0 \tag{4.11}$$

$$\left(D_x^3 + \lambda\right) F \cdot F - D_x G \cdot F = 0 \tag{4.12}$$

(4.12) can be written as

$$\lambda = (G/F)_x - 2(\log F)_{xx} \tag{4.13}$$

If $\lambda = 0$, then

$$G = 2F_x \tag{4.14}$$

is the solution to (4.13). Thus, we have

$$u = (G/F)_x = 2(\log F)_{xx} \tag{4.15}$$

Substituting (4.14) into (4.11) yields

$$(D_t + D_x^3)F_x \cdot F = 0 \tag{4.16}$$

or

$$D_x \left(D_t + D_x^3\right) F \cdot F = 0 \tag{4.17}$$

4.2   Some Properties of the D Operator

We already know in the first section

$$D_t^n D_x^m a \cdot b = \left(\frac{\partial}{\partial t} - \frac{\partial}{\partial t'}\right)^n \left(\frac{\partial}{\partial x} - \frac{\partial}{\partial x'}\right)^m a(x,t)b\left(x',t'\right)|_{x=x',t=t'}$$

Introduce the operator D_z and the differential operator $\dfrac{\partial}{\partial z}$ as follows:

$$D_z = \delta D_t + \varepsilon D_x, \quad \frac{\partial}{\partial z} = \delta\frac{\partial}{\partial t} + \varepsilon\frac{\partial}{\partial x}$$

where δ and ε are constants. The following properties can be easily obtained from the definition of the D operator:

(I) $D_z^m a \cdot 1 = \left(\dfrac{\partial}{\partial z}\right)^m a.$

(II) $D_z^m a \cdot b = (-1)^m D_z^m b \cdot a.$

(II.1) $D_z^m a \cdot a = 0$, for m being an odd number.

(III) $D_z^m a \cdot b = D_z^{m-1}\left(a_z \cdot b - a \cdot b_z\right).$

(III.1) $D_z^m a \cdot a = 2D^{m-1}a_z \cdot a$, for m being an even number.

(III.2) $D_x D_t a \cdot a = 2D_x a_t \cdot a = 2D_t a_x \cdot a.$

(IV) $D_x^m \exp(p_1 x) \cdot \exp(p_2 x) = (p_1 - p_2)^m \exp[(p_1 + p_2)x].$

Let $F(D_t, D_x)$ be a polynomial of D_t and D_x, then

(IV.1) $F(D_t, D_x) \exp(\Omega_1 t + p_1 x) \exp(\Omega_2 t + p_2 x)$

$$= F(\Omega_1 - \Omega_2, p_1 - p_2) / F(\Omega_1 + \Omega_2, p_2 + p_2)$$

$$\cdot F(D_t, D_x) \exp[(\Omega_1 + \Omega_2)t + (p_1 + p_2)x]$$

(V) $\exp(\varepsilon D_x) a(x) \cdot b(x) = a(x + \varepsilon)b(x - \varepsilon).$

(VI) $\exp(\varepsilon D_z) ab \cdot cd = [\exp(\varepsilon D_z) a \cdot c] \cdot [\exp(\varepsilon D_z) b \cdot d]$

$$= [\exp(\varepsilon D_z)a \cdot d][\exp(\varepsilon D_z)b \cdot c]$$

(VI.1) $D_z ab \cdot c = \left(\dfrac{\partial a}{\partial z}\right) bc + a\left(D_z b \cdot c\right).$

(VI.2) $D_z^2 ab \cdot c = \left(\dfrac{\partial^2 a}{\partial z^2}\right) bc + 2\left(\dfrac{\partial a}{\partial z}\right) D_z b \cdot c + a(D_z^2 b \cdot c).$

(VI.3) $D_z^3 ac \cdot bc = \left(D_z^3 a \cdot b\right) c^2 + 3(D_z a \cdot b)D_z^2 b \cdot c.$

(VI.4) $D_x^m \exp(px) a \cdot \exp(px) b = \exp(2px) D_x^m a \cdot b.$

(VII) $\exp(\delta D_t) [\exp(\varepsilon D_x) a \cdot b] \cdot [\exp(\varepsilon D_x) c \cdot d]$

$$= \exp(\varepsilon D_x) [\exp(\delta D_t)a \cdot c] \cdot [\exp(\delta D_t)b \cdot d]$$

$$= [\exp(\delta D_t + \varepsilon D_x)a \cdot d] \cdot [\exp(-\delta D_t + \varepsilon D_x)c \cdot b]$$

The following equations are useful for transforming nonlinear differential equations into bilinear form.

(VIII) $\exp\left(\varepsilon \dfrac{\partial}{\partial z}\right) [a/b] = [\exp(\varepsilon D_z) a \cdot b] / [\cosh(\varepsilon D_z) b \cdot b].$

(VIII.1) $\dfrac{\partial}{\partial z}(a/b) = \dfrac{D_z a \cdot b}{b^2}.$

(VIII.2) $\dfrac{\partial^2}{\partial z^2}(a/b) = \dfrac{D_z^2 a \cdot b}{b^2} - \left(\dfrac{a}{b}\right)\dfrac{D_z^2 b \cdot b}{b^2}.$

(VIII.3) $\dfrac{\partial^3}{\partial z^3}(a/b) = \dfrac{D_z^3 a \cdot b}{b^2} - 3\left[\dfrac{D_z^2 a \cdot b}{b^2} - \dfrac{D_z^2 b \cdot b}{b^2}\right].$

(IX) $2\cosh\left(\varepsilon \dfrac{\partial}{\partial z}\right) \log f = \log[\cosh(\varepsilon D_x) f \cdot f].$

(IX.1) $\dfrac{\partial^2}{\partial z^2} \log f = \dfrac{\partial_z^2 f \cdot f}{2f^2}.$

(IX.2) $\dfrac{\partial^4}{\partial z^4} \log f = \dfrac{D_z^4 f \cdot f}{2f^2} - 6\left[\dfrac{D_z^2 f \cdot f}{2f^2}\right]^2.$

The following equation is useful for transforming bilinear differential equations into original nonlinear equations.

(X) $\exp\left(\varepsilon D_x\right) a \cdot b = \left\{\exp\left[2\cosh\left(\varepsilon\dfrac{\partial}{\partial x}\right)\log b\right]\right\}\left[\exp\left(\varepsilon\dfrac{\partial}{\partial x}\right)(a/b)\right].$

Set $\phi = a/b, \quad u = 2(\log b)_{xx}$, we have

(X.1) $\left(D_x a \cdot b\right)/b^2 = \phi_x.$

(X.2) $\left(D_x^2 a \cdot b\right)/b^2 = \phi_{xx} + u\phi.$

(X.3) $\left(D_x^3 a \cdot b\right)/b^2 = \phi_{xxx} + 3u\phi_x.$

(X.4) $\left(D_x^4 a \cdot b\right)/b^2 = \phi_{xxxx} + 6u\phi_{xx} + (u_{xx} + 3u^2)\phi.$

(XI) $\exp\left(\varepsilon D_x\right) a \cdot b = \exp\left[\sinh\left(\varepsilon\dfrac{\partial}{\partial x}\right)\log(a/b) + \cosh\left(\varepsilon\dfrac{\partial}{\partial x}\right)\cdot\log(ab)\right].$

Let $\varphi = \log(a/b), \quad \rho = \log(ab)$, we have

(XI.1) $\left(D_x a \cdot b\right)/ab = \varphi_x.$

(XI.2) $\left(D_x^2 a \cdot b\right)/a \cdot b = \rho_{xx} + \varphi_x^2.$

(XI.3) $\left(D_x^3 a \cdot b\right)/ab = \varphi_{xxx} + 3\varphi_x\rho_{xx} + \varphi_x^3.$

(XI.4) $\left(D_x^4 a \cdot b\right)/a \cdot b = \rho_{xxxx} + 4\varphi_x\varphi_{xxx} + 3(\rho_{xx})^4 + 6\varphi_x^2\rho_{xx} + \varphi_x^4.$

The properties of the D operator listed above are easy to verify, and we take (X) as an example. Here

$$2\cosh\left(\varepsilon\frac{\partial}{\partial x}\right)\log b = \log b(x+\varepsilon) + \log b(x-\varepsilon)$$

$$\exp\left(\varepsilon\frac{\partial}{\partial x}\right)(a/b) = a(x+\varepsilon)/b(x+\varepsilon)$$

From (V), we have

$$\exp\left(\varepsilon D_x\right) a \cdot b = a(x+\varepsilon)b(x-\varepsilon)$$

Thus

$$\exp\left(\varepsilon D_x\right) a \cdot b = \exp\left[2\cosh\left(\varepsilon\frac{\partial}{\partial x}\right)\log b\right]\cdot\left[\exp\left(\varepsilon\frac{\partial}{\partial x}\right)(a/b)\right]$$

This proves that (X) holds. And equations (X.1 − 4) can be obtained by expanding (X) as a power series of ε and comparing the same power equations of ε.

4.3   Solutions to Bilinear Differential Equations

Now we solve (4.17). Expanding F into a power series with a small parameter ε,

$$F = 1 + \varepsilon f_1 + \varepsilon^2 f_2 + \cdots \tag{4.18}$$

substituting (4.18) into (4.17) and comparing the same power of ε, we have

$$2\frac{\partial}{\partial x}\left(\frac{\partial}{\partial t} + \frac{\partial^3}{\partial x^3}\right)f_1 = 0 \tag{4.19}$$

$$2\frac{\partial}{\partial x}\left(\frac{\partial}{\partial t}+\frac{\partial^3}{\partial x^3}\right)f_2=-D_x\left(D_t+D_x^3\right)f_1\cdot f_1 \tag{4.20}$$

$$2\frac{\partial}{\partial x}\left(\frac{\partial}{\partial t}+\frac{\partial^3}{\partial x^3}\right)f_3=-D_x\left(D_t+D_x^3\right)\left(f_2\cdot f_1+f_1\cdot f_2\right) \tag{4.21}$$

$$\cdots\cdots$$

We seek two types of solutions: I) polynomial solutions, II) solutions in exponential form.

For I), it can be found

$$f_1=a_0+a_1x+a_2x^2+a_3x^3+a_4x^4+bt-24a_4tx \tag{4.22}$$

is a solution of (4.19). When $a_4=0$, $3a_1a_3=a_2^2$, $b=12a_3$, $f_2=0$ can be selected. Therefore, an exact solution for (4.17) can be obtained

$$F=1+\varepsilon\left[a_0+a_1x+(3a_1a_3)^{\frac{1}{2}}x^2+a_3\left(x^3+12t\right)\right] \tag{4.23}$$

If the boundary condition $u|_{x=0}=0$, then $a_1=0$. When selecting $\varepsilon=1$, the expression for F is

$$F=a_3\left[x^3+12(t+\text{const})\right] \tag{4.24}$$

Since $u=2(\log F)_{xx}$, using the bounded condition of the solution, we obtain

$$u=-6x\left(x^3-24t\right)/\left(x^3+12t\right)^2$$

For II), from (4.19) we have

$$f_1=\sum_{j=1}^{N}a_j\exp(\Omega_jt+p_jx) \tag{4.25}$$

here, $\Omega_j+p_j^3=0$, p_j and a_j are all constants.

Substituting (4.25) into (4.20) and utilizing the properties (IV) and (IV.1) of the D operator, we can obtain

$$f_2=\sum_{i>j}^{N}\exp\left(A_{ij}+\eta_i+\eta_j\right) \tag{4.26}$$

where $\exp\left(\eta_j\right)=a_j\exp\left(\Omega_jt+p_jx\right)$, and

$$\exp\left(A_{ij}\right)=-\frac{(p_i-p_j)\left[\Omega_i-\Omega_j+(p_i-p_j)^3\right]}{(p_i+p_j)\left[\Omega_i+\Omega_j+(p_i+p_j)^3\right]}$$

$$= (p_i - p_j)^2 / (p_i + p_j)^2 \tag{4.27}$$

Substituting (4.26) into (4.21) and utilizing the property (VI.4) of D operator and relationship (4.19), we can obtain

$$f_3 = \sum_{i>j>k}^{N} \exp \left(A_{ijk} + \eta_i + \eta_j + \eta_k \right)$$

with

$$\exp \left(A_{ijk} \right) = \exp \left(A_{ij} + A_{ik} + A_{jk} \right)$$

until find f_N, we obtain the exact solution

$$F = \sum_{\mu=0,1} \exp \left(\sum_{i>j}^{(N)} A_{ij} \mu_i \mu_j + \sum_i \mu_i \eta_j \right) \tag{4.28}$$

among them, $\displaystyle\sum_{\mu=0,1}$ represents the sum of all possible combinations of $\mu_1 = 0, 1; \mu_2 = 0, 1; \cdots; \mu_N = 0, 1$. $\displaystyle\sum_{i>j}^{(N)}$ represents the sum of all distinct possible pairs of N elements. The small parameter ε has been absorbed into the constant term a_i. (4.28) together with $u = 2(\log F)_{xx}$ provide the formula for N soliton solutions of the KdV equation.

4.4   Applications in Sine-Gordon Equation and MKdV Equation

We first consider the following sine-Gordon equation

$$\varphi_{xx} - \varphi_{tt} = \sin \varphi \tag{4.29}$$

Suppose $\dfrac{\partial \varphi}{\partial x} \to 0 \ (|x| \to \infty)$. Set

$$\varphi(x, t) = 4 \tan^{-1} [g(x, t) / f(x, t)] \tag{4.30}$$

where

$$f(x, t) = \sum_{n=0}^{\left[\frac{N}{2}\right]} \sum_{N^c 2n} a \left(i_1, i_2, \cdots, i_{2n} \right) \cdot \exp \left(\eta_{i_1} + \eta_{i_2} + \cdots + \eta_{i_{2n}} \right) \tag{4.31}$$

$$g(x, t) = \sum_{m=0}^{[(N-1)/2)]} \sum_{N^c(2m+1)} a \left(j_1, j_2, \cdots, j_{2m+1} \right) \cdot \exp \left(\eta_{j_1} + \eta_{j_2} + \cdots + \eta_{j_{2m+1}} \right)$$

$$\tag{4.32}$$

$$a\left(i_1, i_2, \cdots, i_n\right): \begin{cases} \displaystyle\prod_{k<l}^{(n)} a\left(i_k, i_l\right), & n \geqslant 2 \\[2mm] 1, & n = 0, 1 \end{cases}$$

$$a\left(i_k, i_l\right) = \frac{\left(p_{ik} - p_{il}\right)^2 - \left(\Omega_{ik} - \Omega_{il}\right)^2}{\left(p_{ik} + p_{il}\right)^2 - \left(\Omega_{ik} + \Omega_{il}\right)^2} = -\frac{\left(p_{ik} - p_{il} + \Omega_{ik} - \Omega_{il}\right)^2}{\left(p_{ik} - p_{il} + \Omega_{ik} + \Omega_{il}\right)^2}$$

$$\eta_i = p_i x - \Omega_i t - \eta_i^0, \quad p_i^2 - \Omega_i^2 = 1$$

p_i is an arbitrary finite real constant that determines the amplitude of the i-th soliton, and η_i^0 is an arbitrary finite real constant that determines the phase of the i-th soliton. Assume p_i are different.

For example, $N = 3$, there are

$$f(x, t) = 1 + a(1, 2) \exp\left(\eta_1 + \eta_2\right) + a(1, 3) \exp\left(\eta_1 + \eta_3\right)$$

$$+ a(2, 3) \exp\left(\eta_2 + \eta_3\right)$$

$$g(x, t) = \exp\left(\eta_1\right) + \exp\left(\eta_2\right) + \exp\left(\eta_3\right)$$

$$+ a(1, 2, 3) \exp\left(\eta_1 + \eta_2 + \eta_3\right)$$

$$a(1, 2, 3) = a(1, 2)a(1, 3)a(2, 3),$$

$$\eta_i = p_i x - \Omega_i t, \quad p_i^2 - \Omega_i^2 = 1$$

When $t \to \infty$, keep η_1 finite. For $Q_3/p_3 > Q_2/p_2 > Q_1/p_1 > 0$, $p_i > 0$, $g(x, t)/f(x, t) = \exp\left(\eta_1\right)$, soliton solution can be obtained

$$i(x, t) = -\frac{\partial \varphi}{\partial x} = -2p_1 \mathrm{sech}(\eta_1)$$

We briefly prove that the expressions (4.30), (4.31) and (4.32) are indeed the solutions to equation (4.29).

Therefore, by substituting (4.30) into (4.29), the equation satisfied by f, g can be obtained

$$fg_{xx} - 2f_x g_x + f_{xx}g - (fg_{tt} - 2f_t g_t + f_{tt}g) = fg \tag{4.33}$$

$$f_{xx}f - 2f_x^2 + ff_{xx} - (f_{tt}f - 2f_t^2 + ff_{tt})$$

$$= g_{xx}g - 2g_x^2 + g_{xx}g - (g_{tt}g - 2g_t^2 + gg_{tt}) \tag{4.34}$$

Substituting the expressions (4.31) and (4.32) into (4.33) and (4.34) yields the following relations

$$\sum_{l=0}^{n} \sum_{n^c l} a\left(i_1, i_2, \cdots, i_l\right) a\left(i_{l+1}, i_{l+2}, \cdots, i_n\right)$$

$$\cdot h_1\left(i_1, i_2, \cdots, i_l; i_{l+1}, i_{l+2}, \cdots, i_n\right) = 0$$

$$n = 1, 3, 5, \cdots \leqslant N \tag{4.35}$$

$$\sum_{l=0}^{n} \sum_{n^c l} (-1)^l a\left(i_1, i_2, \cdots, i_l\right) a\left(i_{l+1}, i_{l+2}, \cdots, i_n\right)$$

$$\cdot h_2\left(i_1, i_2, \cdots, i_l; i_{l+1}, i_{l+2}, \cdots, i_n\right) = 0$$

$$n = 2, 4, 6, \cdots \leqslant N \tag{4.36}$$

where

$$h\left(i_1, i_2, \cdots, i_l; i_{l+1}, i_{l+2}, \cdots, i_n\right)$$

$$= \left(p_{i_1} + p_{i_2} + \cdots + p_{i_l} - p_{i_{l+1}} - p_{i_{l+2}} - \cdots - p_{i_n}\right)^2$$

$$- \left(\Omega_{i_1} + \Omega_{i_2} + \cdots + \Omega_{i_l} - \Omega_{i_{l+1}} - \Omega_{i_{l+2}} - \cdots - \Omega_{i_n}\right)^2$$

For a given n, (4.35) and (4.36) can be transformed into the following identity

$$\sum_{\sigma_1, \sigma_2, \cdots \sigma_n = \pm 1} (\Pi_{i=1}^{n} \sigma_i) \hat{b}(\sigma_1 x_1, \sigma_2 x_2, \cdots, \sigma_n x_n)$$

$$\cdot \hat{h}_1(\sigma_1 x_1, \sigma_2 x_2, \cdots, \sigma_n x_n) = 0, \text{ when } n \text{ is an odd number} \tag{4.37}$$

$$\sum_{\sigma_1, \sigma_2, \cdots \sigma_n = \pm 1} (\Pi_{i=1}^{n} \sigma_i) \hat{b}(\sigma_1 x_1, \sigma_2 x_2, \cdots, \sigma_n x_n)$$

$$\cdot \hat{h}_2(\sigma_1 x_1, \sigma_2 x_2, \cdots, \sigma_n x_n) = 0, \text{ when } n \text{ is an even number} \tag{4.38}$$

where

$$\hat{b}(\sigma_1 x_1, \sigma_2 x_2, \cdots, \sigma_n x_n) = \Pi_{k<l}^{(n)}(\sigma_k x_k - \sigma_l x_l)^2$$

$$\hat{h}_1(\sigma_1 x_1, \sigma_2 x_2, \cdots, \sigma_n x_n) = (\Pi_{i=1}^{n} \sigma_i x_i) \left(\sum_{i=1}^{n} \Pi_{l=1, l\neq i}^{n} \sigma_l x_l \right) - \Pi_{i=1}^{n} \sigma_i x_i$$

$$\hat{h}_2(\sigma_1 x_1, \sigma_2 x_2, \cdots, \sigma_n x_n) = \left(\sum_{i=1}^{n} \sigma_i x_i \right) \left(\sum_{i=1}^{n} \Pi_{l=1, l\neq i}^{n} \sigma_l x_l \right)$$

$$x_i = p_i + \Omega_i$$

In fact, note the left of (4.37) as $D_1\left(x_1, x_2, \cdots, x_n\right)$, which has properties

(i) D_1 is a symmetric homogeneous polynomial.

(ii) If $x_1 = \pm x_2$, then

$$D_1\left(x_1, \cdots, x_n\right)_{x_1 = \pm x_2} = 8x_1^4 \prod_{i=3}^{n} \left(x_1^2 - x_i^2\right)^2 D\left(x_3, x_4, \cdots, x_n\right)$$

It is obviously that (4.37) holds for $n = 1$. Assuming (4.37) holds for $n = 2$, then from (i) and (ii), we can see that D_1 is a symmetric homogeneous polynomial of order $2n(n-1)$

$$\prod_{k<l}^{(n)} (x_k^2 - x_l^2)^2$$

On the other hand, it can be directly seen that D_1 is n^2 order polynomial. Therefore, for a given n, $D_1 = 0$. Similarly $D_2 = 0$.

Secondly, we consider the following MKdV equation

$$v_t + 24v^2 v_x + v_{xxx} = 0 \tag{4.39}$$

$$v(x,t) = \frac{\partial \varphi}{\partial x} \tag{4.40}$$

Set

$$\tan \varphi(x,t) = g(x,t)/f(x,t) \tag{4.41}$$

where

$$f(x,t) = \sum_{n=0}^{[N/2]} \sum_{N^c 2n} a(i_1, i_2, \cdots, i_{2n}) \cdot \exp\left(\xi_{i_1} + \xi_{i_2} + \cdots + \xi_{i_{2n}}\right) \tag{4.42}$$

$$g(x,t) = \sum_{n=0}^{[(N-1)/2]} \sum_{N^c 2m+1} a(i_1, i_2, \cdots, i_{2m+1}) \cdot \exp\left(\xi_{i_1} + \xi_{i_2} + \cdots + \xi_{i_{2m+1}}\right) \tag{4.43}$$

$$a(i_1, i_2, \cdots, i_n) = \begin{cases} \prod_{k<l}^{(n)} a(i_k, i_l), & n \geqslant 2 \\ -1, & n = 0,1 \end{cases}$$

$$a(i_k, i_l) = -\frac{(p_{i_k} - p_{i_l})^2}{(p_{i_k} + p_{i_l})^2}$$

$$\xi_i = p_i x - \Omega_i t - \xi_i^0, \quad \Omega_i = p_i^3$$

When $N = 3$, we have

$$f(x,t) = 1 + a(1,2) \exp(\xi_1 + \xi_2) + a(1,3) \exp(\xi_1 + \xi_3) + a(2,3) \exp(\xi_2 + \xi_3)$$

$$g(x,t) = \exp(\xi_1) + \exp(\xi_2) + \exp(\xi_3) + a(1,2,3) \exp(\xi_1 + \xi_2 + \xi_3)$$

$$a(1,2,3) = a(1,2)a(1,3)a(2,3), \quad \xi_i = p_i x - p_i^3 t$$

When $t \to \infty$, ξ_1 is fixed and finite,

$$g/f = \exp(\xi_1), \quad p_3 > p_2 > p_1 > 0$$

there is soliton solution

$$\nu(x,t) = p_1/2 \operatorname{sech} \xi_1$$

f, g satisfy

$$g_t f - g f_t + g_{xxx} f - 3 g_{xx} f_x + 3 g_x f_{xx} - g f_{xxx} = 0 \tag{4.44}$$

$$f f_{xx} - 2 f_x^2 + f_{xx} f + g g_{xx} - 2 g_x^2 + g_{xx} g = 0 \tag{4.45}$$

(4.44), (4.45) can be written in the form

$$\left\{ \left(\frac{\partial}{\partial t} - \frac{\partial}{\partial t'} \right) - \left(\frac{\partial}{\partial x} - \frac{\partial}{\partial x'} \right)^3 \right\} g(x,t) f(x',t') \mid_{t=t',x=x'} = 0$$

$$\left(\frac{\partial}{\partial t} - \frac{\partial}{\partial x'} \right)^2 (f(x,t) f(x',t') + g(x,t) g(x',t')) \mid_{t=t',x=x'} = 0$$

Substituting the expressions (4.42) and (4.43) for f and g into (4.44) and (4.45), we obtain

$$\sum_{l=0}^{n} \sum_{n^c l} \hat{a}(i_1, i_2, \cdots, i_l) \, \hat{a}(i_{l+1}, i_{l+2}, \cdots, i_n)$$

$$\cdot h_1(i_1, i_2, \cdots, i_l, i_{l+1}, \cdots, i_n) = 0$$

$$n = 1, 3, 5 \cdots \leqslant N \tag{4.46}$$

$$\sum_{l=0}^{n} \sum_{n^c l} (-1)^l \hat{a}(i_1, i_2, \cdots, i_l) \, \hat{a}(i_{l+1}, i_{l+2}, \cdots, i_n)$$

$$\cdot h_2(i_1, i_2, \cdots, i_l, i_{l+1}, \cdots, i_n) = 0$$

$$n = 2, 4, 6 \cdots \leqslant N \tag{4.47}$$

$$\hat{a}(i_1, i_2, \cdots, i_n) = \begin{cases} \displaystyle\prod_{k<l}^{(n)} \hat{a}(i_k, i_l), & n \geqslant 2 \\ \\ 1, & n = 0, 1 \end{cases}$$

$$\hat{a}(i_k, i_l) = \frac{(p_{i_k} - p_{i_l})^2}{(p_{i_k} + p_{i_l})^2}$$

$$h_1(i_1, i_2, \cdots, i_l; i_{l+1}, \cdots, i_n)$$

$$= -(p_{i_1}^3 + p_{i_2}^3 + \cdots + p_{i_l}^3 - p_{i_{(l+1)}}^3 - \cdots - p_{i_n}^3)$$

$$+ (p_{i_1} + p_{i_2} + \cdots + p_{i_l} - p_{i_{(l+1)}} - \cdots - p_{i_n})^3$$

$$h_2(i_1, i_2, \cdots, i_l; i_{l+1}, \cdots, i_n)$$

$$= -(p_{i_1} + p_{i_2} + \cdots + p_{i_l} - p_{i_{(l+1)}} - \cdots - p_{i_n})^2$$

(4.46) and (4.47) can be transformed to

$$\sum_{\sigma_1,\sigma_2,\cdots\sigma_n=\pm 1} \hat{b}\left(\sigma_1 p_1, \sigma_2 p_2, \cdots, \sigma_n p_n\right)$$

$$\cdot h_1\left(\sigma_1 p_1, \sigma_2 p_2, \cdots, \sigma_n p_n\right) = 0, \quad n \text{ is odd number} \tag{4.48}$$

$$\sum_{\sigma_1,\sigma_2,\cdots\sigma_n=\pm 1} \left(\prod_{i=1}^n \sigma_i\right) \hat{b}\left(\sigma_1 p_1, \sigma_2 p_2, \cdots, \sigma_n p_n\right)$$

$$\cdot h_2\left(\sigma_1 p_1, \sigma_2 p_2, \cdots, \sigma_n p_n\right) = 0, \quad n \text{ is even number} \tag{4.49}$$

$$\hat{b}\left(\sigma_1 p_1, \cdots, \sigma_n p_n\right) = \prod_{k<l}^n \left(\sigma_k p_k - \sigma_l p_l\right)^2$$

$$h_1\left(\sigma_1 p_1, \cdots, \sigma_n p_n\right) = -\left(\sigma_1 p_1^3 + \sigma_n^3 p_n^3\right) \cdot \left(\sigma_1 p_1 + \sigma_2 p_2 + \cdots + \sigma_n p_n\right)^3$$

$$h_2\left(\sigma_1 p_1, \cdots, \sigma_n p_n\right) = \left(\sigma_1 p_1 + \sigma_2 p_2 + \cdots + \sigma_n p_n\right)^2$$

The left hand of the identity (4.48) is represented as $D_1(p_1, \cdots, p_s)$, which has the following properties:

(i) D_1 is a symmetric homogeneous polynomial,

(ii) D_1 is an even function of $p_1, p_2, \cdots, p_n$

(iii) If $p_1 = p_2$, then

$$D_1(p_1, \cdots, p_n) = 2(2p_1)^2 \prod_{m=2}^n (p_1^2 - p_m^2)^2 \cdot D(p_3, p_4, \cdots, p_n)$$

It is easy to know that (4.48) holds for $n = 1$. If $n = 2$ holds, then D is $2n(n-1)$ order homogeneous symmetric polynomial. On the other hand, it is known that D_1 is $n(n-1)+3$ order polynomial, so for a fixed n, $D_1 = 0$. Similarly, (4.49) holds.

For nonlinear lattice equation

$$m\frac{d^2 r_n}{dt^2} = a\left[e^{-b r_n} - e^{-b r_{n+1}}\right], \quad n = 1, 2 \tag{4.50}$$

where a, b are constants, $r_n = y_n - y_{n-1}$. Using transformations

$$\frac{ab}{m}\left(e^{-br_n} - 1\right) = (\log f_n)_t$$

Hirota obtained the expression for the N soliton solutions of (4.50) as follows

$$f_n(t) = \sum_{\mu=0,1} \exp\left[\sum_{i<j}^{N} B_{ij}\mu_i\mu_j + \sum_{i=1}^{N} \mu_i x_i\right]$$

where

$$x_i = \beta_i t - k_i n + r_i, \quad k_i, r_i \text{ are constants}$$

$$\beta_i = \pm\left(\frac{ab}{m}\right)^{\frac{1}{2}} 2\sin\frac{k_i}{2}$$

$$e^{B_{ij}} = \frac{\dfrac{m}{ab}(\beta_i - \beta_j)^2 - 4\sinh^2\dfrac{k_i + k_j}{2}}{\dfrac{m}{ab}(\beta_i + \beta_j)^2 - 4\sinh^2\dfrac{k_i + k_j}{2}}$$

Here, $\displaystyle\sum_{\mu=0,1}$ represents the sum of all possible combinations of $\mu_1 = 0, 1$; $\mu_2 = 0, 1; \cdots$; $\mu_n = 0, 1$.

For the nonlinear electronic filter equation system

$$\frac{d^2}{dt^2}\log\left(1 + V_n(t)\right) = V_{n+1}(t) - 2V_n(t) + V_{n-1}(t) \tag{4.51}$$

and

$$\begin{cases} \dfrac{dV_n}{dt} = \left(1 + V_n^2\right)(I_n - I_{n-1}) \\[3mm] \dfrac{dI_n}{dt} = \left(1 + I_n^2\right)(V_{n-1} - V_n) \end{cases} \tag{4.52}$$

Hirota obtained N soliton solutions of (4.51) by using the transformation

$$V_n = \left[\tan^{-1} g_n/f_n\right]_t$$

where f_n, g_n have the following form

$$f_n(t) = \sum_{\mu=0,1}^{(l)} \exp\left[\sum_{i<j}^{N} B_{ij}\mu_i\mu_j + \sum_{i=1}^{N} \mu_i x_i\right]$$

$$g_n(t) = \sum_{\mu=0,1}^{(e)} \exp\left[\sum_{i<j}^{N} B_{ij}\mu_i\mu_j + \sum_{i=1}^{N} \mu_i x_i\right]$$

with

$$\alpha_i = \beta_i t - k_i n + \gamma_i$$

$$\beta_i = \pm 2 \sinh \frac{k_i}{2}$$

$$e^{B_{ij}} = -\frac{(\beta_i - \beta_j)^2 - 4\sinh^2 \dfrac{k_i - k_j}{2}}{(\beta_i + \beta_j)^2 - 4\sinh^2 \dfrac{k_i + k_j}{2}}$$

$\overset{(l)}{\underset{\mu=0,1}{\sum}}$, $\overset{(e)}{\underset{\mu=0,1}{\sum}}$ represent the sum of all possible combinations of $\mu_1 = 0,1;\ \mu_2 = 0,1;\cdots;\ \mu_N = 0,1$. But for the former, it is required that

$$\sum_{i=1}^{N(l)} \mu_i = \text{even integer}$$

For the latter, it is required that

$$\sum_{i=1}^{N(e)} \mu_i = \text{odd integer}$$

For Hirota equation

$$i\varphi_t + i3\alpha|\varphi|^2\varphi_x + \rho\varphi_{xx} + i\sigma\varphi_{xxx} + \delta|\varphi|^2\varphi = 0 \qquad (4.53)$$

Its N envelope soliton solutions have the form

$$\varphi = g/f$$

$$f(x,t) = \overset{\prime}{\underset{\mu=0,1}{\sum}} \exp\left[\sum_{i<j}^{2N} B_{ij}\mu_i\mu_j + \sum_{i=1}^{2N} \mu_i x_i\right]$$

$$g(x,t) = \overset{\prime\prime}{\underset{\mu=0,1}{\sum}} \exp\left[\sum_{i<j}^{2N} B_{ij}\mu_i\mu_j + \sum_{i=1}^{2N} \mu_i x_i\right]$$

$$g^*(x,t) = \overset{\prime\prime\prime}{\underset{\mu=0,1}{\sum}} \exp\left[\sum_{i<j}^{2N} B_{ij}\mu_i\mu_j + \sum_{i=1}^{2N} \mu_i x_i\right]$$

$$x_i = k_j x - \beta_j t + \gamma_j, \quad k_j, \gamma_j \text{ are constants}$$

$$\beta_j = -\mathrm{i}\rho k_j^2 + \sigma k_j^3, \ j = 1, 2, \cdots, 2N, \ \mathrm{i} = \sqrt{-1}$$

$$k_{j+N} = k_j^*, \ \beta_{j+N} = \beta_j^*, j = 1, 2, \cdots, 2N$$

Here, the symbol $*$ represents complex conjugation, and

$$B_{ij} = \log\left[\frac{\alpha}{2\sigma}(k_i + k_j)^2\right]$$

for $i = 1, 2, \cdots, N$; $j = N+1, N+2, \cdots, 2N$ or $i = N+1, \cdots, 2N$; $j = 1, 2, \cdots, N$.

$$B_{ij} = -\log\left[\frac{\alpha}{2\sigma}(k_i - k_j)^2\right]$$

for $i = 1, 2, \cdots, N$; $j = 1, 2, \cdots, N$ or $i = N+1, \cdots, 2N$; $j = N+1, \cdots, 2N$.

$\displaystyle\sum_{\mu=0,1}'$ represents the sum of all possible combinations of $\mu_1 = 0, 1$; $\mu_2 = 0, 1; \cdots$; $\mu_{2N} = 0, 1$, and it is required that

$$\sum_{i=1}^{N} \mu_i = \sum_{i=1}^{N} \mu_{i+N}$$

$\displaystyle\sum_{\mu=0,1}''$, $\displaystyle\sum_{\mu=0,1}'''$ represent the sum of all possible combinations of $\mu_1 = 0, 1$; $\mu_2 = 0, 1; \cdots$; $\mu_{2N} = 0, 1$ respectively, and they are required that

$$\sum_{i=1}^{N} \mu_i = 1 + \sum_{i=1}^{N} \mu_{i+N}$$

$$1 + \sum_{i=1}^{N} \mu_i = \sum_{i=1}^{N} \mu_{i+N}$$

4.5   Bäcklund Transform in Bilinear Form

We know that certain nonlinear evolutionary equations can be transformed into bilinear form using the Hirota method. Now we consider the following form of bilinear differential equation

$$F(D_t, D_x)f \cdot f = 0 \tag{4.54}$$

and construct a new differential equation

$$[F(D_t, D_x)f' \cdot f']ff - f'f'[F(D_t, D_x)f \cdot f] = 0 \tag{4.55}$$

Obviously, if f is a solution of (4.54), then it can be inferred from (4.55) that g is another solution of (4.54). Therefore, (4.55) actually provides the Bäcklund transformation of (4.54) regarding to f and g. Let's take the KdV equation as an example to illustrate it. To do this, we also need to use the following formulas, which can be derived from the definition and properties of the D operator mentioned above.

1) $\exp(D_1)[\exp(D_2)a \cdot b] \cdot [\exp(D_3)c \cdot d]$

$$= \exp \frac{1}{2}(D_2 - D_3) \left\{ \exp\left[\frac{1}{2}(D_2 + D_3) + D_1\right] a \cdot d \right\}$$

$$\cdot \left\{ \exp\left[\frac{1}{2}(D_2 + D_3) - D_1\right] c \cdot b \right\}$$

where $D_i = \epsilon_i D_x + \delta_i D_t$, ϵ_i, δ_i are constants, $i = 1, 2, 3$.

2) $\left(D_x^2 a \cdot b\right) cd - ab \left(D_x^2 c \cdot d\right) = D_x \left[(D_x a \cdot d) cb + ad (D_x c \cdot b)\right]$

3) $\left(D_x D_t f' \cdot f'\right) f f - f'f' \left(D_x^2 f \cdot f\right) = 2D_x \left(D_t f' \cdot f\right) \cdot f f'$

4) $\left(D_x^2 f' \cdot f'\right)\right) f f - f'f' \left(D_x^2 f \cdot f\right) = 2D_x \left(D_x f' \cdot f\right) \cdot f f'$

5) $\left(D_x^4 f' \cdot f'\right) f f - f'f'(D_x^4 f \cdot f) = 2D_x \left(D_x^3 f' \cdot f\right) \cdot f f' + 6D_x \left(D_x^2 f' \cdot f\right) \cdot D_x \left(f \cdot f'\right)$

Now consider the KdV equation in bilinear form

$$D_x \left(D_t + c_0 D_x + D_x^3\right) f \cdot f = 0 \tag{4.56}$$

here c_0 is constant. Let f be one solution of (4.56) and f' be the other solution. We write the equation about f, f' as

$$\left[D_x \left(D_t + c_0 D_x + D_x^3\right) f' \cdot f\right] f f - f'f' \left[D_x \left(D_t + c_0 D_x + D_x^3\right) f \cdot f\right] = 0 \tag{4.57}$$

This is the Bäcklund transformation of equation (4.56). From (4.57) we can obtain

$$2D_x \left\{ \left[D_t + (\rho_0 + 3\lambda)D_x + D_x^3\right] f' \cdot f \right\} \cdot (ff')$$

$$+ 6D_x \left[\left(D_x^3 - \mu D_x - \lambda\right) f' \cdot f\right] (D_x f \cdot f') = 0$$

by utilizing properties 2), 3), 4) and 5), where λ and μ are arbitrary constants. If f is a solution of (4.56), then when f' satisfies the following system of equations

$$\left[D_t + (c_0 + 3\lambda) D_x + D_x^3\right] f' \cdot f = 0 \tag{4.58}$$

$$\left(D_x^2 - \mu D_x - \lambda\right) f' \cdot f = 0 \tag{4.59}$$

it is also another solution of (4.56). Equations (4.58) and (4.59) are the Bäcklund transformations of (4.56). Similarly, we can obtain the following nonlinear Bäcklund transformations.

I) Boussinesq equation

$$\left(D_t^2 - D_x^2 - D_x^4\right) f \cdot f = 0$$

$$\text{BT:} \begin{cases} \left(D_t + aD_x^2\right) f' \cdot f = 0 \\ \left(aD_tD_x + D_x + D_x^3\right) f' \cdot f = 0 \end{cases}$$

where $a^2 = -3$.

II) Kadomtsev-Petviashvili equation

$$\left(D_tD_x + D_y^2 + D_x^4\right) f \cdot f = 0$$

$$\text{BT:} \begin{cases} \left(D_y + aD_x^2\right) f' \cdot f = 0 \\ \left(-aD_yD_x + D_t + D_x^3\right) f' \cdot f = 0 \end{cases}$$

where $a^2 = 3$.

III) High-order KdV equation

$$D_x \left(D_t + D_x^5\right) f \cdot f = 0$$

$$\text{BT:} \begin{cases} D_x^3 f' \cdot f = \lambda f' \cdot f \\ \left[D_t - \left(\dfrac{15}{2}\right)\lambda D_x^2 - \left(\dfrac{3}{2}\right) D_x^5\right] f' \cdot f = 0 \end{cases}$$

IV) Shallow water equation

$$D_x \left(D_t - D_tD_x^2 + D_x\right) f \cdot f = 0$$

$$\text{BT:} \begin{cases} \left(D_x^3 - D_x\right) f' \cdot f = \lambda f' \cdot f \\ \left(3D_xD_t - 1\right) f' \cdot f = \mu D_x f' \cdot f \end{cases}$$

References

[1] Hirota R. Direct method of finding exact solutions of nonlinear evolution equations[J]. Bäcklund Transformations, the Inverse Scattering Method, Solitons, and Their Applications, 1976: 40-68.

[2] Satsuma J. N-soliton solution of the two-dimensional Korteweg-deVries equation[J]. Journal of the Physical society of Japan, 1976, 40(1): 286-290.

[3] Ablowitz M J, Kaup D J, Newell A C, et al. The inverse scattering transform-Fourier analysis for nonlinear problems[J]. Studies in applied mathematics, 1974, 53(4): 249-315.

[4] Sawada K, Kotera T. A method for finding N-soliton solutions of the KdV equation and KdV-like equation[J]. Progress of theoretical physics, 1974, 51(5): 1355-1367.

[5] Caudrey P J, Dodd R K, Gibbon J D. A new hierarchy of Korteweg–de Vries equations[J]. Proceedings of the Royal Society of London. A. Mathematical and Physical Sciences, 1976, 351(1666): 407-422.

[6]   GL LAMB J R. Analytical descriptions of ultrashort optical pulse propagation in a resonant medium[J]. Reviews of Modern Physics, 1971, 43(2): 99.

[7]   Wahlquist H D, Estabrook F B. Bäcklund transformation for solutions of the Korteweg-de Vries equation[J]. Physical Review Letters, 1973, 31(23): 1386.

[8]   Satsuma J. Higher conservation laws for the Korteweg-de Vries equation through Bäcklund transformation[J]. Progress of Theoretical Physics, 1974, 52(4): 1396-1397.

[9]   Chen H H. General derivation of Bäcklund transformations from inverse scattering problems[J]. Physical Review Letters, 1974, 33(15): 925.

[10] Wadati M, Sanuki H, Konno K. Relationships among inverse method, Bäcklund transformation and an infinite number of conservation laws[J]. Progress of theoretical physics, 1975, 53(2): 419-436.

[11] Chen H H, Liu C S. Bäcklund transformation solutions of the Toda lattice equation[J]. Journal of Mathematical Physics, 1975, 16(7): 1428-1430.

[12] Wadati M. Wave propagation in nonlinear lattice. I[J]. Journal of the Physical Society of Japan, 1975, 38(3): 673-680.

Chapter 5
Bäcklund Transformation and Infinite Conservation Law

5.1 Sine-Gordon Equation and Bäcklund Transformation

Nonlinear Klein-Gordon equation

$$\varphi_{tt} - \varphi_{xx} + F'(\varphi) = 0 \tag{5.1}$$

When $F'(\varphi) = \varphi$, it is Klein-Gordon equation. When $F'(\varphi) = \sin\varphi$,

$$\varphi_{tt} - \varphi_{xx} + \sin\varphi = 0 \tag{5.2}$$

is the sine-Gordon (SG) equation. If $\sin\varphi \sim \varphi$, we get KG equation. If $\sin\varphi \sim \varphi - \frac{1}{3!}\varphi^3$, that is, $F(\varphi) = \frac{1}{2}\varphi^2 - \frac{1}{24}\varphi^4$, then the φ^4 field equation is obtained

$$\varphi_{tt} - \varphi_{xx} + \varphi - \frac{1}{3!}\varphi^3 = 0 \tag{5.3}$$

If $F'(\varphi) = \sin\varphi + \lambda\sin 2\varphi$, then the equation is

$$\varphi_{tt} - \varphi_{xx} + \sin\varphi + \lambda\sin 2\varphi = 0 \tag{5.4}$$

which is called the double sine-Gordon (DSG) equation.

Under the optical cone coordinate

$$\xi = \frac{(x-t)}{2}, \quad \eta = \frac{(x+t)}{2}$$

equation (5.2) can be reduced to

$$\varphi_{\xi\eta} = \sin\varphi \tag{5.5}$$

SG equation was first obtained from the study of surface geometry for a class of practical problems with Gauss curvature $K = -1$, and later found that it can be reduced to this kind of equation in many physical problems. For example, the propagation of vortex lines of Jesephson junction, in the study of general superconducting junction, Jesephson found that the electric current through the superconducting junction satisfies the following relations:

$$J = J_0 \sin \varphi, \quad \frac{d\varphi}{dt} = \frac{2e}{\hbar} v$$

where v is the voltage and $\varphi = \varphi_1 - \varphi_2$ is the phase difference of the two superconducting wave functions. We can obtain that φ satisfies the following sine-Gordon equation

$$\varphi_{xx} + \varphi_{yy} - LC\varphi_{tt} = \frac{2eLJ_0}{\hbar} \sin \varphi$$

where $L, C, e, J_0, \hbar$ are physical constants. Other related physical problems are: dislocation of crystals, propagation of waves generated by magnetization direction in ferromagnetic material and so on.

We can easily obtain the traveling wave solution of SG equation (5.2). Let $\varphi = \Phi(\xi)$, $\xi = x - Dt$, $D = \text{const} > 0$, then (5.2) is

$$\left(D^2 - 1\right) \Phi_{\xi\xi} + \sin \Phi = 0 \tag{5.6}$$

It is multiplied by Φ_ξ and integrated, there is

$$\frac{1}{2} \left(D^2 - 1\right) \Phi_\xi^2 + 2\sin^2 \frac{1}{2}\Phi = A \tag{5.7}$$

where A is the integral constant. From this we can obtain the soliton solution and the periodic wave train solution of (5.2). When $A = 0$, $D^2 - 1 < 0$, there is

$$\tan \Phi/4 = \pm \exp \left\{ \pm \left(1 - D^2\right)^{-\frac{1}{2}} (\xi - \xi_0) \right\}$$

or

$$\Phi = 4\tan^{-1} \pm \left\{ \pm \left(1 - D^2\right)^{-\frac{1}{2}} (x - Dt) \right\} (\text{ take } \xi_0 = 0) \tag{5.8}$$

It is the soliton solution of (5.2). The periodic wave train is

1) $0 < A < 2$, $D^2 - 1 > 0$, this is the periodic solution, Φ oscillates between $-\Phi_0 < \Phi < \Phi_0$ with respect to $\Phi = 0$, $\Phi_0 = 2\sin^{-1} \left(\frac{A}{2}\right)^{\frac{1}{2}}$.

2) $0 < A < 2$, $D^2 - 1 < 0$, it is still the periodic solution, Φ oscillates between $\pi - \Phi_0 < \Phi < \pi + \Phi_0$ with respect to $\Phi = \pi$.

3) $A < 0$, $D^2 - 1 < 0$, it is the spiral wave

$$\Phi_\xi = \pm \left\{ \frac{2}{1 - D^2} \left(|A| + 2\sin^2 \frac{1}{2}\Phi \right) \right\}^{\frac{1}{2}}$$

4) $A > 2$, $D^2 - 1 > 0$, there is also the spiral wave

$$\Phi_\xi = \pm \left\{ \frac{2}{(D^2 - 1)} \left(A - 2\sin^2 \frac{1}{2}\Phi \right) \right\}^{\frac{1}{2}}$$

5) $A = 2$, $D^2 - 1 > 0$, there is a solution

$$\tan \left(\frac{\Phi + \pi}{4} \right) = \exp \left\{ \pm \left(D^2 - 1 \right)^{-\frac{1}{2}} (\xi - \xi_0) \right\}$$

This represents the kink from $\Phi = -\pi$ to $\Phi = \pi$.

Lamb[9] used the Bäcklund transform (BT) to concretely solve the soliton solution of the SG equation. As we know, the Bäcklund transform is a class of transform relations for unknown function. For (5.5), we introduce the Bäcklund transform

$$\begin{cases} \dfrac{\partial \varphi'}{\partial \xi} = \dfrac{\partial \varphi}{\partial \xi} + 2\lambda \sin \left(\dfrac{\varphi + \varphi'}{2} \right) \\[3mm] \dfrac{\partial \varphi'}{\partial \eta} = -\dfrac{\partial \varphi}{\partial \eta} + \dfrac{2}{\lambda} \sin \left(\dfrac{\varphi' - \varphi}{2} \right) \end{cases} \tag{5.9}$$

where λ is an arbitrary parameter. We take the derivative of the first equation with respect to η and take the derivative of the second equation with respect to ξ in (5.9). Using the consistency of φ' second-order derivative and the original equation $\varphi_{\xi\eta} = \sin\varphi$, it is not difficult to get a new equation $\varphi'_{\xi\eta} = \sin\varphi'$, which is exactly the same form as the old equation, so we can use BT (5.9) to find a new solution. First, the trivial solution $\varphi = 0$ can be obtained, and the other solution φ_1 can be obtained by BT (5.9)

$$\varphi_1 = 4\tan^{-1} \left[\exp \left(\pm \frac{x - Dt}{\sqrt{1 - D^2}} \right) \right], \quad D = \frac{1 - \lambda^2}{1 + \lambda^2} \tag{5.10}$$

φ_1 is just a soliton solution of the SG equation.

From above we see that the soliton solution (5.10) of the SG equation is obtained from BT (5.9) starting from the zero solution. In general, for SG equations, the Bäcklund transformation always causes soliton solutions to come out of nothing, from N soliton solutions to $N + 1$ soliton solutions. Moreover, using the so-called "commutability theorem", starting from the existing sets of solutions, we can get

new solutions only through algebraic operations instead of integrating the system of
equations (5.9). This "commutativity theorem" states: "Starting from the original
solution φ_0 of equation (5.5), whether the transformation (5.9) of the parameter λ_1
is performed first, and then the transformation of the parameter λ_2 is taken, or the
reverse transformation of the parameter λ_2 is performed first, then the transformation
with the parameter λ_1, the resulting solution φ_2 is the same." From the commutativity
theorem, a "nonlinear superposition formula" for the solution of equation (5.5) can
be derived.

$$\tan\left(\frac{\varphi_3 - \varphi_0}{4}\right) = \frac{D_1 + D_2}{D_1 - D_2}\tan\left(\frac{\varphi_1 - \varphi_2}{4}\right) \tag{5.11}$$

Specially taking $\varphi_0 = 0$, φ_1, φ_2 is shown in (5.10). Substituting it into (5.11) gives
the double soliton solution of (5.5) (Perring-Suyrme solution)

$$\tan\varphi/4 = \frac{\text{sh}\left(x/\sqrt{1-D^2}\right)}{\text{ch}\left(Dt/\sqrt{1-D^2}\right)} \tag{5.12}$$

It represents the superposition of two kinks, and

$$\tan\varphi/4 = \frac{\text{sh}\left(Dt/\sqrt{1-D^2}\right)}{D\,\text{ch}\left(x/\sqrt{1-D^2}\right)} \tag{5.13}$$

represents the collision of a positive kink and an anti-kink. In this formula $D = ib$ is
taken, another soliton solution of (5.5) is obtained

$$\tan\varphi/4 = \frac{\sin\left(bt\sqrt{1+b^2}\right)}{\text{ch}\left(x/\sqrt{1+b^2}\right)} \tag{5.14}$$

Its image shows the positive kink and anti-kink cycle to approach and leave, back and
forth together relative vibration. It is called "breather".

We now generalize the above procedure of finding the soliton solution of SG
equation by using the B transformation. Assuming through B transformation

$$\begin{cases} \dfrac{1}{2}\left(\varphi_{jx} - \varphi_{j-1,x}\right) = a_j \sin\dfrac{1}{2}\left(\varphi_j + \varphi_{j-1}\right) \\[3mm] \dfrac{1}{2}\left(\varphi_{jt} - \varphi_{j-1,t}\right) = -\dfrac{1}{a_j}\sin\dfrac{1}{2}\left(\varphi_j - \varphi_{j-1}\right) \end{cases} \tag{5.14'}$$

and $\varphi_0 = 0$, the special solution of (5.5) $\varphi_1, \varphi_2, \cdots, \varphi_k$ can be obtained. Under
appropriate initial conditions, four solutions of (5.5) are linked by two parameters

a_k, a_j:

$$
\begin{cases}
\varphi_{k,j+1} = B_{a_j}\varphi_{kj} \\
\varphi_{k+1,j} = B_{a_k}\varphi_{kj} \\
\varphi_{k+1,j+1} = B_{a_j}B_{a_k}\varphi_{kj} = B_{a_k} \cdot B_{a_j}\varphi_{kj}
\end{cases}
\tag{5.15}
$$

where B_{a_j} is the B transform with parameter a_j. Figure 5-1 shows the commutativity of the B transform.

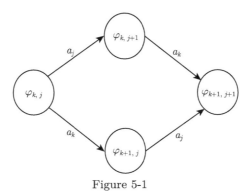

Figure 5-1

Thus N soliton solutions of the sine-Gordon equation can be obtained. The analytical expression of the solution is as follows:

1) When $j > 0$, $\varphi_{j+1,j} = 0$;

$$
\varphi_{j,j} = 4\tan^{-1}\left[e^{-k_j x + \left(\frac{1}{k_j}\right)t + r_{jj}}\right], \gamma_{jj} \text{ is constant}
$$

$$
(-1)^j/k_j < 0
$$

2) When $j > l$,

$$
\varphi_{l,j} = \varphi_{l+1,j-1} + 4\tan^{-1}\left[\frac{\dfrac{1}{k_j} - \dfrac{1}{k_l}}{\dfrac{1}{k_j} + \dfrac{1}{k_l}}\right]\tan\left(\frac{\varphi_{l,j-1} - \varphi_{l+1,j}}{4}\right)
$$

$$
(-1)^l \cdot \frac{1}{k_l} < (-1)^j\frac{1}{k_j}
$$

It can be seen that in order to obtain an expression for N soliton solutions, there must be an expression for all soliton solutions less than N. The solution diagram is shown in Figure 5-2.

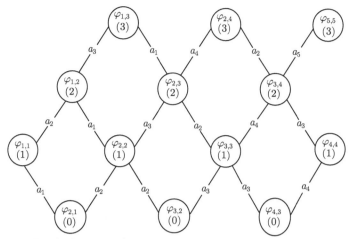

Figure 5-2 Lamb diagram of N soliton. The number in parentheses indicates
the number of solitons

5.2 Bäcklund Transformation of a Class of Nonlinear Evolution Equation

We have introduced the B transformation of the SG equation in the previous section, in fact, not only the SG equation has a B transformation, but also a large number of nonlinear evolution equations have a corresponding B transformation, which is actually a kind of transformation between unknown functions parallel to the contact transformation. Let the transformation $z(x,y) \to z'(x',y')$ satisfy the differential relation

$$\begin{cases} p = f\left(x',y',z',p',q'\right) \\ q = g\left(x',y',z',p',q'\right) \end{cases} \tag{5.16}$$

where $p = \dfrac{\partial z}{\partial x}, q = \dfrac{\partial z}{\partial y}, p' = \dfrac{\partial z'}{\partial x'}, q' = \dfrac{\partial z'}{\partial y'}$ (analogies $r = \dfrac{\partial^2 z}{\partial x^2}, s = \dfrac{\partial^2 z}{\partial x \partial y}, t = \dfrac{\partial^2 z}{\partial y^2}$ and r', s', t' and so on). In the following, we assmue $x = x', y = y'$. From the integrability condition $\dfrac{dp}{dy} = \dfrac{dq}{dx}$ for z, we get

$$Q = f_{y'} - q_{x'} + f_{z'}q' - q_{z'}p' + (f_{p'} - g_{q'}) + f_{q'}t' + g_{p'}r' = 0 \tag{5.17}$$

This integrability condition is satisfied in two cases: or it is identical to zero, i.e.,

$$f_{p'} - g_{q'} = f_{q'} = g_{p'} = 0$$

$$f_{y'} - g_{x'} + f_{z'}q' - q_{z'}p' = 0$$

or $Q = 0$ leads to the second-order Monge-Ampere equation. The former results in a contact transformation, while the latter results in a B transformation. We can obtain the equations of the eigenvalue problem of the scattering inversion of this equation by the B transformation, and in some cases, its infinite conservation law. We give examples as follows:

   1) SG equation

$$S = \sin z \qquad (5.18)$$

Its B transformation is

$$
\begin{cases}
\dfrac{1}{2}(p - p') = a \sin\left[\dfrac{1}{2}(z + z')\right] \\[4mm]
\dfrac{1}{2}(q + q') = a^{-1} \sin\left[\dfrac{1}{2}(z - z')\right]
\end{cases}
\qquad (5.19)
$$

Setting $\Gamma = \tan\left[(z + z')/4\right]$, we can obtain from (5.19)

$$\Gamma_x + a\Gamma - \frac{1}{2}p\left(1 + \Gamma^2\right) = 0 \qquad (5.20)$$

We know the Riccati equation

$$\Gamma_x + 2p\Gamma + Q\Gamma^2 + R = 0 \qquad (5.21)$$

is equivalent to a system of equations

$$
\begin{cases}
w_{1x} + pw_1 = -Rw_2 \\
w_{2x} - pw_2 = Qw_1
\end{cases}
\qquad (5.22)
$$

where $\Gamma = w_1/w_2$. Thus (5.20) is equivalent to the system of equations

$$
\begin{cases}
w_{1,x} + \dfrac{1}{2}aw_1 = \dfrac{1}{2}pw_2 \\[4mm]
w_{2x} - \dfrac{1}{2}aw_2 = -\dfrac{1}{2}pw_1
\end{cases}
\qquad (5.23)
$$

This is the two-component eigenvalue problem equations corresponding to SG equation (5.18) in the scattering inversion method.

   By B transform (5.19) and taking a to be small, we get infinite conservation law of (5.18). In fact, let $z'(x, y, a)$ have the form

$$z'(x, y, a) \approx \sum_{j=0}^{\infty} z'_j(x, y)a^j, \quad a \to 0 \qquad (5.24)$$

Substitute (5.24) into (5.19) to get

$$\sum_{j=0}^{\infty} z'_{jy}a^j = z_y + \frac{2}{a}\sin\left[\frac{1}{2}\left(\sum_{j=0}^{\infty} z'_i a^j - z\right)\right]$$

When $a \to 0$, there is $z'_0 = z$, $z'_1 = 2z_y$, the higher power coefficients of a are equal to

$$\begin{cases} z'_2 = 2z_{yy} \\ z'_3 = 2z_{yyy} + \frac{1}{3}(z_y)^3 \\ z'_4 = 2z_{yyyy} + 2(z_y)^2 z_{yy} \\ z'_5 = 2z_{yyyyy} + 3(z_y)^2 z_{yyy} + 5z_y(z_{yy})^2 + \frac{3}{20}(z_y)^5 \end{cases} \tag{5.25}$$

We have known that equation (5.18) can be written in the energy conservation form

$$\frac{1}{2}\left(z'^2_x\right)_y + (\cos z' - 1)_x = 0 \tag{5.26}$$

Substituting (5.25) into (5.26), an infinite conservation laws are obtained from equal powers of a. The following are several conservation densities

$$T_0 = \frac{1}{2}z_x^2$$

$$T_1 = 2z_{yyyx}z_x + 4z_{yyx}z_{yx} + z_y^2 z_{yx}z_x$$

$$T_2 = 2z_{yyyyyx}z_x + 4z_{yyyyx}z_{yx} + 4z_{yyx}z_{yyyx}$$

$$\quad + 6z_{yyy}z_y z_x z_{yx} + 3z_{xyyy}z_y^2 z_x$$

$$\quad + 10z_{yy_x}z_{yy}z_y z_x + 5z_{yy}^2 z_{yx}z_x + 8z_{yy}z_{yx}^3 z_y$$

$$\quad + 8z_y^2 z_{yx}z_{yy_x} + \frac{3}{4}z_y^4 z_{yx}z_x + \cdots$$

2) KdV equation

$$u_y + 6uu_x + u_{xxx} = 0 \tag{5.27}$$

Letting $z = \int_{-\infty}^{x} u(x', y)\, dx'$, we can obtain the equation for z from (5.27)

$$q + 3p^2 + \alpha = 0, \quad \alpha = z_{xxx} \tag{5.28}$$

Let B transform of (5.27) to be

$$\begin{cases} p = f\left(z, z', p'\right) \\ q = \varphi\left(z, z', q', r, r', p, p'\right) \end{cases} \tag{5.29}$$

Because

$$r = f_z f + f_{z'} p' + f_p r' \tag{5.30}$$

Substituting it into the second formula of (5.29), we have

$$q = \varphi\left(z, z', q', p', r'\right) \tag{5.31}$$

The second-order mixed derivative of z can be written as

$$\frac{dp}{dy} = f_z q + f_{z'} q' + f_{p's'}$$

or

$$\frac{dq}{dx} = \varphi_z p + \varphi_{z'} p' + \varphi_{q'} s' + \varphi_{p'} r' + \varphi_{r'} \alpha'$$

Since these mixed derivatives are required to be equal and z' to satisfy the same equation (5.28), the function $\Omega\left(z, z', p, p', q, q', r', s'\right)$ can be defined so that

$$\Omega = \left(f_{p'} - \varphi_{q'}\right) s' + f_z q + f_{z'} q' - \varphi_z p - \varphi_{z'} p' \\ - \varphi_{p'} r' + \varphi_{r'} \left(q' + 3p'^2\right) = 0 \tag{5.32}$$

Picking $\Omega_{s'} = f_{p'} - \varphi_{q'} = 0$ and f independent of q', r', there is

$$\varphi_{q'q'} = \varphi_{q'r'} = 0, \tag{5.33}$$

$$\Omega_{q'} = f_{p'} f_z + f_{z'} - f f_{p'z} - p' f_{p'z'} - r' f_{pp'} + \varphi_{r'} = 0$$

Due to $\Omega_{q'r'} = -f_{p'p'} + \varphi_{r'r'} = 0$, there is

$$f_{p'p'} = \varphi_{r'r'} = a\left(z, z', r'\right) \tag{5.34}$$

where $a\left(z, z', r'\right)$ is undetermined. Because

$$\Omega_{r'r'r'} = -3f_{p'p'p'} = 0 \tag{5.35}$$

and by (5.33), (5.34) can be written as

$$f\left(z, z', p'\right) = b\left(z, z'\right) p' + c\left(z, z'\right)$$

$$\varphi\left(z, z', q', p', r'\right) = b\left(z, z'\right) q' + \lambda\left(z, z', p'\right) r' + v\left(z, z'p'\right)$$

We determine λ, c, v as follows. From (5.32) there is

$$\Omega_{r'r'} = -2\varphi_{r'p'} = 0$$

Therefore, λ is unrelated to p'. And because $\Omega_{p'p'p'} = 0$, then we have

$$v(z, z', p') = v_2(z, z') p'^2 + v_1(z, z') p' + v_0(z, z') \tag{5.36}$$

If $b(z, z')$ is a constant, we get

$$\begin{cases} p = bp' + c \\ q = bq' + \lambda r' + v_2 p'^2 + v_1 p' + v_0 \end{cases} \tag{5.37}$$

where b is a constant and $c, \lambda, v_i (i = 0, 1, 2)$ are undetermined functions of z, z'. Substituting (2.23) to Ω, since r', p', q' is independent in Ω, the following seven equations are obtained to determine five unknown functions and a constant b.

$$2v_2 = -(b\lambda_z + \lambda_{z'}) \tag{5.38a}$$

$$\lambda = -(bc_z + c_{z'}) \tag{5.38b}$$

$$v_1 = \lambda c_z - c\lambda_z \tag{5.38c}$$

$$v_2 c_z - cv_{2z} + 3\lambda - bv_{1z} - v_{1z'} = 0 \tag{5.38d}$$

$$v_1 c_z - cv_{1z} - v_{0z'} - bv_{0z} = 0 \tag{5.38e}$$

$$bv_{2z} + v_{2z'} = 0 \tag{5.38f}$$

$$v_0 c_z - cv_{0z} = 0 \tag{5.38g}$$

Since z must satisfy the equation (5.28), computing the third derivative from (5.37) gives

$$\alpha = ba' - \lambda r' + 2v_2 p'^2 + p' \left[2bcc_{zz} + 2cc_{zz} \right.$$

$$+ c_z (bc_z + c_{z'}) \left. \right] + c^2 c_{zz} + cc_z^2 = 0 \tag{5.39a}$$

$$v_2 - b + b^2 = 0 \tag{5.39b}$$

$$bc_{zz} + 2b + c_{zz'} = 0 \tag{5.39c}$$

$$c^2 c_{zz} + cc_z^2 + v_0 + 3c^2 = 0 \tag{5.39d}$$

$v_0 = \psi(z') c(z, z')$, is derived from (5.38g), where $\psi(z')$ is undetermined. Integrating (5.39d) once gives

$$c_z^2 + 2c + \psi + Kc^{-2} = 0 \tag{5.40}$$

Taking $K = 0$, (5.40) gives $c_{zz} = -1$ and (5.39c) gives $c_{zz'} = -b$. From (5.39b), (5.38a), (5.38b) , we have

$$c_{z'z'} = 2b + b^2 \qquad (5.41)$$

Integrate (5.41) to get

$$c(z, z') = m - \frac{1}{2}\left[z^2 + 2bzz' - b(2 + b)z'^2\right] + kz + lz'$$

where k, l, m are integral constants. Let $m \neq 0, k = l = 0$, (5.38a) to (5.38g) give

$$\lambda = 2b\left(z - z'\right)$$

$$v_1 = -2bm - b\left(z^2 - 2zz' + b^2 z'^2\right)$$

$$v_2 = b - b^2$$

$$\psi = -2m - 2b(1 + b)z'^2$$

From $b = -1$, the B transformation of the KdV equation (5.27) is

$$\begin{cases} p + p' = m - \dfrac{1}{2}\left(z - z'\right)^2 \\ q + q' = \left(z - z'\right)\left(r - r'\right) - 2\left(p^2 + pp' + p'^2\right) \end{cases} \qquad (5.42)$$

From the solution of $z' = 0$ to the KdV equation, we can obtain from (5.42)

$$\begin{cases} p = m - \dfrac{1}{2}z^2 \\ q = zr - 2p^2 = -2mp \end{cases}$$

The solution to this equation is

$$z = (2m)^{\frac{1}{2}} \tanh\left[\left(\frac{m}{2}\right)^{\frac{1}{2}}(x - 2my)\right]$$

$$u = p = m\,\mathrm{sech}^2\left[\left(\frac{m}{2}\right)^{\frac{1}{2}}(x - 2my)\right] \qquad (5.43)$$

which is the solution of the KdV equation (5.27). If $\Gamma = z - z'$, we get from (5.42)

$$\Gamma_x - \frac{1}{2}\Gamma^2 + (m - 2p) = 0$$

which is equivalent to

$$\begin{cases} v_{1x} = (2p - m)v_2 \\ v_{2x} = -\dfrac{1}{2}v_1 \end{cases} \qquad (5.44)$$

3) MKdV equation

$$u_y + 6u^2 u_x + u_{xxx} = 0 \tag{5.45}$$

Integrating it yields

$$q + 3p^2 + \alpha = 0$$

Similar to the steps performed for equation KdV above, the B transformation of MKdV equation (5.45) is given as

$$\begin{cases} p = bp' + a \sin v \\ q = bq' - 2a \left[br' \cos v + p'^2 \sin v + \dfrac{1}{2} a \left(p + bp' \right) \right] \end{cases} \tag{5.46}$$

where $b = \pm 1$.

If $\Gamma = \tan \left[\dfrac{1}{2} \left(z + bz' \right) \right]$, we get

$$\Gamma_x + a\Gamma - p \left(1 + \Gamma^2 \right) = 0$$

which is equivalent to eigen set of equations of scattering problem (5.45)

$$\begin{cases} w_{1x} + \dfrac{1}{2} a w_1 = p w_2 \\ w_{2x} - \dfrac{1}{2} a w_2 = p w_1 \end{cases} \tag{5.47}$$

4) Nonlinear Schrödinger equation

The complex conjugate form of the nonlinear Schrödinger equation is

$$\begin{cases} iq + r + z^2 \bar{z} = 0 \\ -i\bar{q} + \bar{r} + \bar{z}^2 z = 0 \end{cases} \tag{5.48}$$

where "$-$" means conjugate of complex numbers, then its B transformation is

$$\begin{cases} p = p' - \dfrac{1}{2} i w \tau + ikv \\ q = q' + \dfrac{1}{2} \tau \left(p + p' \right) - k_n + \dfrac{1}{4} iv \left(|w|^2 + |v|^2 \right) \end{cases} \tag{5.49}$$

where $w = z + z', v = z - z', \tau = \pm i \left(b - 2|v|^2 \right)^{\frac{1}{2}}$, n, b and k are arbitrary real constants.

Let $\Gamma = \left(b - 2|v|^2 \right)^{\frac{1}{2}} / \sqrt{2} y$, and from (5.49) we get

$$z \left[\Gamma_x + ik\Gamma + \tau^{-\frac{1}{2}} \left(z\Gamma^2 + \bar{z} \right) \right] = z' \left[\Gamma_x + ik\Gamma + \tau^{-\frac{1}{2}} \left(z'\Gamma^2 + z^{-1} \right) \right]$$

which is equivalent to

$$\begin{cases} w_{1x} + \dfrac{1}{2}ikw_1 = -\tau^{-\frac{1}{2}}\bar{z}w_2 \\[3mm] w_{2x} - \dfrac{1}{2}ikw_1 = \tau^{-\frac{1}{2}}zw_1 \end{cases} \tag{5.50}$$

5.3   B Transformation Commutability of the KdV Equation

The KdV equation

$$u_y + 6uu_x + u_{xxx} = 0 \tag{5.51}$$

is invariant under BT

$$B_\beta u' : \begin{cases} u_x = \beta - u_x' - \dfrac{1}{2}\left(u - u'\right)^2 \\[3mm] u_y = -u_y' + \left(u - u'\right)\left(u_{xx} - u_{xx}'\right) \\[2mm] \quad\quad -2\left[u_x^2 + u_x u_x' + u_x'^2\right] \end{cases} \tag{5.52}$$

where β is an arbitrary Bäcklund transform parameter. From (5.51), it can be easily concluded that

$$u_y + 3u_x^2 + u_{xxx} = 0 \tag{5.53}$$

We have the following theorem.

Theorem 5.1   *If $u_{\beta_i} = B_{\beta_i}u_0(i = 1, 2)$ is the solution to equation (5.53), and it is generated from the known solution u_0 and the BT with parameter β_i, then we can obtain a new solution φ to equation (5.53),*

$$\varphi = u_0 + 2\left(\beta_1 - \beta_2\right)/\left(u_{\beta_1} - u_{\beta_2}\right) \tag{5.54}$$

where $\varphi = B_{\beta_1}B_{\beta_2}u_0 = B_{\beta_2}B_{\beta_1}u_0$.

Proof   In fact, we have

$$u_{0x} + u_{\beta_1,x} = \beta_1 - \dfrac{1}{2}\left(u_0 - u_{\beta_1}\right)^2 \tag{5.55a}$$

$$u_{0x} + u_{\beta_2,x} = \beta_2 - \dfrac{1}{2}\left(u_0 - u_{\beta_2}\right)^2 \tag{5.55b}$$

$$u_{\beta_1,x} + u_{\beta_1\beta_2,x} = \beta_2 - \dfrac{1}{2}\left(u_{\beta_1} - u_{\beta_1\beta_2}\right)^2 \tag{5.55c}$$

$$u_{\beta_2,x} + u_{\beta_2\beta_1,x} = \beta_1 - \dfrac{1}{2}\left(u_{\beta_2} - u_{\beta_2\beta_1}\right)^2 \tag{5.55d}$$

where $u_{\beta_1\beta_2} = B_{\beta_2}B_{\beta_1}u_0, u_{\beta_2\beta_1} = B_{\beta_1}B_{\beta_2}u_0$. If $\varphi = u_{\beta_1\beta_2} = u_{\beta_2\beta_1}$, then from (5.55a), (5.55b), (5.55c) and (5.55d), it can be obtained that

$$u_{\beta_1x} - u_{\beta_2x} = \beta_1 - \beta_2 + \dfrac{1}{2}\left(u_{\beta_1} - u_{\beta_2}\right)\left(2u_0 - u_{\beta_1} - u_{\beta_2}\right) \tag{5.56a}$$

$$u_{\beta_1 x} - u_{\beta_2 x} = \beta_2 - \beta_1 + \frac{1}{2}\left(u_{\beta_2} - u_{\beta_1}\right)\left(u_{\beta_1} + u_{\beta_2} - 2\varphi\right) \qquad (5.56\text{b})$$

So subtracting (5.56b) from (5.56a) yields

$$\varphi = u_0 + 2\left(\beta_1 - \beta_2\right)/\left(u_{\beta_1} - u_{\beta_2}\right)$$

It is easy to verify that (5.54) is indeed the solution to (5.53).

Similarly, the MKdV equation

$$v_y + 6v^2 v_x + v_{xxx} = 0 \qquad (5.57)$$

is invariant under the Bäcklund transformation

$$B_\beta u' : \begin{cases} u_x = \alpha u'_x + \beta \sin\left(u + \alpha u'\right) \\ u_y = \alpha u'_y - \beta\left[2\alpha u'_{xx}\cos\left(u + \alpha u'\right) + 2u'^2_x\sin(u \\ \qquad + \alpha u') + \beta\left(u_x + \alpha u'_x\right)\right], \alpha = \pm 1 \end{cases} \qquad (5.58)$$

where β is any Bäcklund parameter. We have a theorem:

Theorem 5.2 *If $u_{\beta_i}(i = 1, 2)$ represents the solution of (5.57) and is generated from the Bäcklund transformation (5.58), $\beta = \beta_i(i = 1, 2)$ starting from the solution u_0, then the new solution can be expressed as*

$$\tan\left(\frac{\varphi - u_0}{2}\right) = \alpha\left(\frac{\beta_1 + \beta_2}{\beta_1 - \beta_2}\right)\tan\left(\frac{u_{\beta_1} - u_{\beta_2}}{2}\right) \qquad (5.59)$$

where $\varphi = B_{\beta_1} B_{\beta_2} u_0 = B_{\beta_2} B_{\beta_1} u_0$.

Example 5.1 *Assuming $u_0 = 0$, it can be obtained from (5.58) that*

$$u_{\beta_i} = 2\tan^{-1} e^{\mu_i} \qquad (5.60)$$

where

$$\mu_i = \beta_i x - \beta_i^3 y + \gamma_i \qquad (5.61)$$

$\gamma_i(i = 1, 2)$ are integral constants, a new solution can be obtained from the commutativity theorem (5.59)

$$\varphi = \pm 2\tan^{-1}\left[\left(\frac{\beta_1 + \beta_2}{\beta_1 - \beta_2}\right)\frac{\sinh\left\{\frac{1}{2}\left(\mu_1 - \mu_2\right)\right\}}{\cosh\left\{-\frac{1}{2}\left(\mu_1 + \mu_2\right)\right\}}\right] \qquad (5.62)$$

5.4   Bäcklund Transformations for High-Order KdV Equation and High-Dimensional Sine-Gordon Equation

Sawada and Kotera [22] proposed the following high-order KdV equation

$$u_t + 180u^2 u_x + 30 \left(uu_{xxx} + u_x u_{xx}\right) + u_{xxxxx} = 0 \tag{5.63}$$

The bilinear form of equation (5.63) is easily obtained as

$$D_x \left(D_t + D_x^5\right) f \cdot f = 0 \tag{5.64}$$

The BT of (5.64) constructed by Sawada and Kaup[22] is

$$\left(D_t - \frac{15}{2}\beta D_x^2 - \frac{3}{2}D_x^5\right) f' \cdot f = 0 \tag{5.65a}$$

$$\left(D_x^3 - \beta\right) f' \cdot f = 0 \tag{5.65b}$$

where β is the Bäcklund transformation parameter. To prove it, we only need to prove that when (5.65a) and (5.65b) hold, there is

$$P \equiv f' \cdot f' D_x \left(D_t + D_x^5\right) f \cdot f - ff D_x \left(D_t + D_x^5\right) f' \cdot f' = 0 \tag{5.66}$$

In fact, from the properties of the D operator in Chapter 4, it is easy to know

$$P = D_x \left[2 \left(f' \cdot f\right)\left(D_t f' \cdot f\right) + \frac{3}{4} \left(f'f\right) \cdot \left(D_x^5 f' \cdot f\right)\right.$$
$$\left. - \frac{15}{4} \left(D_x f' \cdot f\right) \cdot \left(D_x^4 f' \cdot f\right) + \frac{15}{2} \left(D_x^2 f' \cdot f\right) \cdot \left(D_x^3 f' \cdot f\right)\right]$$
$$+ \frac{5}{4} D_x^3 \left[\left(f'f\right) \cdot \left(D_x^3 f' \cdot f\right) - 3 \left(D_x f' \cdot f\right) \cdot \left(D_x^2 f' \cdot f\right)\right]$$

Further, it can be transformed into

$$P = D_x \left[2 \left(f' \cdot f\right)\left(D_t f' \cdot f\right) - 3 \left(f'f\right) \cdot \left(D_x^5 f' \cdot f\right)\right.$$
$$\left. + 15 \left(D_x^2 f' \cdot f\right) \cdot \left(D_x^3 f' \cdot f\right)\right] + 5 D_x^3 \left(f' \cdot f\right) \cdot \left(D_x^3 f' \cdot f\right) \tag{5.67}$$

Substitute the Bäcklund relationship (5.65a) and (5.65b) into (5.67) to obtain

$$P = D_x \left[15\beta \left(f' \cdot f\right) \cdot \left(D_x^2 f' \cdot f\right) + 15\beta \left(D_x^2 f' \cdot f\right) \cdot \left(f'f\right)\right] = 0 \tag{5.68}$$

Now we consider the following three-dimensional sine-Gordon equation

$$\left(\sum_{i=1}^{3} \frac{\partial^2}{\partial x_i^2} - \frac{\partial^2}{\partial t^2}\right) u = \sin u \tag{5.69}$$

Liebbrandt and Christiansen pointed out that equation (5.69) has the Bäcklund transformation

$$\left\{ I\frac{\partial}{\partial x^1} + i\sigma_1\frac{\partial}{\partial x^2} + i\sigma_3\frac{\partial}{\partial x^3} + \sigma_2\frac{\partial}{\partial t} \right\} \left\{ \frac{\alpha - i\beta}{2} \right\}$$

$$= \exp\left[i\theta\sigma_1 \exp\left[(-i\varphi\sigma_2) \exp\left(-\tau\sigma_1 \right) \right] \right] \cdot \sin\left\{ \frac{\alpha + i\beta}{2} \right\} \tag{5.70}$$

where σ_1, σ_2 and σ_3 are the pauli spin matrix, and I is the 2×2 identity matrix. The real Bäcklund parameter (θ, φ, τ) are confined to the region: $0 \leqslant \theta \leqslant 2\pi, 0 \leqslant \varphi \leqslant 2\pi, -\infty < \tau < \infty$. And the real functions α, β satisfy

$$\left\{ \sum_{i=1}^{3} \frac{\partial^2}{\partial x^{i2}} - \frac{\partial^2}{\partial t^2} \right\}$$

$$\alpha\left(x^1, x^2, x^3, t\right) = \sin\alpha\left(x^1, x^2, x^3, t\right) \tag{5.71a}$$

$$\beta\left(x^1, x^2, x^3, t\right) = \sinh\beta\left(x^1, x^2, x^3, t\right) \tag{5.71b}$$

respectively. Equation (5.70) can be rewritten as

$$\left\{ I\frac{\partial}{\partial x^1} + iP \right\} \left\{ \frac{\alpha - i\beta}{2} \right\}$$

$$= \left[A_1 + iA_2 \right] \sin\left\{ \frac{\alpha + i\beta}{2} \right\} \tag{5.72}$$

where

$$\begin{cases} P = \sigma_1\dfrac{\partial}{\partial x^2} + \sigma_3\dfrac{\partial}{\partial x^3} - i\sigma_2\dfrac{\partial}{\partial t} \\[2mm] \sigma_1 = \begin{pmatrix} 0 & 1 \\ 1 & 0 \end{pmatrix}, \sigma_2 = \begin{pmatrix} 0 & -i \\ i & 0 \end{pmatrix} \\[3mm] \sigma_3 = \begin{pmatrix} 1 & 0 \\ 0 & -1 \end{pmatrix} \end{cases} \tag{5.73}$$

and

$$\begin{cases} A_1 = l\cos\theta \\[2mm] A_2 = \begin{pmatrix} \sin\theta\sin\varphi\cosh\tau & (\cos\varphi - \sin\varphi\sinh\tau)\sin\theta \\ (\cos\varphi + \sin\varphi\sin h\tau)\sin\theta & -\sin\theta\sin\varphi\cosh\tau \end{pmatrix} \end{cases} \tag{5.74}$$

Dividing equation (5.72) into real and imaginary parts yields

$$I\frac{\partial}{\partial x}\left\{ \frac{\alpha}{2} \right\} + P\left\{ \frac{\beta}{2} \right\} = A_1\sin\left(\frac{\alpha}{2} \right)\cosh\left(\frac{\beta}{2} \right)$$

$$- A_2 \cos\left(\frac{\alpha}{2}\right) \sinh\left(\frac{\beta}{2}\right) \tag{5.75a}$$

$$P\left\{\frac{\alpha}{2}\right\} - I\frac{\partial}{\partial x^1}\left\{\frac{\beta}{2}\right\} = A_1 \cos\left(\frac{\alpha}{2}\right) \sinh\left(\frac{\beta}{2}\right)$$

$$+ A_2 \sin\left(\frac{\alpha}{2}\right) \cosh\left(\frac{\beta}{2}\right) \tag{5.75b}$$

We can construct the simplest nontrivial solutions for (5.71a) and (5.71b). Taking $\beta = \beta_0 = 0$ first, and then taking $\alpha = \alpha_0 = 0$ in (5.75a) and (5.75b), we can obtain

$$\alpha_1\left(x^1, x^2, x^3; \theta, \varphi, \tau\right) = 4\tan^{-1}\left\{a_0 \exp R\right\} \tag{5.76a}$$

$$\beta_1\left(x^1, x^2, x^3; \theta, \varphi, \tau\right)$$

$$= \begin{cases} 4\tan^{-1}\left\{a_1 \exp R\right\}, & R \leqslant 0 \\ 4\cosh^{-1}\left\{a_1 \exp R\right\}, & R > 0 \end{cases} \tag{5.76b}$$

where

$$R = x^1 \cos\theta + x^2 \sin\theta \cos\varphi + \sin\theta \sin\varphi$$

$$\cdot \left(x^3 \cosh\tau + t \sinh\tau\right) \tag{5.77}$$

a_i $(i = 0, 1)$ are all integral constants. The solution (5.76a) represents a single soliton solution to equation (5.71a), while the β solution (5.76b) lacks the characteristics of solitons.

5.5   Bäcklund Transformation of Benjamin-Ono Equation

The Bäcklund transformation of the Benjamin-Ono equation describing the propagation of deep water waves is easily obtained

$$u_t + 2uu_x + H\left[u_{xx}\right] = 0 \tag{5.78}$$

where H is the Hilbert transform operator, which is determined by Cauchy principal value integration

$$Hf(x) = \frac{1}{\pi}P\int_{-\infty}^{\infty}\frac{f(z)}{z - x}dz$$

We first find the bilinear representation of BO equation (5.78). Set

$$u(x, t) = i\frac{\partial}{\partial x}\left(\log\left[f'/f\right]\right) \tag{5.79}$$

where

$$f \propto \prod_{n=1}^{N} (x - z_n(t)), \quad f' \propto \prod_{n=1}^{N} (x - z'_n(t)) \tag{5.80}$$

z_n, z'_n are complex, and $\mathrm{Im}\, z_n > 0, \mathrm{Im}\, z'_n < 0, \forall n, N \in Z^+$. Successively

$$u = i \left(\frac{f'_x}{f'} - \frac{f_x}{f} \right) = i \sum_{n=1}^{N} \left\{ \frac{1}{x - z'_n} - \frac{1}{x - z_n} \right\} \tag{5.81}$$

Therefore

$$Hu = \frac{i}{\pi} P \int_{-\infty}^{\infty} \frac{1}{(z - x)} \cdot \sum_{n=1}^{N} \left\{ \frac{1}{z - z'_n} - \frac{1}{z - z_n} \right\} dz \tag{5.82}$$

To calculate (5.82), the contour C is taken as shown in Figure 5-3.

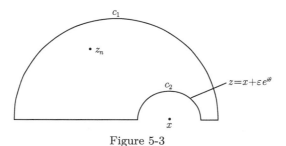

Figure 5-3

According to the residue theorem, we get

$$\frac{1}{2\pi i} \oint_c \frac{1}{(z - x)} \left[\frac{1}{z - z'_n} - \frac{1}{z - z_n} \right] dz = \mathrm{res}\,(z = z_n)$$

Therefore

$$\lim_{\varepsilon \to 0} \frac{1}{2\pi i} \int_{-\infty}^{x-\varepsilon} \frac{1}{z - x} \left[\frac{1}{z - z'_n} - \frac{1}{z - z_n} \right] dz$$

$$+ \lim_{\varepsilon \to 0} \frac{1}{2\pi i} \int_{c_2} \frac{1}{z - x} \left[-\frac{1}{z - z'_n} - \frac{1}{z - z_n} \right] dz$$

$$+ \lim_{\varepsilon \to 0} \frac{1}{2\pi i} \int_{x+\varepsilon}^{\infty} \frac{1}{z - x} \left[\frac{1}{z - z'_n} - \frac{1}{z - z_n} \right] dz$$

$$= \frac{1}{x - z_n} \tag{5.83}$$

$$\frac{1}{2\pi i} P \int_{-\infty}^{\infty} \frac{1}{z-x} \left[\frac{1}{z-z_n'} - \frac{1}{z-z_n} \right] dz$$

$$= \frac{1}{x-z_n} - \lim_{\varepsilon \to 0} \frac{1}{2\pi i}$$

$$\cdot \int_{c_2} \frac{1}{z-x} \left[\frac{1}{z-z_n'} - \frac{1}{z-z_n} \right] dz$$

$$= \frac{1}{x-z_n} - \lim_{\varepsilon \to 0} \left\{ \frac{1}{2\pi i} \int_{\pi}^{0} \varepsilon^{-1} e^{-i\theta} \right.$$

$$\cdot \left. \left[\frac{1}{x+\varepsilon e^{i\theta} - z_n'} - \frac{1}{x+\varepsilon e^{i\theta} - z_n} \right] \varepsilon i e^{i\theta} d\theta \right\}$$

$$= \frac{1}{x-z_n} + \frac{1}{2} \left[\frac{1}{x-z_n'} - \frac{1}{x-z_n} \right]$$

$$= \frac{1}{2} \left[\frac{1}{x-z_n'} + \frac{1}{x-z_n} \right] \tag{5.84}$$

It can be inferred from this that

$$H_u = \frac{i}{\pi} \sum_{n=1}^{N} \pi i \left(\frac{1}{x-z_n'} + \frac{1}{x-z_n} \right)$$

$$= -\sum_{n=1}^{N} \left(\frac{1}{x-z_n'} + \frac{1}{x-z_n} \right) \tag{5.85}$$

$$= -\left[\frac{f_x'}{f'} + \frac{f_x}{f} \right] = -\frac{\partial}{\partial x} (\log [f'f])$$

Substituting (5.79) and (5.85) into BO equation (5.78) yields

$$\frac{\partial}{\partial x} \left[i\frac{\partial}{\partial t} (\log [f'/f]) - \frac{\partial}{\partial x} (\log [f'/f])^2 \right.$$

$$\left. - \frac{\partial^2}{\partial x^2} (\log [f'f]) \right] = 0$$

Integrating with respect to x and selecting any integration function on t as 0 yields

$$i\frac{\partial}{\partial t} (\log [f'/f]) - \left[\frac{\partial}{\partial x} (\log [f'/f]) \right]^2$$

$$- \frac{\partial^2}{\partial x^2} (\log [f'f]) = 0$$

Therefore, there are

$$i (ff_t' - f'f_t) - ff_{xx}' + 2f_x'f_x - f'f_{xx} = 0 \tag{5.86}$$

Finally, the equation can be written as

$$[i\,(f'_t f - f' f_{t'}) - (f'_{xx}f - 2f'_x f_{x'} + f' f_{x'x'})]_{x'=x,t'=t} = 0 \tag{5.87}$$

or

$$\left[i\,(f'_t f - f' f_{t'}) - \left(\frac{\partial}{\partial x} - \frac{\partial}{\partial x'}\right)(f'_x f - f' f_{x'})\right]_{x'=x,t'=t} = 0$$

Then

$$\left[i\left(\frac{\partial}{\partial t} - \frac{\partial}{\partial t'}\right) - \left(\frac{\partial}{\partial x} - \frac{\partial}{\partial x'}\right)^2\right]$$

$$\cdot f'(x,t) f\,(x',t')|_{x'=x,t'=t} = 0$$

namely

$$\left(iD_t - D_x^2\right) f' \cdot f = 0 \tag{5.88}$$

So, we deduce that if f', f is determined by (5.80) and (5.88) holds, then the function u defined by (5.79) is the solution to the BO equation.

Let (f, f') be a pair of solutions to (5.88), and (g, g') be determined by the Bäcklund relationship introduced by a similar method

$$\left(iD_t - 2i\lambda D_x - D_x^2 - \mu\right) f \cdot g = 0 \tag{5.89a}$$

$$\left(iD_t - 2i\lambda D_x - D_x^2 - \mu\right) f' \cdot g' = 0 \tag{5.89b}$$

$$\left(D_x + i\lambda\right) f \cdot g' = i\nu f' g \tag{5.89c}$$

where λ, μ, ν are any parameters. It can be proven that g and g' satisfy the equation

$$\left(iD_t - D_x^2\right) g' \cdot g = 0 \tag{5.90}$$

In fact, it only needs to be proven that

$$P = g'g\left(iD_t - D_x^2\right) f' \cdot f - f' \cdot f\left(iD_t - D_x^2\right) g' \cdot g = 0 \tag{5.91}$$

is derived from equations (5.88)-(5.89c). Now, (5.90) is naturally satisfied. Due to

$$g'g\left(D_t f' \cdot f\right) - f' f\left(D_t g' \cdot g\right) = fg\left(D_t f' \cdot g'\right) - f'g'\left(D_t f \cdot g\right)$$

$$P = fg\left(iD_t f' \cdot g'\right) - f'g'\left(D_t f \cdot g\right) - gg'\left(D_x^2 f' \cdot f\right) + f'f\left(D_x^2 g' \cdot g\right)$$

$$= 2i\lambda\left[fgD_x f' \cdot g' - f'\cdot g' D_x f \cdot g\right] + \left[fgD_x^2 f' \cdot g' - f'g'D_x^2 f \cdot g\right]$$

$$- g'gD_x^2 f' \cdot f + f'f\left(D_x^2 g' \cdot g\right)$$

from the properties of the D operator, it can be concluded that

$$fgD_xf' \cdot g' - f'g'D_xf \cdot g = D_xf'g \cdot fg' \qquad (5.92a)$$

$$fgD_x^2f' \cdot g' - f'g'D_x^2f \cdot g$$
$$= D_x\left[(D_xf' \cdot g) \cdot fg' + f'g \cdot (D_xf \cdot g')\right] \qquad (5.92b)$$

$$f'fD_x^2g' \cdot g - g' \cdot gD_x^2f' \cdot f$$
$$= D_x\left[(D_xg' \cdot f) \cdot f'g + fg' \cdot (D_xf' \cdot g')\right] \qquad (5.92c)$$

Adding (5.92b) and (5.92c) and using the anti symmetry of the D operator, there is

$$fgD_x^2f' \cdot g' - f'g'D_x^2f \cdot g + f'fD_x^2g' \cdot g - g'gD_x^2f' \cdot f$$
$$= 2D_x\left[f'g \cdot (D_xf \cdot g')\right] \qquad (5.93)$$

Furthermore, based on the properties of the D operator

$$cd\,(D_xa \cdot b) - ab\,(D_xc \cdot d) = D_xad \cdot bc \qquad (5.94)$$

it can be obtained that

$$P = 2i\lambda D_xf'g \cdot fg' + 2D_x\left[f'g \cdot (D_xf \cdot g')\right]$$
$$= 2D_x\left[f'g \cdot (D_x + i\lambda)f \cdot g'\right] \qquad (5.95)$$

From the Bäcklund relationship (5.89c), it can be inferred that

$$P = 2D_x\left[f'g \cdot (i\nu f'g)\right] = 2i\nu D_x\left[f'g \cdot f'g\right] = 0$$

this is what we need to prove. Assuming

$$v = i\frac{\partial}{\partial x}\left(\log\left[g'/g\right]\right) \qquad (5.96)$$

this provides a new solution to the BO equation.

For Nakamura's modified BO equation

$$u_t - 2\lambda u_x + 2\nu e^u u_x + Hu_{xx} + u_xHu_x = 0 \qquad (5.97)$$

where H is a Hilbert transform operator, λ, ν are constants. Similarly, setting

$$u(x,t) = u_0 + \log\left[\frac{f'g}{fg'}\right] \qquad (5.98)$$

it can be proven that $u(x,t)$ defined by (5.98) is the solution to MBO equation (5.97) if f', f, g, g' satisfies the condition

$$\left(iD_t - 2i\lambda D_x - D_x^2 - \mu\right) f \cdot g = 0 \tag{5.99a}$$

$$\left(iD_t - 2i\lambda D_x - D_x^2 - \mu\right) f' \cdot g' = 0 \tag{5.99b}$$

$$\left(D_x + i\lambda\right) f \cdot g' = i\nu' f' g \left(\nu' = \nu e^{u_0}\right) \tag{5.99c}$$

Now consider the following equtaion which describes the finite depth laminar wave propagation

$$u_t + 2uu_x + G\left[u_{xx}\right] = 0 \tag{5.100}$$

where G is the integration operator,

$$G[u(x,t)] = \frac{1}{2}\lambda \int_{-\infty}^{\infty} \left[\coth \frac{\pi}{2}\lambda \left(x' - x\right) \right.$$
$$\left. - \operatorname{sgn}\left(x' - x\right) \right] u\left(x', t\right) dx' \tag{5.101}$$

λ^{-1} is a parameter that characterizes the depth of the fluid. For shallow water waves, it is the KdV equation when $\lambda \to \infty$. For deep water waves, it can be reduced to the BO equation when $\lambda = 0$. Set

$$u(x,t) = i\frac{\partial}{\partial x}\left(\log\left[\frac{\bar{f}}{f}\right]\right) \tag{5.102}$$

where

$$f(x,t) = \prod_{n=1}^{N} \left[1 + \exp\left\{\lambda\left[\lambda\left(\operatorname{Im} z_n\right)\left(x - \lambda t\right) - \bar{z}_n\right]\right\}\right] \tag{5.103}$$

$z_n(n = 1, 2, \cdots, N)$ is complex, $0 < \lambda \operatorname{Im} z_n < \pi$. $\bar{f}$ is a complex conjugate of f. For equation (5.100), its Bäcklund relationship can be obtained

$$\left(iD_t + i\left(\lambda - 2\lambda'\right) D_x - D_x^2 - \mu'\right) f \cdot g = 0 \tag{5.104a}$$

$$\left(iD_t + i\left(\lambda - 2\lambda'\right) D_x - D_x^2 - \mu'\right) \bar{f} \cdot \bar{g} = 0 \tag{5.104b}$$

$$\left(D_x + i\lambda'\right) f \cdot \bar{g} = i\nu' fg \tag{5.104c}$$

where λ', μ', ν' are any parameters.

5.6   Infinite Conservation Laws for the KdV Equation

As is well known, there are three important conservation laws in physics, namely the law of mass conservation, the law of momentum conservation, and the law of energy conservation. In mathematics, when a physical problem can be described by a differential equation in the form of

$$u_t = K(u) \tag{5.105}$$

The conservation law corresponding to this equation refers to the divergence form that can be written as follows

$$\frac{\partial T}{\partial t} + \frac{\partial X}{\partial x} = 0 \tag{5.106}$$

Here T and X are both related to the unknown function $u(x,t)$. T is generally called conservation density, and X is called the conserved flow. When X is zero at the boundary of the region, it can be inferred from (5.106) that $I = \int T dx$ is an invariant independent of time.

   The existence of solitons and infinite conservation laws is closely related. More and more facts now indicate that nonlinear equations with soliton solutions often have infinite conservation laws. On the other hand, conservation integrals are important mathematical tools, and with them, we can make prior estimates of the solutions to differential equations. These prior estimates are the core and key to the existence and uniqueness theorem of solutions to differential equations. As Lax pointed out, having an infinite number of conservation laws is the most important characteristic that distinguishes the KdV equation from other evolutionary equations.

   For the KdV equation

$$u_t - 6uu_x + u_{xxx} = 0 \tag{5.107}$$

Its corresponding first form of conservation law is

$$u_t - \left(3u^2 + u_{xx}\right)_x = 0 \tag{5.108}$$

From this, it can be concluded that

$$\int_{-\infty}^{\infty} u(x,t)dx$$

$$= \int_{-\infty}^{\infty} u(x,0)dx = M_0 \text{ (conservation of momentum)} \tag{5.109}$$

We have assumed that u and its derivatives tend to zero at $|x| \to \infty$.

Multiplying u by the KdV equation yields the second conservation law

$$\left(\frac{1}{2}u^2\right)_t + \left(-2u^3 + uu_{xx} - \frac{1}{2}u_x^2\right)_x = 0 \tag{5.110}$$

From this, it can be concluded that

$$E = \int_{-\infty}^{\infty} \frac{1}{2}u^2(x,t)dx$$

$$= \int_{-\infty}^{\infty} \frac{1}{2}u^2(x,0)dx = E_0(\text{conservation of energy}) \tag{5.111}$$

The third conservation law can be expressed as

$$\left(u^3 + \frac{1}{2}u_x^2\right)_t + \left(-\frac{9}{2}u^4 + 3u^2 u_{xx} - 6uu_x^2\right.$$

$$\left. +u_x u_{xxx} - \frac{1}{2}u_{xx}^2\right)_x = 0 \tag{5.112}$$

From this, it can also be inferred that

$$\int_{-\infty}^{\infty} \left(u^3 + \frac{1}{2}u_x^3\right) dx$$

$$= \int_{-\infty}^{\infty} \left(u^3(x,0) + \frac{1}{2}u_x^2(x,0)\right) dx \tag{5.113}$$

These conservation laws were first obtained by Whitham[1], and then Miura[2], and others. They specifically wrote out the forms of these conservation laws. Miura also utilized a skillful function transformation to obtain infinite conservation laws for the KdV equation.

Now we consider the following MKdV equation

$$Qv \equiv v_t - 6vv_x + v_{xxx} = 0 \tag{5.114}$$

It has the following relationship with KdV equation (5.107):

Theorem 5.3 *If v satisfies (5.114), then let $u = v^2 + v_x$, which satisfies KdV equation (5.107), that is*

$$Pu \equiv u_t - 6uu_x + u_{xxx} = 0$$

Proof In fact, $Pu = \left(2u + \frac{\partial}{\partial x}\right) Qv$, so

$$Qv = 0 \Rightarrow Pu = 0$$

On the contrary, it may not necessarily correct.   □

It is not difficult to verify that the KdV equation (5.107) preserves its form with respect to transformations

$$t \rightarrow t', x \rightarrow x' - 6ct', u \rightarrow u' + c \ (c \text{ is any constant})$$

We introduce the following transformation

$$t' = t, \quad x' = x + \frac{3}{2\varepsilon^2}t$$

$$u(x,t) = u\left(x', t'\right) + \frac{1}{4\varepsilon^2}, \quad \varepsilon > 0$$

then (5.107) remains unchanged. If $v(x,t) = \varepsilon w\left(x', t'\right) + \frac{1}{2\varepsilon}$, then transform $u = v^2 + v_x$ is

$$u\left(x', t'\right) = w\left(x', t'\right) + \varepsilon w_{x'}\left(x', t'\right) + \varepsilon^2 w^2\left(x', t'\right)$$

If the "$\prime$" symbol is omitted, there is

$$0 = Pu = u_t - 6uu_x + u_{xxx}$$

$$= \left(1 + \varepsilon \frac{\partial}{\partial x} + 2\varepsilon^2 w\right)\left[w_t - 6\left(w + \varepsilon^2 w^2\right)w_x\right.$$

$$\left. + w_{xxx}\right] \equiv LRw$$

where

$$Rw \equiv w_t - 6\left(w + \varepsilon^2 w^2\right)w_x + w_{xxx}$$

Since the KdV equation does not contain ε, it can be assumed that u is independent of ε, and $w(x,t)$ depends on ε. According to the relationship $u = w + \varepsilon w_x + \varepsilon^2 w^2 (\varepsilon \ll 1)$, we know that w is a function of u. Expand it about ε

$$w = w_0 + \varepsilon w_1 + \varepsilon^2 w_2 + \cdots$$

$$= u - \varepsilon u_x - \varepsilon^2\left(u^2 - u_{xx}\right) + \cdots$$

where $w_i(i = 0, 1, 2, \cdots)$ is a polynomial of $u, u_x, u_{xx}, \cdots$. Substituting w into

$$Rw = w_t + \left(-3w^2 - 2\varepsilon^2 w^3 + w_{xx}\right)_x = 0$$

and setting the high power coefficient of ε to 0, we can obtain infinite conservation laws for the KdV equation.

5.7   Infinite Conserved Quantities of AKNS Equation

According to the scattering inversion method, for more general Zakharov Shabat equations

$$\begin{cases} v_{1x} = -i\zeta v_1 + q v_2 \\ v_{2x} = i\zeta v_2 + r v_1 \end{cases} \tag{5.115}$$

The defined eigenfunctions $\varphi, \bar{\varphi}, \psi, \bar{\psi}$ have boundary conditions

$$\begin{matrix} \varphi & \sim \begin{pmatrix} 1 \\ 0 \end{pmatrix} e^{-i\xi x}, & \psi & \sim \begin{pmatrix} 0 \\ 1 \end{pmatrix} e^{i\xi x} \\ x & \to -\infty, & x & \to +\infty \\ \bar{\varphi} & \sim \begin{pmatrix} 0 \\ -1 \end{pmatrix} e^{i\xi x}, & \bar{\psi} & \sim \begin{pmatrix} 1 \\ 0 \end{pmatrix} e^{-i\xi x} \end{matrix} \tag{5.116}$$

Here $q, r \to 0$ when $|x| \to \infty$. $\zeta = \xi + i\eta$ is the eigenvalue. By using the WKB method, it can be obtained that

$$\psi e^{-i\zeta x} \sim \begin{pmatrix} 0 \\ 1 \end{pmatrix} + \frac{1}{2i\zeta} \begin{bmatrix} q \\ -\int_x^\infty qr dx' \end{bmatrix} + \cdots$$

$$\bar{\psi} e^{i\zeta x} \sim \begin{pmatrix} 1 \\ 0 \end{pmatrix} + \frac{1}{3i\zeta} \begin{bmatrix} \int_x^\infty qr dx' \\ -r \end{bmatrix} + \cdots$$

$$\varphi e^{i\zeta x} \sim \begin{pmatrix} 1 \\ 0 \end{pmatrix} - \frac{1}{2i\zeta} \begin{bmatrix} \int_{-\infty}^x qr dx' \\ r \end{bmatrix} + \cdots \tag{5.117}$$

$$\bar{\varphi} e^{-i\zeta x} \sim \begin{pmatrix} 0 \\ -1 \end{pmatrix} - \frac{1}{2i\zeta} \begin{bmatrix} q \\ \int_{-\infty}^x qr dx' \end{bmatrix} + \cdots$$

$$a(\zeta) = w(\varphi, \psi) \sim 1 - \frac{1}{2i\zeta} \int_{-\infty}^\infty qr dx' + \cdots$$

$$\bar{a}(\zeta) = w(\bar{\varphi}, \bar{\psi}) \sim 1 + \frac{1}{2i\zeta} \int_{-\infty}^\infty qr dx' + \cdots \tag{5.118}$$

where $w(u, v) = u_1 v_2 - v_1 u_2$. When $x \to \infty$, there is

$$\psi \sim \begin{pmatrix} 0 \\ 1 \end{pmatrix} e^{i\zeta x} \tag{5.119}$$

Therefore, there is

$$a(\zeta) = \lim_{x \to \infty} \left(\varphi_1 e^{i\zeta x} \right) \tag{5.120}$$

Set $e^{\hat{\varphi}} = \varphi_1 e^{i\zeta x}$, from (5.115), we have

$$\left(\varphi_1 e^{i\zeta x} \right)_x = q\varphi_2 e^{i\zeta x}$$
$$\left(\varphi_2 e^{-i\zeta x} \right)_x = r\varphi_1 e^{-i\zeta x} \tag{5.121}$$

Eliminating φ_2 and using the definition of $\hat{\varphi}$ mentioned above, as well as (5.121), we have

$$\left(\frac{1}{qe^{2i\zeta x}} \left(e^{\hat{\varphi}} \right)_x \right)_x = re^{-2i\zeta x} e^{\hat{\varphi}} \tag{5.122}$$

or

$$\hat{\varphi}_x = \frac{1}{2i\zeta} \left[-qr + \hat{\varphi}_x^2 + q \left(\frac{\hat{\varphi}_x}{q} \right)_x \right] \tag{5.123}$$

From (5.117) we have (when $\zeta \to \infty$)

$$\varphi e^{i\zeta x} \sim \begin{pmatrix} 1 \\ 0 \end{pmatrix} + O(1/\zeta) \tag{5.124}$$

$\hat{\varphi}$ expands on ζ

$$\hat{\varphi} = \frac{\hat{\varphi}_1}{2i\zeta} + \frac{\hat{\varphi}_2}{(2i\zeta)^2} + \frac{\hat{\varphi}_3}{(2i\zeta)^3} + \cdots$$
$$= \sum \frac{\hat{\varphi}_n}{(2i\zeta)^n} \tag{5.125}$$

At the same time, a series of solvable equations were generated by (5.123), and the first few were

$$\hat{\varphi}_{1x} = -qr \Rightarrow \hat{\varphi}_1 = -\int_{-\infty}^{x} qr\,dy$$
$$\hat{\varphi}_{2x} = q \left(\frac{\hat{\varphi}_{0x}}{q} \right)_x = qr_x \Rightarrow \hat{\varphi}_2 = -\int_{-\infty}^{x} qr_y\,dy \tag{5.126}$$
$$\hat{\varphi}_3 = -\int_{-\infty}^{x} \left(qr_{yy} - q^2 r^2 \right) dy$$

Generally $\hat{\varphi}_n$ has the following circular formula

$$\hat{\varphi}_{n+1} = q \left(\frac{\hat{\varphi}_n}{q} \right)_x + \sum_{k=1}^{n-1} \hat{\varphi}_k \varphi_{n-k} \quad (n \geqslant 1) \tag{5.127}$$

where $\hat{\varphi}_0 = 0$, $\hat{\varphi}_1 = -qr$. Note that $a(\zeta)$ is independent of t, and thus the conserved quantity can be obtained as

$$\ln a\,(\zeta) = \lim_{x \to \infty} \ln\left(\varphi_1 e^{i\zeta x}\right)$$

$$= \lim_{x \to \infty} \hat{\varphi} = \sum_1^{\infty} \lim_{x \to \infty} \hat{\varphi}_n / (2i\zeta)^n \qquad (5.128)$$

The conserved quantity $C_n = \lim_{x \to \infty} \hat{\varphi}_n, n = 1, 2, \cdots$. From (5.126), it can be obtained that

$$\begin{cases} c_1 = \displaystyle\int_{-\infty}^{\infty} qr dy \\[2mm] c_2 = \displaystyle\int_{-\infty}^{\infty} qr_y dy \\[2mm] c_3 = \displaystyle\int_{-\infty}^{\infty} \left(qr_{yy} - q^2 r^2\right) dy \end{cases} \qquad (5.129)$$

For a class of high-order KdV equations

$$u_t + u^q u_x + u_x^p = 0 \qquad (5.130)$$

where p, q are non-negative integers $p \geqslant 2$. In 1970, the founders of soliton theory, Kruskal, Miura [3] guessed the numbers of conservation laws of it are shown in the table.

p \ q		0	1	2	$\geqslant 3$
even		1	1	1	1
odd	3	∞	∞	∞	∞
	$\geqslant 5$	∞	3	3	3

For this conjecture, Tu Guizhang et al.[4] successfully and completely proved it using the method of symmetric functions.

For a broader class of KdV equations

$$u_t + f(u)_x = \beta u_{xxx} \qquad (5.131)$$

when $f(u)$ is a polynomial of u, its three conservation laws were specifically identified in reference:

$$
\begin{cases}
T_1 = u, X_1 = f(u) - \beta u_{xx} \\[2mm]
T_2 = \dfrac{1}{2}u^2 \\[2mm]
X_2 = \displaystyle\int_0^u f'(u)u\,du - \beta u u_{xx} + \dfrac{1}{2}\beta u_x^2 \\[2mm]
T_3 = \dfrac{\beta}{2}u_x^2 + \displaystyle\int_0^u f(u)\,du \\[2mm]
X_3 = \beta f'(u)u_x^2 + \beta^2 u_x u_{xxx} - \dfrac{\beta^2}{2}u_{xx}^2 \\[2mm]
\qquad + \dfrac{1}{2}f^2(u) - \beta f(u)u_{xx}
\end{cases}
\tag{5.132}
$$

In reference [6], for the following nonlinear evolution equations

$$
u_t = H\left(u, u_1, \cdots, u_n\right)\left(u_i = \mathscr{D}^i u, \quad \mathscr{D} = \frac{d}{dx}\right)
\tag{5.133}
$$

where $H\left(u, u_1, \cdots, u_n\right)$ represents the constant coefficient polynomial of u_i, they used infinitesimal transformations and the relationship between symmetry and conservation laws to prove that if $H = \mathscr{D}g$ (g is a gradient polynomial), then (5.133) has at least three conservation laws.

In reference [7], for the Boussinesq equation

$$
u_u = u_{xx} + \left(3u^2\right)_{xx} + u_{xxxx}
\tag{5.134}
$$

its infinite multiple conservation laws were obtained by utilizing its Bäcklund transformation.

References

[1]   Whitham G B. Non-linear dispersive waves[J]. Proceedings of the Royal Society of London. Series A. Mathematical and Physical Sciences, 1965, 283(1393): 238-261.

[2]   Miura R M, Gardner C S, Kruskal M D. Korteweg-de Vries equation and generalizations. II. Existence of conservation laws and constants of motion[J]. Journal of Mathematical physics, 1968, 9(8): 1204-1209.

[3]   Kruskal M D, Miura R M, Gardner C S, et al. Korteweg-deVries equation and generalizations. V. Uniqueness and nonexistence of polynomial conservation laws[J]. Journal of Mathematical Physics, 1970, 11(3): 952.

[4]   屠规彰, 秦孟兆. 非线性演化方程的不变群与守恒律——对称函数方法 [J]. 中国科学 A 辑, 1980(5):3-14.

[5] Tu G Z. On the similarity solution of evolution equation $ut = H(x, t, u, ux, uxx, ...)$[J]. Letters in Mathematical Physics, 1980, 4(4): 347-355.

[6] 秦孟兆, 屠规彰. 一类演化方程的三个基本守恒律 [J]. 应用数学学报, 1982, 5(2):155-164.

[7] 屠规彰. Boussinesq 方程的 Bäcklund 变换与守恒律 [J]. 应用数学学报, 1981(1):6.

[8] Lamb Jr G L. Bäcklund transformations for certain nonlinear evolution equations[J]. Journal of Mathematical Physics, 1974, 15(12): 2157-2165.

[9] Payne D A. Bäcklund transformations in several variables[J]. Journal of Mathematical Physics, 1980, 21(7): 1593-1602.

[10] Tu G Z. Proceedings of the 1980 Beijing Symposium on Differential Geometry and Differential Equations[M], Science Press, 1982, 3: 1465.

[11] Satsuma J, J. Kaup D. A Bäcklund transformation for a higher order Korteweg-de Vries equation[J]. Journal of the Physical Society of Japan, 1977, 43(2): 692-697.

[12] Tu G Z. A commutativity theorem of partial differential operators[J]. Communications in Mathematical Physics, 1980, 77: 289-297.

[13] Hirota R. Exact solution of the Korteweg—de Vries equation for multiple collisions of solitons[J]. Physical Review Letters, 1971, 27(18): 1192.

[14] Hirota R. Exact solution of the sine-Gordon equation for multiple collisions of solitons[J]. Journal of the Physical Society of Japan, 1972, 33(5): 1459-1463.

[15] Hirota R. Exact solution of the modified Korteweg-de Vries equation for multiple collisions of solitons[J]. Journal of the Physical Society of Japan, 1972, 33(5): 1456-1458.

[16] Hirota R. Exact N-soliton solution of a nonlinear lumped network equation[J]. Journal of the Physical Society of Japan, 1973, 35(1): 286-288.

[17] Hirota R. Exact N-soliton solution of nonlinear lumped self-dual network equations[J]. Journal of the Physical Society of Japan, 1973, 35(1): 289-294.

[18] Hirota R. Exact three-soliton solution of the two-dimensional sine-Gordon equation[J]. Journal of the Physical Society of Japan, 1973, 35(5): 1566-1566.

[19] Hirota R. A new form of Bäcklund transformations and its relation to the inverse scattering problem[J]. Progress of Theoretical Physics, 1974, 52(5): 1498-1512.

[20] Hirota R. The Bäcklund and inverse scattering transform of the K-dV equation with nonuniformities[J]. Journal of the Physical Society of Japan, 1979, 46(5): 1681-1682.

[21] Xun-Cheng H. Relations connecting the scale transformation and Backlund transformation for the cylindrical Korteweg-de Vries equation[J]. Journal of Physics A: Mathematical and General, 1982, 15(7): L347.

[22] Sawada K, Kotera T. A method for finding N-soliton solutions of the KdV equation and KdV-like equation[J]. Progress of theoretical physics, 1974, 51(5): 1355-1367.

Chapter 6
Multidimensional Solitons and Their Stability

6.1 Introduction

After encountering a large number of one-dimensional soliton problems, we inevitably raise the question: Does multidimensional soliton exist? If it exists, what is its behavior status? This is indeed a widely concerned and very important issue. A lot of work has been carried out on the problem of multidimensional solitons, and some meaningful results have been achieved. Although this is relatively small in number, from the perspective of the need to carry out work, the current work is only the beginning. Of course, the problem of multidimensional solitons is indeed a very complex and difficult problem, involving a series of problems that must be solved. From the current situation, there are at least the following questions: ① Do solutions for multidimensional "Solitary waves" and "Standing waves" exist? From a mathematical perspective, this problem is related to the existence of non-zero solutions for a class of nonlinear elliptic equations and certain boundary value problems of equation systems. ② Are the solutions of these "solitary waves" and "standing waves" stable? Will it collapse or collapse within a limited time? This is one of the most concerning issues in physics. ③ Are these solutions "soliton" solutions? Does its waveform and amplitude remain unchanged or change little under the interaction? Partial answers have been provided to some specific questions in these questions. It has already been pointed out in [1] and [2] that a class of nonlinear wave equations for real (uncharged) scalar fields do not have multi-dimensional steady solutions, that is, if a steady solution exists, it is only stable in plane geometry. In [3] and [4], certain sufficient criteria for the existence of solitary wave and standing wave solutions to the nonlinear Klein-Gordon equation were studied. The existence conditions for multidimensional nonlinear Langmuir solitary waves and periodic traveling wave solutions were discussed in [5]. In [7], the conditions for the existence of a class of solitons formed by three-dimensional quantity fields were studied in detail, a general theorem was established for their stability and numerical results were given for some special problems. In [8], the existence and stability of three-dimensional ion acoustic solitons in low-pressure magnetized plasmas were proved. The expressions for three soliton solutions of the two-dimensional sine-Gordon equation were given in [9]. In [10] and [11], the solitary wave problem of multidimensional nonlinear Schrödinger equations was discussed. In [12], cylindrical solitons in water waves were taken into account and numerical calculations were presented. Guo Boling et al. derived the two-dimensional Boussinesq equation and the two-dimensional KdV equation, and discussed their solitary wave solutions. In [13], the stability problem of solitons in nonlinear Klein-Gordon equations was discussed, and it was pointed out that the soliton in equations having third-order nonlinear term is unstable, while the soliton in equations having fifth order nonlinear term is stable. The existence and collapse of plasma multidimensional solitons were studied and discussed in [14] and [15]. From the current research work, considering the existence problem of multi-dimensional soliton, a large number of articles were devoted to the study of fully symmetric stable solutions, which maked the problem mathematically one-dimensional, that is, examining spherical or cylindrical symmetric models. As for the dynamic problems of "multidimensional" solitons, which examine the process of forming interactions, numerical calculations using computers are in a prominent position. From this perspective, it has promoted the development of computational mathematics for a new type of evolutionary equation and system calculation methods. This chapter mainly introduces the main results of the existence of multidimensional solitons in several important nonlinear evolution equations, and briefly introduces and comments on their stability and collapse.

6.2   The Existence Problem of Multidimensional Solitons

We define solitary waves as a solution to the wave equation, its maximum amplitude $\sup\limits_{x}|\varphi(x,t)|$ does not tend to zero at $t \to \infty$, but for each t, it tends to zero at $|x| \to \infty$. Physically speaking, some physical quantities such as charge and energy are concentrated in a finite region of space (i.e. non diffusive) at all times. Solitary waves generally have two special forms: (1) Traveling waves, $\varphi = u(x - ct)$, where c is a constant vector. (2) Standing wave $\varphi = \exp(i\omega t)u(x)$, where ω is a real number, $i = \sqrt{-1}$. The solitary waves commonly referred to the traveling wave (1), but in recent years, standing wave (2) with oscillation factor is also called soliton. For example, traveling wave solutions with oscillation factors for nonlinear Schrödinger equations have been referred to as envelope solitary waves. In some literature, "solitary waves" are sometimes confused with "solitons". However, correctly speaking, solitons should be understood as solitary waves with a certain "safety factor", that is, in the interaction of these solitary waves, their amplitude and shape remain unchanged or only slightly change. In the following, we will discuss the existence of solitary waves, solitons in several important nonlinear wave equations and equation systems, respectively.

(I) Nonlinear Klein-Gordon equation (real NLKG equation)

$$\varphi_{tt} - \Delta\varphi + m^2\varphi + f(\varphi) = 0 \tag{6.1}$$

where $x = (x_1, x_2, \cdots, x_n) \in R^n$, Δ is the Laplace operator, $m > 0$. Let $f(0) = 0$, $f\left(re^{i\theta}\right) = f(r)e^{i\theta}$, if φ has a standing wave form solution (2), then equation (6.1) can be reduced to

$$-\Delta u + \left(m^2 - \omega^2\right)u + f(u) = 0 \tag{6.2}$$

We will prove that if $f(u)$ satisfies certain growth conditions, $|\omega| < m$, then (6.2) has a non trivial solution with an exponent approaching zero at $|x| \to \infty$. If there is a traveling wave solution (1) in (6.1), then there is

$$-\sum_{ij=1}^{n} a_{ij}\frac{\partial^2 u}{\partial x_i \partial x_j} + m^2 u + f(u) = 0 \tag{6.3}$$

where $a_{ij} = \delta_{ij} + c_i c_j$. If $|c| < 1$, then (a_{ij}) is a positive definite matrix. In fact, $\sum_{ij} a_{ij}\xi_i\xi_j = |\xi|^2 - (c.\xi)^2 \geqslant \left(1 - (c)^2\right)|\xi|^2$ holds for $\xi = (\xi_1, \xi_2, \cdots, \xi_n) \in R^n$.

Through some transformations such as rotation, (6.3) and (6.2) can be simplified into the following equations

$$-\Delta u + F(u) = 0, x \in R^n \tag{6.4}$$

where $F(u) = f(u) + (\text{const })u$. We assume that $F(0) = 0$, which means that equation (6.4) always has a trivial solution $u = 0$. Letting F be a real continuous function and $G' = F$, $G(0) = 0$, we can easily obtain some necessary conditions for equation (6.4) to have a solution.

Theorem 6.1   *If $u(x, t)$ is the solution to (6.4) and tends to zero when $|x| \to \infty$, then there is*

$$(n-2)\int |\nabla u|^2 dx = -(n-2)\int uF(u)dx$$

$$= -2n\int G(u)dx \tag{6.5}$$

Therefore, if $sF(s)$, $G(s)(n \neq 1)$ or $H(s) = (n-2)sF(s) - 2nG(s)$, or $-H(s)$ is positive $(s \neq 0)$, then (6.4) only has one trivial solution. For any non trivial solution, the energy is positive:

$$E(t) = \int \left[\frac{1}{2}|\nabla u|^2 - G(u)\right]dx$$

$$= \frac{1}{n}\int |\nabla u|^2 dx > 0$$

Proof   We prove that equation (6.5) holds. Let $\bar{u}$ represent the complex conjugate of u, we have

$$-(\Delta u)\bar{u} = \nabla(\nabla u \cdot \bar{u}) + |\nabla u|^2$$

Then multiplying $\bar{u}$ by (6.4) and integrating with respect to x, and letting u and its derivative approach zero when $|x| \to \infty$, we obtain

$$\int \left[|\nabla u|^2 + \operatorname{Re} \bar{u} F(u) \right] dx = 0$$

on the other hand, $r\dfrac{\partial \bar{u}}{\partial r} = \sum x_i \bar{u}_i$ and identity equation

$$-\operatorname{Re} u_{ij}\bar{u}_i = -\operatorname{Re}(u_j x_i \bar{u}_i)_j + \left(\frac{1}{2} x_i |u_j|^2 \right)_i + \left(1 - \frac{n}{2} \right) |u_j|^2$$

$$\operatorname{Re} F(u) x_i \bar{u}_i = (x_i G(u))_i - nG(u)$$

$$\int \left[(n-2)|\nabla u|^2 + 2nG(u) \right] dx = 0$$

From these, it can be concluded that (6.5) holds. Let $L \equiv -\sum a_{ij}\dfrac{\partial^2}{\partial x_i \partial x_j} + a_0$, where the constant matrix a_{ij} is positive and a_0 is a positive number. Let $F_1(s), F_2(s)$ be a real continuous function, where $s \in [0, \infty), G_1(s), G_2(s)$ are indefinite integrals of F_1, F_2, respectively. Assuming the following conditions are satisfied:

$$F_1(s) \geqslant 0, \ F_2(s) > 0, s > 0 \tag{6.6a}$$

$$\text{when } s \to 0, F_1(s) = O(s), F_2(s) = o(s) \tag{6.6b}$$

$$F_2(s) = o\left(s^l + F_1(s)\right), \ s \to \infty \tag{6.6c}$$

$$F_2(s) = O\left(s^l + G_1(s)/s\right), s \to \infty \tag{6.6d}$$

where $l = \dfrac{n+2}{n-2}$ and $n > 3$.   □

Theorem 6.2   *If conditions (6.6a), (6.6b), (6.6c), and (6.6d) are satisfied, then there exists $\lambda > 0$ and solution $u \in H^1$ of*

$$Lu + F_1(u) = \lambda F_2(u) \tag{6.7}$$

is non negative. When $|x| \to \infty$, u exponentially decays to zero and $\int G_1(u(x))dx < \infty$.

Annotation 6.1   *When $n = 1$ or $n = 2$, Theorem 6.1 is still correct, and it can hold under weaker conditions. Here are a few examples to illustrate how to utilize the results of Theorem 6.1 and Theorem 6.2.*

Example 6.1

$$-\Delta u + u - |u|^{q-1}u = 0, \ x \in R^n, n \geqslant 3, q > 1$$

Applying Theorem 6.1 gives

$$F(s) = s - |s|^{q-1}s, \quad G(s) = \frac{s^2}{2} - \frac{|s|^{q+1}}{q+1}$$

Let $\alpha^{-1} = 2^{-1} - (q+1)^{-1}$, therefore when the coefficients of the following equation

$$\left(\frac{n-2}{2}\right) sF(s) - nG(s) = -s^2 + \left(1 - \alpha^{-1}n\right)|s|^{q+1}$$

are the same sign, i.e. $\alpha \leqslant n$, or $q \geqslant \dfrac{n+2}{n-2}$, there is no non trivial solution. Therefore, let $1 < q < \dfrac{n+2}{n-2}$. Any solution must satisfy the identity (6.5), that is

$$\alpha(n-2)\int |\nabla u|^2 dx = \frac{n\alpha}{\alpha - n}\int |u|^2 dx$$

$$= n\int |u|^{q+1} dx$$

If $F_1(s) = 0, F_2(s) = |s|^{q-1}s, L = -\Delta + I, \lambda = 1$, according to Theorem 6.2, it can be inferred that non negative solutions exist.

Example 6.2 $-\Delta u + \left(m^2 - w^2\right)u + |u|^{p-1}u - \lambda|u|^{q-1}u = 0$, *where $x \in R^n, m^2 - w^2 > 0$, and p, q are mutually different numbers greater than 1. We will discuss in four situations.*

Case A : $1 < q < \max\left(p, \dfrac{n+2}{n-2}\right)$. Theorem 6.2 asserts that for a $\lambda > 0$, there exists a non trivial solution. Note that

$$G(s) = \frac{1}{2}\left(m^2 - w^2\right)s^2 + \frac{1}{p+1}|s|^{p+1} - \frac{\lambda}{q+1}|s|^{q+1}$$

is bounded and there exists λ_*, such that $G(s)$ is non negative when $\lambda \geqslant \lambda_*$, and according to Theorem 6.1, there are only trivial solutions. According to Theorem 6.1, if a non trivial solution exists, its energy integral must be positive. Now there is an *energy density*

$$\frac{1}{2}|\varphi_t|^2 + \frac{1}{2}|\nabla\varphi|^2 + G(\varphi)$$

$$= \frac{1}{2}|\nabla u|^2 + w^2 u^2 + G(u)$$

If $w > 0$ and λ is slightly greater than λ_*, it is easy to verify that the above equation is positive. In [14], these solutions have been calculated when $n = 3$, $p = 5$, and $q = 3$, and noteworthy results have been obtained: when the energy density is positive, the disturbance of the positive solution with respect to the initial conditions is stable. Under the selection of p, q, n mentioned above, and taking $m^2 - w^2 = 1$, there is an inequality

$$\lambda s^4 = (2s)\left(\frac{1}{2}\lambda s^3\right) \leqslant \frac{1}{2}(2s)^2 + \frac{1}{2}\left(\frac{1}{2}\lambda s^3\right)^2$$

$$= 2s^2 + \frac{1}{8}\lambda^2 s^6$$

$$G(s) = \frac{1}{2}s^2 + \frac{1}{6}s^6 - \frac{\lambda}{4}s^4 \geqslant \left(\frac{1}{6} - \frac{1}{32}\lambda^2\right)s^6$$

We can obtain $\lambda_* = (4.3)^{-\frac{1}{2}}$.

Case B : $p < q < \dfrac{n+2}{n-2}$. By applying Theorem 6.2, we can prove that for every $\lambda > 0$, there exists an infinite sequence of non trivial solutions.

Case C: $p \leqslant (n+2)/(n-2) \leqslant q$. If $\alpha^{-1} = 2^{-1} + (q+1)^{-1}$, $\beta^{-1} = 2^{-1} + (p+1)^{-1}$, then $\alpha \leqslant n \leqslant \beta$, and

$$\frac{n-2}{2}sF(s) - nG(s)$$

$$= -s^2 - \left(1 - \frac{n}{\beta}\right)|s|^{p+1} + \left(1 - \frac{n}{\alpha}\right)\lambda|s|^{\alpha+1}$$

According to Theorem 6.1, there exists a non trivial solution.

Case D: $(n+2)/(n-2) < p < q$, we don't know if there is a non trivial solution in this situation.

Annotation6.2 *The existence of non trivial solutions to equation $-\Delta u + F(u) = 0$ essentially requires $F'(0) \geqslant 0$. In fact, let $-\alpha = F'(0) < 0$ and $f(s) = F(s)/s + \alpha$, assume that $u(x)$ is a non trivial solution and very small at infinity, let $q(x) = f(u(x))$. Now, the equation can be rewritten as $-\Delta u + qu = \alpha u$. Let $u(x)$ be sufficiently small at infinity, such that $q(x) = O|x|^{-1}$. We can easily understand that the operator $-\Delta + q$ does not have positive eigenvalues, so we have a contradiction. Therefore,*

$$F'(0) \geqslant 0$$

We are now considering the solution to the axisymmetric problem. Axial solution $u(r)$: For $r = |x| \neq 0$, it is continuous and satisfies equation

$$u_{rr} + \frac{n-1}{r}u_r - F(u) = 0, \ 0 < r < \infty$$

where $F(u) = u + F_1(u) - \lambda F_2(u)$. So

$$r^{1-n}\left(r^{n-1}u_r\right)_r = u_{rr} + ((n-1)/r)u_r$$

is continuous, therefore $u \in c^2, r \neq 0$. Let

$$q(r) = \frac{F(u(r))}{u(r)} = 1 + F_1'(0) + p(r)$$

from (6.6a) and (6.6b), it can be seen that $p(r) \to 0, \ r \to \infty$. Therefore, for a sufficiently large r, there is $q(r) \geqslant \frac{1}{2}$. Let $v = r^{(n-1)/2}u$, which satisfies the equation

$$v_{rr} - \left[q(r) + \frac{(n-1)(n-3)}{4r^2}\right]v = 0$$

$$\left(\frac{1}{2}v^2\right)_{rr} = v_r^2 + \left[q(r) + \frac{(n-1)(n-3)}{4r^2}\right]v^2$$

So for a sufficiently large $r, w = v^2$, satisfy the inequality $w_{rr} \geqslant w$. w and u as exponential decay from this inequality.

In fact, for a large r, it can be deduced that $Q = e^{-r}(w_r + w)$ is non decreasing. If Q remains non positive for a large r, then $(e^r w)_r = e^{2r}Q \leqslant 0$, which leads to $w = O\left(e^{-r}\right)(r \to \infty)$. If $Q \geqslant 2\delta > 0$, then $w_r + w$ must not be integrable near infinity. However, due to $u \in H^1$, the functions v^2, v_r^2, w, and w_r are integrable on the interval $k < r < \infty$, which leads to a contradiction and proves the exponential decay of the solution.

The relationship between the axial solution and the solution to a certain type of definite solution problem is as follows:

Theorem 6.3 *Assuming $L = I - \Delta$, $F_1 = 0$, F_2 is a continuous real function, such that*

(i) $sF_2(s) > 0, s \neq 0$

(ii) $F_2(s) = O\left(|s|^p\right), |s| \to \infty$, *and* $p < (n+2)/(n-2) = l$

(iii) F_2 *is an odd function, and* $F_2(0) = 0$

then for any $\gamma > 0$, there exists an infinite pair of axial solutions $(\lambda_k, \pm u_k), k = 0, 1, 2, \cdots$ for $Lu = \lambda F_2(u)$, and

$$(Lu_k, u_k) = \gamma$$

Theorem 6.4 *Let F be a real continuous function that satisfies*

(i) $F(s)/s \to -\infty, s \to +\infty$

(ii) $sF(s) \geqslant \alpha G(s), \alpha > 2$

(iii) $F(s) = o(s), s \to 0$

(iv) $F_2(s) = O\left(|s|^p\right), |s| \to \infty, p < (n+2)/(n-2) = l$

then the equation $Lu + F(u) = 0$ has at least one non trivial solution. If F is an odd function, then there exist infinite pairs of distinct solutions $\pm u_k (k = 0, 1, 2, \cdots)$.

Example 6.3 $-\Delta u + u - |u|^{q-1} u = 0, 1 < q < \dfrac{n+2}{n-2}$. *Theorem th6.2.3 and Theorem th6.2.4 state that there are multiple axial solutions $u_0, u_1, u_2, \cdots$.*

Example 6.4 $-\Delta u + u + |u|^{p-1} u - \lambda |u|^{q-1} u = 0, \ 1 < p < q < \dfrac{n+2}{n-2}$. *For any $\lambda > 0$, Theorem 6.2.4 can be applied to derive at least one non trivial solution. For large p, q, the problem remains unresolved.*

(II) Multidimensional Nonlinear Langmuir Waves

$$\begin{cases} i\dfrac{\partial \boldsymbol{E}}{\partial t} = -\nabla^2 \boldsymbol{E} + n\boldsymbol{E} \\[2mm] \dfrac{\partial^2 n}{\partial t^2} = \nabla^2 \left[n + g\left(|\boldsymbol{E}|^2\right) \right] \end{cases} \tag{6.8}$$

where, $i = \sqrt{-1}$, $\boldsymbol{E} = (E_1, E_2, \cdots, E_N)$ is the complex amplitude of the high-frequency electric field, n is the low-frequency disturbance of the ion density with respect to its constant equilibrium state, and g is the given function of $|\boldsymbol{E}|^2$. When $g\left(|\boldsymbol{E}|^2\right) = |\boldsymbol{E}|^2$, (6.8) is the Zakharov equations. When $g\left(|\boldsymbol{E}|^2\right) = \chi\left(1 - \exp\left(-|\boldsymbol{E}|^2\right)\right)$, where χ is a normal number, it corresponds to the saturation state of ion density. Assuming $\boldsymbol{k}$ is the unit vector on R^n and v is the traveling wave velocity, we seek the following traveling wave solution of (6.8)

$$\boldsymbol{E}(\boldsymbol{x}, t) = \boldsymbol{h}(\boldsymbol{k} \cdot \boldsymbol{x} - vt), n(\boldsymbol{x}, t) = s(\boldsymbol{k} \cdot \boldsymbol{x} - vt) \tag{6.9}$$

where $\boldsymbol{h}$ and s are the vector function to be solved and the function to be solved, respectively. Let $|\boldsymbol{h}(\xi)|, |s(\xi)|$ be uniformly bounded, where $\xi = \boldsymbol{k} \cdot \boldsymbol{x} - vt$, and assume

$$\boldsymbol{h}(\xi) |, s(\xi) \to 0(|\xi| \to \infty)$$

Replace (6.9) with (6.8) to obtain the equation that $\boldsymbol{h} = (h_1, h_2, \cdots, h_N)$ and s satisfy

$$-iv\dfrac{d\boldsymbol{h}}{d\xi} + \dfrac{d^2\boldsymbol{h}}{d\xi^2} = s(\xi)\boldsymbol{h}(\xi) \tag{6.10a}$$

$$\left(v^2 - 1\right) \dfrac{d^2 s}{d\xi^2} = \dfrac{d^2}{d\xi^2} g\left(|\boldsymbol{h}|^2\right) \tag{6.10b}$$

Integrate equation (6.10b) to obtain

$$\left(v^2 - 1\right) s(\xi) = g\left(\left(|\boldsymbol{h}(\xi)|^2\right) + \hat{c}\xi + c \right. \tag{6.11}$$

where $\hat{c}$ and c are both integral constants. Assume $v^2 \neq 1$, from the boundedness requirement of $s(\xi)$, let $\hat{c} = 0$, solve $s(\xi)$ from (2.15) and substitute it into (2.13) to obtain a complex equation that $\boldsymbol{h}$ satisfies

$$\frac{d^2 \boldsymbol{h}}{d\xi^2} - iv\frac{d\boldsymbol{h}}{d\xi} = \left(v^2 - 1\right)^{-1}\left[g\left(|\boldsymbol{h}(\xi)|^2\right) + c\right]\boldsymbol{h} \tag{6.11'}$$

For convenience, $((6.11'))$ is written in polar coordinate form. If

$$h_i(\xi) = A_i(\xi)\exp[i\theta(\xi)], \quad j = 1, 2, \cdots, N$$

then there are

$$\frac{d^2 A_j}{d\xi^2} + A_j\theta_j'(\xi)\left[v - \theta_j'(\xi)\right]$$
$$= \left(v^2 - 1\right)^{-1} A_j\left[g\left(\|A\|^2\right) + c\right] \tag{6.12a}$$

$$\frac{d^2\theta}{d\xi^2} = \left(v - 2\theta_j'(\xi)\right)\frac{d}{d\xi}\ln A_j, j = 1, 2, \cdots, N \tag{6.12b}$$

where $A = (A_1, A_2, \cdots, A_N)$, $\|A\| = |\boldsymbol{h}|$, $\theta_j' = \dfrac{d\theta_j}{d\xi}$. Integrating (6.12b) yields

$$\theta_j'(\xi) = \left(v - \mu_j A_j^{-2}(\xi)\right)/2$$
$$\mu_j = A_j^2(0)\left(v - 2\theta_j'(0)\right) \tag{6.13}$$

Substituting (6.13) into (6.12a) yields a system of differential equations satisfying A_j

$$\frac{d^2 A_j}{d\xi^2} = f\left(\mu_j, c, A\right) A_j, j = 1, 2, \cdots, N \tag{6.14}$$

where

$$f\left(\mu_j, c, A\right) = \left(\mu_j^2 A_j^{-4} - v^2\right)/4$$
$$+ \left(v^2 - 1\right)^{-1}\left[g\left(\|A\|^2\right) + c\right] \tag{6.15}$$

(6.14) can be written in the following form

$$\frac{d^2 A_j}{d\xi^2} = \frac{\partial U}{\partial A_j}, j = 1, 2\cdots, N \tag{6.16}$$

where

$$U(\boldsymbol{A}, \boldsymbol{\mu}, c) = U_1\left(\|\boldsymbol{A}\|^2, c\right) - \sum_{i=1}^{N} \mu_j^2 / 8A_j^2 \tag{6.17a}$$

$$2U_1\left(\|\boldsymbol{A}\|^2, c\right) = \int_0^{\|\boldsymbol{A}\|^2} [Kg(\eta) + \gamma] d\eta \tag{6.17b}$$

$$K = \left(v^2 - 1\right)^{-1}, \gamma = \left(v^2 - 1\right)^{-1} c - v^2/4 \tag{6.17c}$$

and $\boldsymbol{\mu} = (\mu_1, \mu_2, \cdots, \mu_N)$. The first integral of (6.16) is

$$I\left(\boldsymbol{A}(\xi), \boldsymbol{A}'(\xi)\right) = \|\boldsymbol{A}'(\xi)\|^2 - 2U_1\left(\|\boldsymbol{A}(\xi)\|^2, c\right)$$
$$- \sum_{j=1}^{N} \mu_j^2 A_j^{-2}(\xi)/4 = c_1 \tag{6.18}$$

where $\|\boldsymbol{A}'(\xi)\|^2 = \sum_{j=1}^{N} \left[\dfrac{dA_j(\xi)}{d\xi}\right]^2$, and

$$c_1 = \|\boldsymbol{A}'(0)\|^2 - 2U_1\left(\|\boldsymbol{A}(0)\|^2, c\right) - \sum_{j=1}^{N} \left[v - 2\theta_j'(0)\right] \tag{6.19}$$

Obviously, for a certain $j, \mu_j \neq 0$, and since c_1 is finite, then when

$$\|\boldsymbol{A}'(\xi)\| \to 0, \ \|\boldsymbol{A}(\xi)\| \to 0$$

$I\left(\boldsymbol{A}(\xi), \boldsymbol{A}'(\xi)\right) \to -\infty$. Therefore, there are no solitary wave solutions of (6.16) or (6.12a) that result in $\|\boldsymbol{A}(\xi)\| \to 0, \|\boldsymbol{A}'(\xi)\| \to 0(|\xi| \to \infty)$. When $\mu = 0$, we have

$$\theta_j(\xi) = \theta_j(0) + \frac{1}{2}v\xi \quad j = 1, 2, \cdots, N \tag{6.20}$$

Now (6.16) becomes

$$\frac{d^2 A_j}{d\xi^2} = \frac{\partial U_1}{\partial A_j}, \ j = 1, 2, \cdots, N \tag{6.21}$$

The equilibrium point of (6.21) is $(A_c, 0) \in R^{2N}$, such that A_c is the steady point of U_1 or equation $f(0, c, \boldsymbol{A})\boldsymbol{A} = 0$. Obviously, A_c includes $A = 0$ and A satisfies $g\left(\|\boldsymbol{A}\|^2\right) = v^2\left(v^2 - 1\right)/4c$. For (6.21), let $u(\xi) = \|\boldsymbol{A}(\xi)\|^2$ direct calculations lead to

$$\frac{d^2 u}{d\xi^2} = 2\|\boldsymbol{A}'(\xi)\|^2 + 2\boldsymbol{A}(\xi)\frac{d^2 \boldsymbol{A}}{d\xi^2}$$
$$= 2\|\boldsymbol{A}'(\xi)\|^2 + 2u\tilde{f}(c, u) \tag{6.22}$$

where $\tilde{f}\left(c, \|\boldsymbol{A}\|^2\right) = f(0, c, A)$. For the fixed c_1 and $\mu = 0$, (6.22) becomes

$$\frac{d^2 u}{d\xi^2} = 2\left[u\tilde{f}(c, u) + c_1 + 2U_1(u, c)\right] = p\left(u, c, c_1\right) \tag{6.23}$$

The initial conditions are

$$u(0) = \|\boldsymbol{A}(0)\|^2, u'(0) = 2\boldsymbol{A}(0) \cdot \boldsymbol{A}'(0) \tag{6.24a}$$

$$\|\boldsymbol{A}'(0)\|^2 = c_1 + 2U_1\left(\|\boldsymbol{A}(0)\|^2, c\right) \geqslant 0 \tag{6.24b}$$

The first integral of (6.23) is

$$[u'(\xi)]^2 = [u'(0)]^2 + \int_{u(0)}^{u(\xi)} QP\left(\eta, c, c_1\right) d\eta$$

$$= Q\left(u, c, c_1, u'(0)\right) \tag{6.25}$$

where $u' = \dfrac{du}{d\xi}$. Equation (6.25) is correct only when its right side is non negative.

Implicit expression of $\|A(\xi)\|^2$ can be obtained by integrating (6.25)

$$\int_{\|A(0)\|^2}^{\|A(\xi)\|^2} Q\left(\eta, c, c_1, u'(0)\right)^{-\frac{1}{2}} d\eta = \pm\xi \tag{6.26}$$

We now discuss the existence of solutions with $\|A(\xi)\| \to 0(|\xi| \to \infty)$ for equation (6.21).

Theorem 6.5   *If*

$$Kg(u) + \gamma \geqslant 0, \quad \forall u \geqslant 0 \tag{6.27}$$

then there is no solution for (6.21), such that

$$\|\boldsymbol{A}(0)\| > 0, \quad \|\boldsymbol{A}(\xi)\| \to 0(|\xi| \to \infty)$$

Proof   For $u \geqslant 0$, condition (6.27) is equivalent to $\tilde{f}(c, u) \geqslant 0$. From (6.22) we have $\dfrac{d^2 u}{d\xi^2} \geqslant 0$. Further, we deduce that $\|\boldsymbol{A}(\xi)\|^2$ (corresponding to any solution of (6.21)) is a convex function of ξ. Therefore, there cannot be solutions with $\|A(0)\| > 0$, $\|\boldsymbol{A}(\xi)\| \to 0(|\xi| \to \infty)$. $\qquad\square$

Theorem 6.6   *If the following conditions are satisfied:*

(i) $v^2\left(v^2 - 1\right) > 4c, v^2 < 1$

(ii) $g(s)$ *is a strict monotone increasing function,* $g(0) = 0$, *and there exists a positive number* $u_1 < \infty$, *such that*

$$\int_0^{u_1} g(\eta)d\eta = \left[v^2\left(v^2 - 1\right)/4 - c\right] u_1 \tag{6.28a}$$

$$\int_0^u g(\eta)d\eta > \left[v^2 \left(v^2 - 1\right)/4 - c\right] u_1, \quad \forall u > u_1 \qquad (6.28b)$$

then (6.21) has a solution, $\boldsymbol{A}(\xi) \geqslant 0, \forall \xi \in R$. If $\|\boldsymbol{A}(0)\| > 0, \|\boldsymbol{A}'(0)\| = 0$, then we get $\|\boldsymbol{A}(\xi)\| \to 0, \|\boldsymbol{A}'(\xi)\| \to 0(|\xi| \to \infty)$.

For equation (6.21), in order to seek the periodic function solution $\boldsymbol{A}(\xi)$ with modulus ξ, we have

Theorem 6.7 *If the consitions for Theorem 6.6 are satisfied and there exists a real number $\gamma_e > 0$ satisfies*

$$Kg\left(r_e^2\right) + \gamma = 0, \quad or \quad g\left(r_e^2\right) = v^2 \left(v^2 - 1\right)/4 - c \qquad (6.29)$$

then there exists a solution $\boldsymbol{A}(\xi)$ for (6.21), where $\|\boldsymbol{A}(\xi)\|$ is a periodic function of ξ.

Theorem 6.8 *Assuming the following conditions are satisfied*
(i) *$g(u)$ is a real monotonically increasing function, and $g(0) = 0$*
(ii) *$v^2 \left(v^2 - 1\right) < 4c, v^2 > 1$*
(iii) *If the initial condition $\boldsymbol{A}(0), \boldsymbol{A}'(0)$ satisfies $\|\boldsymbol{A}(0)\| > 0$ and*

$$\tilde{c}_1 = \|\boldsymbol{A}'(0)\|^2 - 2U_1 \left(\|\boldsymbol{A}(0)\|^2, c\right) \geqslant 0$$

then there exists a solution $\boldsymbol{A}(\xi)$ of (6.21), $\|\boldsymbol{A}(\xi)\|$ is definitely not a periodic function of ξ.

(III) Three dimensional FDS nonlinear wave equation

$$\Box\varphi + \alpha^2 \chi^2 \varphi = 0 \qquad (6.30a)$$

$$\Box\chi + \alpha^2 \chi|\varphi|^2 + \frac{1}{2}\chi \left(\chi^2 - 1\right) = 0 \qquad (6.30b)$$

where $\Box \equiv \dfrac{\partial^2}{\partial t^2} - \Delta, \ \Delta \equiv \dfrac{\partial^2}{\partial x_1^2} + \dfrac{\partial^2}{\partial x_2^2} + \dfrac{\partial^2}{\partial x_3^2}$. For complex fields, we consider the following traveling wave solutions with oscillation factors

$$\varphi(r, t) = \frac{1}{\sqrt{2}}\psi(r)e^{-i\omega t} \qquad (6.31)$$

From (6.30a), (6.30b), we can obtain

$$\nabla^2\psi - \alpha^2 \chi^2 \psi + \omega^2 \psi = 0 \qquad (6.32a)$$

$$\nabla^2\chi - \alpha^2 \psi^2 \chi - \frac{1}{2} \left(\chi^2 - 1\right) \chi = 0 \qquad (6.32b)$$

Charge $Q = \omega \displaystyle\int \dot{\psi}^2 d^3x$, system energy $E = \displaystyle\int \varepsilon d^3\chi$, where

$$\varepsilon = \frac{1}{2}(\nabla\chi)^2 + (\nabla\psi)^3 + \frac{1}{2}\left(\omega^2 + \alpha^2\chi^2\right)\psi^2$$

$$+ \frac{1}{4} \left(\chi^2 - 1 \right)^2$$

Given

$$\xi = \left(\alpha^2 - \omega^2 \right)^{\frac{1}{2}}, \ \chi = 1 - \frac{1}{2} (\xi/\alpha)^2 x$$

$$\omega = 2^{-\frac{1}{2}} \frac{\xi}{\alpha} y \tag{6.33}$$

we consider the spherically symmetric solution and x, y as a function of r. Substitute (6.33) into (6.32b) and compare the lowest order terms of ξ to obtain

$$x = y^2$$

Furthermore, from (6.32a), we can obtain

$$\frac{1}{r^2} \frac{d}{dr} \left(r^2 \frac{dy}{dr} \right) - y + y^2 = 0 \tag{6.34}$$

Boundary condition

$$\begin{cases} \dfrac{dy}{dr} = 0, r = \infty \\ y = 0, r \to \infty \end{cases} \tag{6.35}$$

It is easy to determine that (6.34) and (6.35) have an infinite number of solutions, while corresponding to problems (6.32a) and (6.32b), the solution with the lowest energy and no intersection with the radial direction is an soliton solution, which is stable.

(IV) Multidimensional nonlinear Schrödinger equation

$$i u_t + \nabla^2 u + q \left(|u|^2 \right) u = 0 \tag{6.36}$$

If

$$u = f \left(x_1', x_2', \cdots, x_n' \right) e^{i\theta}, \quad x_j' = x_j - c_j t$$

$$\theta = \sum_{j=1}^{N} k_j x_j - wt \tag{6.37}$$

then

$$u_t = \left[f(-i\omega) - \sum_{j=1}^{N} f_{x_j'} c_j \right] e^{i\theta}$$

$$u_{x_i x_j} = \left[-k_i k_j f + i \left(k_i f_{x_j'} + k_j f_{x_i'} \right) + f_{x_i' x_j'} \right] e^{i\theta}$$

$$(j = 1, 2, \cdots, N)$$

Substituting them into (6.36) yields

$$\omega f - i \sum_{j=1}^{N} f'_{xj} c_j + 2i \sum_{j=1}^{N} k_j f_{x'_j}$$

$$- \sum_{j=1}^{N} k_j^2 f + \nabla'^2 f + q\left(f^2\right) f = 0$$

Taking $c_j = 2k_j$ and eliminating the imaginary part yields

$$\nabla'^2 f + \left(\omega - \sum_{j=1}^{N} k_j^2\right) f + q\left(f^2\right) f = 0 \tag{6.38}$$

especially for spherical symmetry, there is

$$\frac{1}{\rho^{n-1}} \frac{\partial}{\partial \rho}\left(\rho^{n-1} \frac{\partial f}{\partial \rho}\right)$$
$$+ \left(\omega - \sum_{j=1}^{n} k_j^2\right) f + q\left(f^2\right) f = 0 \tag{6.39}$$

When $n = 1$, $\omega = k_i^2 - \eta^2$, we obtain the soliton solution of the one-dimensional nonlinear Schrödinger equation, which is stable. For solitary wave solution with $n > 1$, it is unstable.

(V) In low-pressure magnetized plasma, the three-dimensional ion acoustic equation is

$$\begin{cases} \dfrac{\partial n}{\partial t} + \operatorname{div} n\boldsymbol{V} = 0 \\[2mm] \dfrac{\partial \boldsymbol{V}}{\partial t} + (\boldsymbol{V} \cdot \nabla)\boldsymbol{V} = -e\nabla\varphi/M + [\boldsymbol{V}, W_{H_i}] \\[2mm] \Delta\varphi = -4\pi e\left(n - n_0\right)\exp\left(e\varphi/T_c\right) \end{cases} \tag{6.40}$$

Simplify it to dimensionless form

$$\frac{\partial u}{\partial \tau} + \frac{\partial}{\partial \xi_z}\left(\Delta_{\xi\xi} + u\right) u = 0 \tag{6.41}$$

where $\tau = \dfrac{1}{2}\omega_{p_i} t, u = \dfrac{v_z}{2c_s}$. Letting $u = u\left(\xi_x - \lambda\tau\right)$ gives

$$\Delta_{\xi\xi} u - (\lambda - u)u = 0 \tag{6.42}$$

When $\lambda = c^2 > 0$, it has an exponentially decaying solution when $|\xi| \to \infty$. The simplest case is ball symmetry:

$$\frac{1}{\xi^2}\frac{d}{d\xi}\xi^2\frac{du}{d\xi} - \left(c^2 - u\right)u = 0 \tag{6.43}$$

The solution of (6.43) exists as a three-dimensional soliton solution, which is stable. By numerical calculation, the graph can be obtained as shown in Figure 6-1.

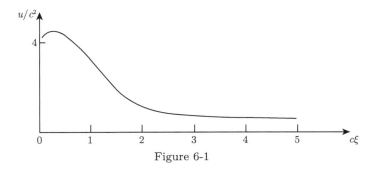

Figure 6-1

(VI) Two dimensional sine-Gordon equation

$$\varphi_{xx} - \varphi_{yy} - \varphi_{tt} = \sin\varphi \tag{6.44}$$

Its three soliton solutions were considered in [9]. The formal solution of (6.44) is

$$\varphi(x, y, t) = 4\tan^{-1}[g(x, y, t)/f(x, y, t)]$$

where

$$f = 1 + a(1,2)e^{\eta_1+\eta_2} + a(1,3)e^{\eta_1+\eta_3} + a(2,3)e^{\eta_2+\eta_3}$$

$$g = e^{\eta_1} + e^{\eta_2} + e^{\eta_3} + a(1,2)a(1,3)a(2,3)e^{\eta_1+\eta_2+\eta_3}$$

$$a(i,j) = \frac{\left(p_i - p_j\right)^2 + \left(q_i - q_j\right)^2 - \left(\Omega_i - \Omega_j\right)^2}{\left(p_i + p_j\right)^2 + \left(q_i + q_j\right)^2 - \left(\Omega_i + \Omega_j\right)^2}$$

$$\eta_1 = p_i x + q_i y - \Omega_i t - \eta_i^0 \ (\eta_i^0 \text{ is a constant })$$

$$p_i^2 + q_i^2 - \Omega_i^2 = 1, \ i = 1, 2, 3$$

and

$$\begin{vmatrix} p_1 & q_1 & \Omega_1 \\ p_2 & q_2 & \Omega_2 \\ p_3 & q_3 & \Omega_3 \end{vmatrix} = 0$$

(VII) Complex nonlinear field equation

$$\nabla^2\psi - c^{-2}\frac{\partial^2\psi}{\partial t^2} = k^2\psi - \mu^2|\psi|^2\psi \tag{6.45}$$

Assuming $\psi = \varphi(r)e^{i\omega t}$, where $\varphi(r)$ is real and spherically symmetric, then (6.45) becomes

$$\frac{d^2\varphi}{dr^2} + \frac{2}{r}\frac{d\varphi}{dr} = \left(k^2 - \frac{w^2}{c^2}\right)\varphi - \mu^2\varphi^3 \tag{6.46a}$$

$$\frac{d\varphi}{dr}\bigg|_{r=0} = 0, \ \varphi \to 0, r \to \infty \tag{6.46b}$$

Setting $k^2 - w^2/c^2 > 0$ and making the transformation

$$r' = r\left(k^2 - w^2/c^2\right)^{\frac{1}{2}}, \ \varphi' = \mu\varphi\left(k^2 - w^2/c^2\right)^{-\frac{1}{2}}$$

then φ' satisfies equation

$$\frac{d^2\varphi'}{dr'^2} + \frac{2}{r'}\frac{d\varphi'}{dr'} = \varphi' - \varphi'^3$$

It can be proven that solution $\psi = \varphi(r)e^{i\omega t}$ of (6.45) is unstable for small perturbations.

6.3　Stability and Collapse of Multidimensional Solitons

One of the most important and natural requirements for solitons in plasma physics and various field models is that they must be stable, that is from a "process" perspective, solitons must have a sufficiently long lifetime. In other words, the lifetime of solitons must be much longer than the characteristic interaction time of solitons. The stability in this regard varies depending on the direction of disturbance, including "longitudinal" stability and "transversal" stability. In terms of stability analysis and handling, there are two types: linear stability and nonlinear stability. Nonlinear stability generally refers to the stability based on a certain functional. The stability typically considered in physics means that the energy of the system is minimized. In multidimensional situations, many solitary waves are unstable.

1. We first start with the simplest case, a real (uncharged) scalar field, described by the following nonlinear wave equation

$$\Box F(p) + F'(\varphi) = 0 \quad \left(\Box \equiv \frac{\partial^2}{\partial t^2} - \nabla^2, \ F'(\varphi) = \frac{dF}{d\varphi}\right) \tag{6.47}$$

The Hamiltonian of a steady field is

$$E = \int \left[\frac{1}{2}(\nabla\varphi)^2 + F(\varphi) \right] dx = K + V \tag{6.48}$$

After calibration transformation $\varphi_\alpha = \varphi(\alpha x)$, we can obtain

$$E\left[\varphi_\alpha\right] = \alpha^{2-n}K + \alpha^{-n}V \tag{6.49a}$$

$$\left.\frac{dE}{d\alpha}\right|_{\alpha=1} = 0, \quad V = \frac{2-n}{n}K \tag{6.49b}$$

$$\left.\frac{d^2E}{d\alpha^2}\right|_{\alpha=1} = -2(n-2)K \tag{6.49c}$$

From (6.49b) and (6.49c), it can be seen that $n = 1$ gives the minimum value of E, which is stable. $n > 2$ gives the maximum value of $E[\varphi]$, which is unstable. $n = 2$ gives a inflection point.

2. If an electric field exists ($Q \neq 0$), then the above situation undergoes a qualitative change. For the three-dimensional FDS nonlinear wave equation (6.30a) and (6.30b), when

$$Q > Q_s = \frac{1}{2}\left(\frac{4\pi}{3\alpha}\right)^4$$

it can be proven that $E_{\min} < Q_m$ (corresponding to the free meson solution), its soliton solution is absolutely stable.

3. φ^5 nonlinear wave equation

$$\nabla^2\psi - \frac{1}{c^2}\frac{\partial^2\psi}{\partial t^2} = k^2\psi - \mu^2|\psi|^2\psi + \lambda|\psi|^4\psi \tag{6.50}$$

where λ is real and usually positive. Considering its spherically symmetric solution, (6.50) becomes

$$\frac{d^2\varphi'}{dr'^2} + \frac{2}{r'}\frac{d\varphi'}{dr'} = \varphi' - \varphi'^3 + \beta\varphi'^5 \tag{6.51}$$

where

$$\varphi' = \mu\rho\left(1 - \omega'^2\right)^{-\frac{1}{2}}, r' = kr\left(1 - \omega'^2\right)^{\frac{1}{2}}$$
$$\beta = \lambda k^2\left(1 - \omega'^2\right)/\mu^4 \tag{6.52}$$

Both first-order perturbation theory and direct perturbation method can be used to prove the existence of the steady solution of (6.50).

4. Three dimensional ion acoustic wave equation (6.42) in low-pressure magnetized plasma. Its energy is

$$\mathscr{H} = \int \left[\frac{1}{2} \left(D_\xi u \right)^2 - \frac{1}{3} u^3 \right] d\xi$$

By using Hölder inequality

$$\int u^3 d\xi \leqslant \left(\int u^2 d\xi \right)^{\frac{1}{2}} \left(\int u^4 d\xi \right)^{\frac{1}{2}}$$

and interpolation inequality

$$\int u^4 dx \leqslant 4 \left(\int u^2 d\xi \right)^{\frac{1}{2}} \left(\int |\nabla u|^2 d\xi \right)^{\frac{3}{2}} \tag{6.53}$$

of $\int u^4 d\xi$, we can obtain

$$\mathscr{H} \geqslant \int \frac{(\nabla u)^2}{2} d\xi - \frac{2}{3} \left(\int u^2 d\xi \right)^{\frac{3}{2}} \left(\int (\nabla u)^2 d\xi \right)^{\frac{3}{4}}$$

$$\geqslant -\frac{1}{6} \left(\int u^2 d\xi \right)^3$$

From this, it can be inferred that the functional $\mathscr{H}$ has a lower bound, so the three-dimensional spherically symmetric soliton solution reaches an absolute minimum and is therefore stable.

5. Collapse of Langmuir waves

Similar to the mechanical phenomenon of the focal effect of spherical shock waves, a similar phenomenon occurs in the dissipation mechanism of "Langmuir condensation" - the condensation of turbulent energy in the long wavelength region of the spectrum. This aggregation indicates the instability of multidimensional Langmuir solitons.

Example 6.5 *In the φ^3 approximation, CLW is described by equation*

$$\nabla^2 \left(i\psi_t + \nabla^2 \psi \right) - \operatorname{div} \left(|\nabla \psi|^2 \nabla \psi \right) = 0 \tag{6.54}$$

where ψ is the high-frequency potential envelope. Considering the case of spherical symmetry, (6.54) becomes

$$i\varphi_t + \nabla_{rr}^2 \cdot \varphi - \frac{n-1}{r^2} \varphi + |\varphi|^2 \varphi = 0 \tag{6.55}$$

where $\varphi = -\nabla\psi$ *and* $\varphi(0) = 0$.   (6.54) *and* (6.55) *respectively have conserved*
quantities

$$s = \int |\nabla\psi|^2 d^3r, \quad s_2 = \int \left[|\nabla^2\psi|^2 - \frac{1}{2} |\nabla\psi|^4 \right] d^3r \tag{6.56a}$$

$$s = \int_0^\infty |\varphi|^2 r^2 dr \tag{6.56b}$$

$$s_2 = \int_0^\infty \left[|(r\varphi)_r|^2 + 2|\varphi|^2 - \frac{1}{2} r^2 |\varphi|^4 \right] dr \tag{6.56c}$$

Consider accelerating the quasi planar soliton of (6.55) *towards the origin* (*see Figure*
6-2).

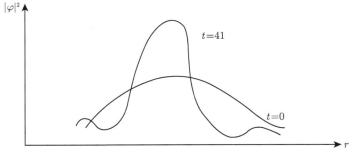

Figure 6-2

Let $D = \langle r^2 \rangle_\varphi = \int_0^\infty |\varphi|^2 r^4 dr$, *and from* (6.55), *we can obtain*

$$\frac{d^2 D}{dt^2} = 6s_2 - 2 \int_0^\infty |(r\varphi)_r|^2 \, dr$$

$$- 4 \int_0^\infty |\varphi|^4 \cdot r^2 dr < 6s_2$$

Assuming $s_2 \leqslant 0$, *the above equation integrates with respect to* t *and yields*

$$D \leqslant 3s_2 t^2 + c_1 t + c_2$$

When $t \to t_0 = \dfrac{c_1 + \left(c_1^2 + 12 c_2 \, |s_2| \right)^{\frac{1}{2}}}{6 \, |s_2|}$, *the local solution of the initial value problem*
causes singularity. If $c_1 > 0$, *small* t-*wave packets will be dispersed. If* $c_1 < 0$, *it will*
cause contraction.

Example 6.6 *Equation system*

$$\text{div}\left(-2i\nabla\psi_t - \nabla\nabla^2\psi + \Phi\nabla\psi\right) = 0 \tag{6.57a}$$

$$\left(\frac{\partial^2}{\partial t^2} - \nabla^2\right)\Phi = \nabla^2\left(|\nabla\psi|^2\right) \tag{6.57b}$$

introducing low-frequency potential u,

$$u_t = \Phi + |\varphi|^2 = \Phi + |\nabla\psi|^2 \tag{6.58a}$$

$$\nabla^2 u = \Phi_t \tag{6.58b}$$

$$\nabla^2\left(i\psi_t + \nabla^2\psi\right) = \text{div}(\Phi\nabla\psi) \tag{6.58c}$$

easily yields

$$s_2 = \int\left[|\nabla^2\psi|^2 + \Phi|\nabla\psi|^2 + \frac{1}{2}\Phi^2 + \frac{1}{2}(\nabla u)^2\right]d^3r \tag{6.59}$$

Assuming $s_2 \leqslant 0$, for the system of equations (6.58a), (6.58b) and (6.58c), we can obtain the self transformation in two extreme cases. At the quasi-static limit, it transforms into (6.54); at the ultrasonic limit, Φ in the right side of (6.58a) can be ignored. In the first case, the self transformation of (6.54) is

$$\psi = \exp\left\{-i\mu^2\ln\left(t_0 - t\right)\right\}\chi(\boldsymbol{\xi}), \quad \boldsymbol{\xi} = \frac{\boldsymbol{x}}{\sqrt{t_0 - t}} \tag{6.60}$$

$\chi(\boldsymbol{\xi})$ satisfies equation

$$\nabla^2\left(-\mu^2\chi + \frac{1}{2}i\boldsymbol{\xi}\cdot\nabla\chi + \nabla^2\chi\right) + \text{div}\left(|\nabla\chi|^2\nabla\chi\right) = 0 \tag{6.61}$$

Zakharov obtains a spherically symmetric solution of (6.58c) in the region of $|\xi| \geqslant \dfrac{1}{\mu}$ in [15], where χ satisfies equation

$$i\boldsymbol{\xi}\cdot\nabla\chi = 2\mu^2\chi$$

therefore there are

$$\chi \approx |\xi|^{-2i\mu^2}\chi_0$$

From (6.58b), there is

$$s_2(t) = s_2(0)/\sqrt{t_0 - t}$$

At the ultrasound limit, the transformation $i\psi_t \to -\mu^2(t)\psi$ is performed, and from (6.58a), (6.58b) and (6.58c), we have

$$\nabla^2\left(-\mu^2(t)\chi + \nabla^2\chi\right) - \text{div}(\Phi\nabla\chi) = 0 \tag{6.62a}$$

$$\Phi_{tt} = \nabla^2 |\nabla \chi|^2 \tag{6.62b}$$

This system of equations allows for the following transformation

$$\mu^2(t) = \frac{\mu_0^2}{t_0 - t}, \quad \chi = \frac{\eta(\xi)}{(t_0 - t)^{1 - \frac{2}{n}}}$$

$$\Phi = \frac{D(\xi)}{(t_0 - t)}, \quad \xi = r(t_0 - t)^{-\frac{2}{n}} \tag{6.63}$$

where n is the dimension of the space. The solution of (6.63) has the following properties:

(i) *(6.63) leads to $s_2 = 0$.*

(ii) $\dfrac{\nabla^2 \Phi}{\Phi_{tt}} \approx \dfrac{(t_0 - t)^2}{r^2} \approx \dfrac{(t_0 - t)^{2 - \frac{4}{n}}}{|\xi|^2}.$

(iii) $|\varphi(0,t)|^2 = |\nabla \chi(0,t)|^2 = f(t) = \dfrac{f_0}{(t_0 - t)^2}.$

(iv) $|\Phi(0,t)| = \varphi_0 (t_0 - t)^{\frac{4}{3}}.$

From (3.24), it can be concluded that for planar solitons, when $t \to t_0$, $\dfrac{\nabla^2 \Phi}{\Phi_{tt}} \to \infty$,

for two-dimensional collapse $\nabla^2 \Phi / \Phi_{tt} \to const$, and finally in the three-dimensional case $\Delta^2 \Phi / \Phi_{tt} \to 0$.

If the initial selection is made in the three-dimensional space $(r, z$ coordinates)

$$\rho = \nabla^2 \psi = \begin{cases} \rho_0 \sqrt{w} \sin \dfrac{\pi z}{2}, & w > 0 \\ 0, & w \leqslant 0 \end{cases} \tag{6.64}$$

$$w = 1 - \frac{1}{4}\left(r^2 + z^2\right)$$

For the system of equations (6.58a), (6.58b) and (6.58c), we also present

$$\Phi(r, z, 0) = -|\nabla \psi|^2, \quad \Phi_t(r, z, 0) = 0$$

(6.64) describes the dipole type charge along the z axis when $t = 0$.

The calculation results indicate its collapse behavior over time, and in fact, it is well described by self transformation (6.63) as shown in Figures 6-3 and 6-4.

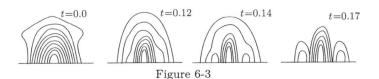

$t=0.0$ $t=0.12$ $t=0.14$ $t=0.17$

Figure 6-3

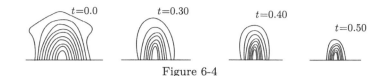

Figure 6-4

References

[1] Derrick G H. Comments on nonlinear wave equations as models for elementary particles[J]. Journal of Mathematical Physics, 1964, 5(9): 1252.

[2] Hobart R H. Non-linear field equilibria[J]. Proceedings of the Physical Society, 1965, 85(3): 610.

[3] Berger M S. On the existence and structure of stationary states for a nonlinear Klein-Gordan equation[J]. Journal of Functional Analysis, 1972, 9(3): 249-261.

[4] Strauss, W.A. Existence of solitary waves in higher dimensions[J]. Commun. Math. Phys. 55, 149–162 (1977).

[5] Gibbons J, Thornhill S G, Wardrop M J, et al. On the theory of Langmuir solitons[J]. Journal of Plasma Physics, 1977, 17(2): 153-170.

[6] Manakov S V. On the theory of two-dimensional stationary self-focusing of electromagnetic waves[J]. Soviet Physics-JETP, 1974, 38(2): 248-253.

[7] Friedberg R, Lee T D, Sirlin A. Class of scalar-field soliton solutions in three space dimensions[J]. Physical Review D, 1976, 13(10): 2739.

[8] Zakharov V E, Kuznetsov E A. On three dimensional solitons[J]. Zhurnal Eksp. Teoret. Fiz, 1974, 66: 594-597.

[9] Hirota R. Exact three-soliton solution of the two-dimensional sine-Gordon equation[J]. Journal of the Physical Society of Japan, 1973, 35(5): 1566-1566.

[10] Ablowitz M J. Lectures on the inverse scattering transform[J]. Studies in Applied Mathematics, 1978, 58(1): 17-94.

[11] Makhankov V G. Dynamics of classical solitons (in non-integrable systems)[J]. Physics reports, 1978, 35(1): 1-128.

[12] Dell'Antonio G, Doplicher S, Jona-Lasinio G. Lecture Notes in Physics[J]. Presented at the capitolium, 1977: 2.

[13] Anderson D L T, Derrick G H. Stability of Time-Dependent Particlelike Solutions in Nonlinear Field Theories. I[J]. Journal of Mathematical Physics, 1970, 11(4): 1336-1346.

[14] Shabat A, Zakharov V. Exact theory of two-dimensional self-focusing and one-dimensional self-modulation of waves in nonlinear media[J]. Sov. Phys. JETP, 1972, 34(1): 62.

[15] Degtiarev L M, Zakharov V E. Dipole character of the collapse of Langmuir waves[J]. JETP Letters, 1974, 20: 164-166.

[16] Ablowitz M J, Kaup D J. AC Newell and H. Segur[J]. Stud. Appl. Math, 1974, 53: 249.

[17] Shabat A, Zakharov V. Exact theory of two-dimensional self-focusing and one-dimensional self-modulation of waves in nonlinear media[J]. Sov. Phys. JETP, 1972, 34(1): 62.

[18] Shabat A B. Inverse-scattering problem for a system of differential equations[J]. Functional Analysis and Its Applications, 1975, 9(3): 244-247.

Chapter 7
Numerical Calculation Methods for Some Nonlinear Evolution Equations

7.1 Introduction

With the further development of research on soliton problems, a large number of numerical calculation methods for nonlinear evolutionary equations (usually with dispersion terms) with soliton solutions have emerged and flourished. In fact, when the soliton problem began to achieve remarkable results, numerical calculations of nonlinear equations had already played an important role. For example, for the KdV equation, although Korteweg de Vries had analyzed solitary waves in 1895, the rich content of its nonlinear phenomena was unknown. It was not until 1965 that Zabusky and Kruskal discovered the great stability of the waveform after soliton interactions through numerical calculations of the KdV equation, which was obtained from harmonic lattice models, that people became interested in and valued soliton phenomena. Other calculations such as the FPU problem and Perring and Syryme's calculation of the two soliton solutions (Kink) of the s, sine-Gordon equation provide important basis for physically analyzing the existence of solitons. With the deepening and complexity of the study of soliton problems, especially the qualitative and quantitative research on the interaction of multiple solitons, quasi solitons, and the existence and interaction of multidimensional solitons, numerical computation plays an increasingly important role. It is no exaggeration to say that numerical calculations have become the main tool for examining the stability of soliton problems in laser and plasma physics.

For the numerical calculation of nonlinear evolutionary equations with soliton solutions, it is generally required that the calculation is stable, able to adapt to large gradients of soliton solution changes, and the calculation format meets the characteristics of conservation law. There are two commonly used numerical methods: the finite difference method and the function approximation method, namely the Finite Element method and the Collocation method.

Now we consider the general evolution equation

$$u_i = L(u) \tag{7.1}$$

where $L(u)$ is a general nonlinear differential operator. For the finite difference method, we use the difference operator $L_h(u_m^n)$ to approximate $L(u)$, where $u_m^n = u(x_m, t_n)$, $x_m = mh$, $t_n = nk$. We usually use the following methods to discretize the time derivative:

$$u_m^{n+1} - u_m^n = kL_h\left(u_m^n\right) \tag{7.2}$$

$$u_m^n - u_m^{n-1} = kL_h\left(u_m^n\right) \tag{7.3}$$

$$u_m^{n+1} - u_m^{n-1} = 2kL_h\left(u_m^n\right) \tag{7.4}$$

We know that (7.2) is a simple explicit format, (7.3) is a simple implicit format, and (7.4) is a "Leafrog" format. There are two more complex and important formats, one is the Crack-Nicholson format, which is the sum of (7.2) and (7.3), and the other is the postscript format (the Hopscotch format), which calculates using the (7.2) format when $n + m$ is odd and the (7.3) format when $n + m$ is even, then the result of the simultaneous calculation becomes explicit, as shown in Fig.7-1. To make (7.3) explicit,

Figure 7-1

the nonlinear part of $L_h(u_m^n)$ must be averaged spatially

$$L_h\left(\frac{1}{2}\left(u_{m+1}^n + u_{m-1}^n\right)\right)$$

All these methods must meet the stability condition, otherwise the calculation cannot proceed. For linear equations, the Crack-Nicholson scheme may be the most effective, but for nonlinear equations, it is cumbersome because a large number of nonlinear simultaneous equations need to be solved at each step. As is well known, the Leapfrog scheme (7.4) is unstable for linear heat conduction equations, but it is a very suitable method for second-order hyperbolic equations. As for the Hopscotch scheme, it is simple, fast, and stable, but for parabolic equations, the time step must be limited by $k \approx h^2$ to ensure reasonable accuracy.

The function approximation method, as the name suggests, approximates the exact solution using an approximate solution

$$u(x,t) \approx \tilde{u}(x,t) = \sum_{i=1}^{N} c_i(t)\varphi_i(x) \tag{7.5}$$

which is defined in a finite dimensional subspace, where $\varphi_i(x)$ is the basis function for the approximation space. For example, it is usually chosen as a trigonometric function, which leads to the finite F-transform or the pseudospectral method. Using local basis for piecewise polynomials, we obtain the finite element method. Let φ_i satisfy the boundary conditions, and let

$$r(x,t) = \tilde{u}_t - L(\tilde{u}) = \sum_{i=1}^{N} c_i(t)\varphi_i(x) - L(\tilde{u}) \tag{7.6}$$

and we require that the remaining $r(x,t)$ be small in some sense, such as

$$\int_0^1 r(x,t)\varphi_i(x)dx = 0, \quad j = 1, 2, \cdots, N \tag{7.7}$$

Then this is the Galerkin method, and (7.7) leads to a series of ordinary differential equations.

If it is required to strictly satisfy

$$r(x_j, t) = 0, \quad j = 1, 2, \cdots, N \tag{7.8}$$

on a given set of points (such as the Gaussian points), that is, the Collocation method is obtained.

Next we will discuss in detail the numerical methods used to calculate certain special nonlinear evolution equations.

7.2   The Finite Difference Method and Galerkin Finite Element Method for KdV Equations

In [1], numerical calculations were performed on the initial and boundary problem of the KdV equation

$$u_t + uu_x + \delta^2 u_{xxx} = 0 \tag{7.9}$$

$$u|_{t=0} = \cos \pi x \tag{7.10}$$

$$u(x+2, t) = u(x,t) \tag{7.11}$$

where using the following difference scheme

$$u_m^{n+1} = u_m^{n-1} - \frac{1}{3}\frac{k}{h}\left(u_{m+1}^n + u_m^n + u_{m-1}^n\right)\left(u_{m+1}^n - u_{m-1}^n\right)$$

$$- \left(\frac{\delta^2 k}{h^3}\right)\left(u_{m+2}^n - 2u_{m+1}^n + 2u_{m-1}^n - u_{m-2}^n\right)$$

$$(m = 0, 1, 2, \cdots, 2N - 1) \tag{7.12}$$

$$u_m^0 = \cos \pi x_m \qquad (7.13)$$

$$u_m^n = u_{m+2N}^n \qquad (7.14)$$

here, k is the time step, $h = \dfrac{1}{N}$ is the spatial step, $u_m^n = u(mh, nk)$. This differential scheme has conserved momentum $\displaystyle\sum_{m=0}^{2N-1} (u_m^n)$ and almost conserved energy $\displaystyle\sum_{m=0}^{2n-1} \frac{1}{2} (u_m^n)^2$, with $\delta = 0.022$. In this case, the initial dispersion is relatively small compared to the nonlinear term because

$$\left\{ \max \left| \delta^2 u_{xxx} \right| / \max \left| u u_x \right| \right\}_{t=0} = 0.004$$

The calculation results are divided into three time stages:

(i) At first, the first and second terms of equation (7.39) dominate, leading to the usual catch-up phenomenon. Then, the solution is essentially determined by the hyperbolic equation $u_t + u u_x = 0$, where $u \approx \cos \pi (x - ut)$.

(ii) When u is sufficiently steep, the third term becomes important and disrupts the formation of discontinuous solutions. Then, small wavelength vibrations develop on the left, and the amplitude of dispersion vibrations increases, eventually forms a series of individual solitons.

(iii) Each soliton moves at a uniform velocity, which is proportional to the amplitude. Due to periodicity, two or more solitons overlap in space, resulting in nonlinear interactions. After a short interaction, they exhibit no effect on their size and shape. In Fig.7-2, curve A represents the initial value of (7.10) at $t = 0$. The curve B represents the graph of the solution (7.9) when $u = \cos(\pi x - ut)$ generates multiple values at $x = \dfrac{1}{2}, t = t_B = \dfrac{1}{\pi}$, and the curve C represents the graph of the dispersion structure fully developing into a series of solitons at $t = 3.6 t_B$. Then the position of the maximum value of the soliton forms a straight line.

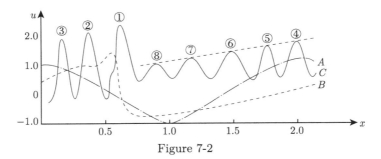

Figure 7-2

The Gauss initial function was calculated in [2]. Considering the initial value problem

$$v_t + vv_x + \frac{1}{\sigma^2}v_{xxx} = 0 \quad (-\infty < x < \infty, t > 0) \tag{7.15}$$

$$v|_{t=0} = \varphi(x) = e^{-x^2} \quad (-\infty < x < \infty) \tag{7.16}$$

It was found that when $4 < \sigma < 7$, two solitons were formed. When $7 < \sigma < 11$, there were three solitons. When $\sigma \approx 11$, there were four solitons. When $\sigma \approx 16$, there were six solitons. However, when $\sigma \ll \sigma_3 = \sqrt{12}$, there were no solitons and only dispersive vibration waves. For some intermediate values of σ, there were both solitons and dispersive vibrations, as shown in Fig.7-3. Then

$$\sigma_c = 6\sigma_s^2 \int_{-\infty}^{\infty} [\varphi(\xi)]^2 d\xi / \left(\int_{-\infty}^{\infty} \varphi(\xi)d\xi \right)^3, \varphi(\xi) = e^{-\xi^2}, \sigma_s = \sqrt{12}$$

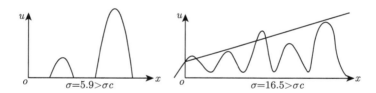

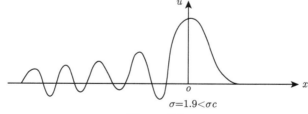

Figure 7-3

For the initial value problem of the KdV equation

$$u_t + \varepsilon u u_x + \mu u_{xxx} = 0 \tag{7.17}$$

$$u|_{t=0} = \frac{1}{2}[1 - \tanh(x - 25)/25] \tag{7.18}$$

using the (7.12) format, the calculation results are shown in Fig.7-4.

Equation (2.4) is a three-layer format, and the first step can be taken as (non-central-format)

$$u_m^1 = u_m^0 - \frac{1}{6}\varepsilon\frac{k}{h}\left(u_{m+1}^0 + u_m^0 + u_{m-1}^0\right)\left(u_{m+1}^0 - u_{m-1}^0\right)$$

$$-\frac{1}{2}\mu\frac{k}{h^3}\left[u_{m+2}^0 - 2u_{m+1}^0 + 2u_{m-1}^0 - u_{m-2}^0\right]$$

The truncation error of this format and the corresponding equation is $O\left[k^3 + kh^2\right]$. Its linear stability condition is

$$\frac{k}{h}\left(\varepsilon\left|u_0\right| + \left(4r/h^2\right)\right) \leqslant 1, \quad \left|u_0\right| \leqslant \max\left|u\right| \tag{7.19}$$

In order to reduce excessive storage usage, the Hopscotch method is used to switch to a two-layer format.

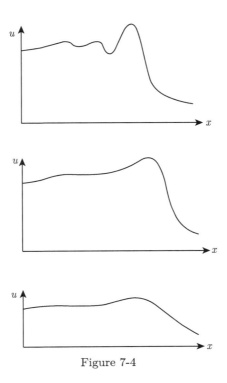

Figure 7-4

For equation (7.17), its difference scheme is as follows:

$$v_m^{n+1} = v_m^n - \frac{1}{2}\frac{k}{h}\varepsilon(f_{m+1}^n - f_{m-1}^n)$$

$$- \frac{k\mu}{2h^3}(v_{m+2}^n - 2v_{m+1}^n + 2v_{m-1}^n - v_{m-2}^n)$$

$$m + n \text{ is even}$$

$$v_m^{n+1} = u_m^n - \frac{1}{2}\frac{k}{h}\varepsilon(f_{m+1}^{n+1} - f_{m-1}^{n+1})$$

$$- \frac{k\mu}{2h^3}(v_{m+2}^{n+1} - 2v_{m+1}^{n+1} + 2V_{m-1}^{n+1} - v_{m-2}^{n+1})$$

$$m + n \text{ is odd}$$

(7.20)

It is not difficult to verify that the truncation error between this and equation (7.21) is

$$kO\left(k^2 + \left(\frac{k}{h}\right)^2 + h^2\right)$$

and its linear stability condition is

$$\frac{k}{h}|\varepsilon|u_0| - (2\mu/h^2)| \leqslant 1 \tag{7.21}$$

From this, it can be seen that although the H-method has the advantage of low storage capacity, but the time step size is smaller than that required by the Zabusky-Kruskal format.

For the KdV equation, another numerical method is used, which is the function approximation method. We consider the periodic initial value problem of the KdV equation as follows

$$\begin{cases} u_t + uu_x + u_{xxx} = 0, \ 0 < t \leqslant T, x \in R \\ u(x,0) = u_0(x), \ x \in R \\ u(x+1,t) = u(x,t), \ \forall x,t \end{cases} \tag{7.22}$$

Let $u_0(x)$ be a sufficiently smooth function with a period of 1, and the solution of (7.22) exists and sufficiently smooth. Then we take the finite dimensional subspace

$$S^\mu = \{\chi(x), \chi \in [0,1];$$

The periodic expansion of χ is $C^k(R), \chi(x)$ is a polynomial in the interval $[ih, (i+1)h](i = 0, 1, \cdots, h^{-1})$ of order less than $\mu - i\}$

and let μ, k are integers, $\mu - 1 > k \geqslant 0$, $k \geqslant 2$. The Galerkin approximation for problem (7.22) is now defined as

$$(U_t + U_{xxx} + UU_x, \chi + h^3\chi_{xxx}) = 0, \ \chi \in S^\mu$$

$$0 \leqslant t \leqslant T, \quad U(0) \in S^{\mu} \tag{7.23}$$

We have the following result.

Theorem 7.1   *Assuming $k \geqslant 2$, the initial value $U(0)$ satisfies*

$$\|U(0) - u_0\|_{L_2[0,1]} \leqslant C_1 h^{\mu}$$

then there exists a constant C, h_0 that depends on T, u_0 and C_1, such that for $0 \leqslant t \leqslant T$, $0 \leqslant h \leqslant h_0$, the Galerkin approximate solution to problem (7.23) exists and has an estimate

$$\|U(t) - u(\cdot, t)\|_{L_2[0,1]} \leqslant Ch^{\mu} \tag{7.24}$$

The research on numerical calculation and methods of KdV equations can be found in $[1] - [5]$ and $[19]$.

7.3   The Finite Difference Method for Nonlinear Schrödinger Equations

We consider the initial and boundary problem of a class of the Nonlinear Schrödinger Equations as follows

$$iu_t - [a(x)u_x]_x + \beta|u|^2 u + f(x)u = 0$$
$$0 < x < 1, t > 0 \tag{7.25}$$

$$u|_{x=0} = u|_{x=1} = 0, \ t \geqslant 0 \tag{7.26}$$

$$u|_{t=0} = u_0(x), \ 0 \leqslant x \leqslant 1 \tag{7.27}$$

where $i = \sqrt{-1}$, $\beta > 0, a(x), f(x)$ are known functions, $a(x) \geqslant \alpha > 0$, $u_0(x)$ is a known complex valued function, and $u(x,t)$ is a complex valued function to be solved. Let $Q = [0,1] \times [0,T]$ be a rectangular region, and we partition the region into many small grids using the straight line $t = mk, \chi = ph$, where m is an integer, $m \in [0, [T/h]]$, p is an integer, and $p \in [0, [h^{-1}]]$, as shown in Fig.7-5. Let the entire interior grid be Q_h, and the remaining grids with boundary points be s_h, represented by $\Omega(t)$ as $(t = \text{const})\cap Q_h$.

Let

$$\varphi_x(x,t) = \frac{1}{h}[\varphi(x+h,t) - \varphi(x,t)] = D_+\varphi$$

$$\varphi_{\bar{x}}(x,t) = \frac{1}{h}[\varphi(x,t) - \varphi(x-h,t)]$$

$$\varphi_{\hat{x}}(x,t) = \frac{1}{2h}[\varphi(x+h,t) - \varphi(x-h,t)]$$

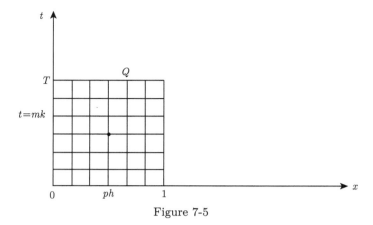

Figure 7-5

Similarly, $\varphi_t, \varphi_{\bar{t}}$ can be defined. We define the discrete modulus as follows:

$$\|\varphi\|_{\Omega(t)}^2 = h \sum_{\Omega} |\varphi|^2(x, t)$$

$$\|\varphi\|_{Q_h}^2 = kh \sum_{Q_h} |\varphi(x, t)|^2$$

$$\|\varphi\|_{l,Q}^2 = \|\varphi\|_{\Omega(t)}^2 + \sum_{|s| \leqslant l} \|D_t^s \varphi\|_{\Omega(t)}^2$$

$$\|\varphi\|_{L_\infty(\Omega)} = \sup_{x_i \in \Omega} |\varphi(x_i)|, \ \|D_t^l \varphi\|_{L_\infty} = \sup_{x_i \in \Omega} |D_t^l \varphi|$$

We consider the following initial and boundary problem of the four implicit difference equation

$$i\varphi_{\bar{t}} - [b(x)\varphi_x]_{\bar{x}} + \beta|\varphi|^2\varphi + f(x)\varphi = 0 \tag{7.28}$$

$$\varphi|_{s_h} = 0, \ \varphi|_{t=0} = u_0(x) \tag{7.29}$$

For the solutions of (7.28) and (7.29), we have the following estimates.

Lemma 7.1 *If the following conditions are met*
(i) $\beta > 0, f(x), b(x)$ are real functions,
(ii) $u_0(x) \in C^0$,
then there is

$$\|\varphi\|_{Q(T)}^2 \leqslant 2\|u_0(x)\|_{L_2}^2 = E_0 \tag{7.30}$$

Proof   Multiplying (7.28) by $\bar{\varphi}$ yields

$$i\varphi_{\bar{t}}\bar{\varphi} - \bar{\varphi}[b(x)\varphi_x]_{\bar{x}} + \beta|\varphi|^4 + f(x)|\varphi|^2 = 0 \tag{7.31}$$

By summation by parts and utilizing boundary conditions, and taking the imaginary part of Q from (7.31), we can obtain

$$\sum_Q \left(|\varphi|_{\bar{t}}^2 + k\, |\varphi_{\bar{t}}|^2 \right) = 0$$

This leads to (7.30). $\square$

Lemma 7.2   *If the condition of Lemma 7.1 is satisfied and $u_0(x) \in H^1, 0 \leqslant \alpha \leqslant b(x) \leqslant M, |u_0|^2 \in C^0, f(x) \geqslant 0$, then we have estimate:*

$$\frac{\alpha}{2} \left\| \varphi_x \right\|_{\Omega(T)}^2 + \frac{\beta}{4} \left\| \varphi^2 \right\|_{\Omega(T)}^2 + \frac{1}{2} \left\| f^{\frac{1}{2}} \varphi \right\|_{\Omega(T)}^2$$

$$\leqslant M \left\| u_0(x) \right\|_{L_2}^2 + \frac{\beta}{2} \left\| u_0^2 \right\|_{L_2}^2 + \left\| f(x) u_0^2 \right\|_{L_1} \tag{7.32}$$

Proof   Multiplying (7.28) by $\bar{\varphi}_t$ yields

$$i|\varphi_i|^2 - \bar{\varphi}_t \left[b\varphi_x \right]_{\bar{x}} + \beta |\varphi|^2 \varphi \varphi_{\bar{t}} + f(x) \varphi \varphi_{\bar{t}} = 0 \tag{7.33}$$

By taking the imaginary part of (7.33), multiplying it by kh, and summing up Q, we can obtain

$$\frac{\alpha h}{2} \sum_{\Omega(T)} |\varphi_x(T)|^2 + \frac{\beta h}{4} \sum_{\Omega(T)} |\varphi(T)|^4 + \frac{h}{2} \sum_{\Omega(T)} f(x)|\varphi(T)|^2$$

$$\leqslant \frac{Mh}{2} \sum_{\Omega(0)} |\varphi_x(0)|^2 + \frac{\beta h}{4} \sum_{\Omega(0)} |\varphi(0)|^4 + \frac{h}{2} \sum_{\Omega(0)} f(x)|\varphi(0)|^2$$

Thus, when $h \leqslant h_0$, we obtain (7.32). $\square$

Now examining the following differential initial and boundary problem

$$i\varphi_{\bar{t}} - [b\varphi_x]_{\bar{x}} + \beta |\varphi|^2 \varphi + C(x,t,\varphi) + g(x,t) = 0 \tag{7.34}$$

$$\varphi|_{s_h=0}, \ \varphi|_{t=0} = u_0(x) \tag{7.35}$$

and considering the solutions φ of (7.34)and (7.35), we have the following estimate.

Lemma 7.3   *If the conditions are satisfied*
(i) β is a real number, $b(x)$ is a real function
(ii) $|C(x,t,\varphi)\bar{\varphi}| \leqslant M|\varphi|^2$, where M is a normal number
(iii) $u_0(x) \in C^0, g(x,t) \in C^0(Q)$
then there are

$$\|\varphi\|_{Q(T)}^2 \leqslant 2 \left(\|u_0\|_{L_2}^2 + \|g\|_{L_2(\rho)}^2 \right) e^{(2M+1)T} = C_2 \tag{7.36}$$

Proof   Similar to Lemma 7.3.

We ssume that there exists the smooth solutions of problems (7.25), (7.26), and (7.27) on the region $Q = [0,1] \times [0,T]$. Now we consider the difference initial and boundary problem corresponding to (7.25) - (7.27)

$$i\varphi_{\bar{t}} - [b(x)\varphi_x]_{\bar{x}} + \beta|\varphi|^2\varphi + f(x)\varphi = 0 \tag{7.37}$$

$$\varphi|_{s_h=0}, \varphi|_{t=0} = u_0(x) \tag{7.38}$$

where $b(x) = a\left(x + \dfrac{h}{2}\right)$. We have the following convergence theorem.

Theorem 7.2   *If the conditions of Lemma 7.2 are satisfied, and let* $u(x,t)$, $\varphi(x,t)$ *are the solutions to problems (7.25) - (7.27) and problems (7.37) and (7.38), respectively, then there are*

$$\|u - \varphi\|_{\Omega(T)} = O\left(k + h^2\right) \tag{7.39}$$

Proof   Since $u(x,t)$ is a smooth solution to the initial and boundary solution problems (7.25) - (7.27), according to Taylor expansion we have

$$iu_t - [b(x)u_x]_{\bar{x}} + \beta|u|^2 u + fu = O\left(k + h^2\right)$$

where $b(x) = a(x + \dfrac{h}{2})$. Let $\varepsilon(x,t) = u(x,t) - \varphi(x,t)$, then

$$i\varepsilon_{\bar{t}} - [b(x)\varepsilon_x]_{\bar{x}} + \beta\left(|u|^2 u - |\varphi|^2\varphi\right) + f\varepsilon = O\left(k + h^2\right) \tag{7.40}$$

$$\varepsilon|_{s_h} = 0, \varepsilon|_{t=0} = 0 \tag{7.41}$$

$$\beta(|u|^2 u - |\varphi|^2\varphi) = \beta|u|^2(u - \varphi) + \beta\varphi(|u|^2 - |\varphi|^2)$$

$$= \beta|u|^2\varepsilon + \beta\varphi(|u| + |\varphi|)(|u| - |\varphi|)$$

Due to the assumption that the solutions of (7.25) - (7.27) are smooth and bounded, and the consistent bounded estimate of the differential solution φ can be obtained from Lemma 7.2, so

$$\left|\bar{\varepsilon}\beta\left(|u|^2 u - |\varphi|^2\varphi\right)\right| \leqslant M|\varepsilon|^2$$

where $M = |\beta|[\|u\|_\infty^2 + \|\varphi\|_{L_\infty}(\|u\|_{L_\infty} + \|\varphi\|_{L_\infty})]$, using Lemma 7.3 (note that when $f(x)$ is a real function, estimator is independent of it), then we obtain

$$\|\varepsilon\|_{\Omega(T)}^2 \leqslant 2\left\|O\left(k + h^2\right)\right\|_\Omega^2 e^{(2M+1)T}$$

$\square$

Theorem 7.3   *The solutions φ of the difference equation (7.37) and (7.38) is stable depend on the modulus $\|\cdot\|_\Omega$ based on its initial value.*

Proof   Similar to the proof of Theorem 7.1.

For the six point symmetric format (Crank-Nicalson format):

$$i\varphi_{\bar{t}} - \frac{1}{2}\left[(b\varphi_x)_{\bar{x}} + (b\varphi_x(t-k))_{\bar{x}}\right]$$

$$+ \frac{\beta}{2}\left[|\varphi(t)|^2\varphi(t) + |\varphi(t-k)|^2\varphi(t-k)\right] \tag{7.42}$$

$$+ f(x)\frac{1}{2}[\varphi(t) + \varphi(t-k)] = 0$$

$$\varphi_{s_h} = 0, \quad \varphi|_{t=0} = u_0(x) \tag{7.43}$$

where $b(x) = a\left(x + \dfrac{h}{2}\right)$, and we have the following result.

Theorem 7.4   *If $u(x,t)$, $\varphi(x,t)$ are the solutions to problems (7.25) - (7.27) and problems (7.42), (7.43), respectively, then there are*

$$\|u - \varphi\|_{Q(T)} = O\left(k^2 + h^2\right) \tag{7.44}$$

Proof   Similar to the proof of Theorem 7.1.

Theorem 7.5   *The difference equations (7.42) and (7.43) are stable with respect to their initial values according to the modulus $\|\cdot\|_\Omega$.*

For the solution of nonlinear algebraic equations (7.37) and (7.38), the chasing iteration method can generally be adopted.

Numerical calculations show that the conservation schemes for the problems (7.25) - (7.27)

$$i\varphi_{\bar{t}} - \frac{1}{2}(b\varphi_x)_{\bar{x}} + (b\varphi_x(t-k)_{\bar{x}}] + \frac{\beta}{4}[|\varphi(t)|^2 + |\varphi(t-k)|^2]$$

$$\cdot (\varphi(t) + \varphi(t-k)) + f(x)\frac{1}{2}[\varphi(t) + \varphi(t-k)] = 0 \tag{7.45}$$

$$\varphi|_{s_h} = 0, \quad \varphi|_{t=0} = u_0(x) \tag{7.46}$$

have better conservation properties than the six point symmetry scheme. Therefore, the calculation results are good. For example, using the format (3.21), the following initial and boundary problem

$$\begin{cases} iu_t + u_{xx} + 2|u|^2 u = 0 \\ u|_{t=0} = \text{sech}(x+10)\exp(2i(x+10)) \\ u|_{x=\pm 15} = 0 \end{cases}$$

was calculated in [11]. *It was found that it has better accuracy compared to the exact solution*

$$u(x,t) = \text{sech}(x + 10 - 4t) \exp[2i(x+10) - 3it]$$

The numerical calculation methods for nonlinear Schrödinger equations and their systems (including multidimensional ones) can be found in [7] − [8].

7.4   Numerical Calculation of the RLW Equation

For the regularized long wave equation (RLWE)

$$u_t + u_x + uu_x - u_{xxt} = 0 \tag{7.47}$$

BBM has proven the existence and uniqueness of its solutions in [16]. We can use the following three-layer differential format to approximate it:

$$w_{m-1}^{n+1} - \left(2 + h^2\right) w_m^{n+1} + w_{m+1}^{n+1}$$

$$= w_{m-1}^{n-1} - \left(2 + h^2\right) w_m^{n-1} + w_{m+1}^{n-1} - kh\left(1 + w_m^n\right) \tag{7.48}$$

$$\cdot \left(w_{m+1}^n - w_{m-1}^n\right)$$

Obviously, the truncation error of format (7.48) and equation (7.47) is

$$\frac{h^2}{6} u_{xxx}(1 + u) + \left(k^2/6\right) u_{ttt}$$

In practical calculations, if $u \ll 1$ and $u_t \sim u_x$, then the two terms may cancel each other out when $h = k$. The interaction between two and three solitons was calculated using this format, as shown in Fig.7-6.

A more accurate difference scheme calculation indicates that the two stronger solitons are inelastic and have a small vibration tail.

For (7.47), there are other differential formats can be used, such as:

$$u_{\bar{t}} - u_{\hat{x}} + \frac{1}{2}(u^2)_{\hat{x}} - u_{x\bar{x}\bar{t}} = 0 \tag{7.49}$$

and

$$u_{\bar{t}} - u_{\hat{x}} + \frac{1}{3}((u^2)_{\hat{x}} + uu_{\hat{x}}) - u_{x\bar{x}\bar{t}} = 0 \tag{7.50}$$

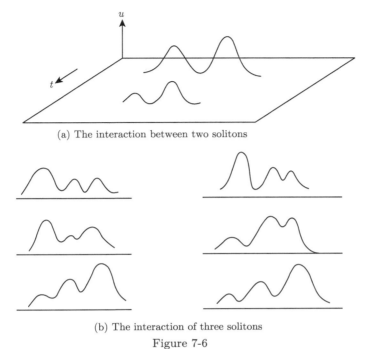

(a) The interaction between two solitons

(b) The interaction of three solitons

Figure 7-6

7.5   Numerical Computation of the Nonlinear Klein–Gordon Equation

For nonlinear wave equations

$$u_{tt} - u_{xx} + F'(u) = 0 \tag{7.51}$$

The selection of different forms of $F'(u)$ plays an important role in the study of soliton, such as:

$$F'(u, \lambda) = \sin u + \lambda \sin 2u \qquad \text{(Double sine-Gordon)} \tag{7.52}$$

$$F'(u, \lambda) = \sin u \qquad \text{(sinc-Gordon)} \tag{7.53}$$

$$F'(u) = -u + u^3 \qquad (\varphi_-^4) \tag{7.54}$$

$$F'(u) = u - u^3 \qquad (\varphi_+^4) \tag{7.55}$$

$$F'(u) = \begin{cases} \dfrac{\pi}{4}, & 2n\pi < u \leqslant (2n+1)\pi \\ 0, & u = n\pi \\ -\dfrac{\pi}{4}, & (2n+1)\pi < u < (2n+2)\pi \end{cases}$$

$$n = 0, \pm 1, \pm 2, \cdots \tag{7.56}$$

For SG, two simple calculation methods have been considered, one of which is the simple Leapfrog format.

$$u_m^{n+1} = - u_m^{n-1} + \frac{k^2}{h^2} \left[u_{m+1}^n + u_{m-1}^n \right]$$
$$+ 2 \left[1 - \frac{k^2}{h^2} \right] u_m^n - k^2 \sin u_m^n \tag{7.57}$$

Through linear stability analysis, it can be concluded that this format is unstable when $k = h$, and can overcome instability when $k = 0.95h$. It applies to the case of two Kinks. Numerical calculations were performed. Another format is to transform the original equation into the first-order system of equations

$$\begin{cases} u_x + u_t = v \\ v_x - v_t = \sin u \end{cases} \tag{7.58}$$

then introduce $\xi = t - x, \eta = t + x$ to convert (7.58) into

$$u_\eta = \frac{1}{2} v, v_\xi = -\frac{1}{2} \sin u \tag{7.59}$$

It is called the characteristic form, where the characteristic line is a straight line. The ordinary differential equation is solved using a pre corrected format, which is more accurate but time-consuming to iterate.

In [16], Ablowitz et al. proposed a new format. For equation (7.51), we set $u_m^n = u(mh, nh), v_m^n = u\left(\left(m + \frac{1}{2} \right) h, \left(n + \frac{1}{2} \right) h \right), \omega_m = u_t(mh, 0),$

$$v_m^0 = \frac{1}{2} \left(u_m^0 + u_{m+1}^0 \right) + \frac{h}{4} \left(v_m + w_{m+1} \right)$$
$$- \frac{h^2}{8} F' \left(\frac{u_m^0 + u_{m+1}^0}{2} \right) + O\left(h^3 \right) \tag{7.60}$$

$$u_m^{n+1} = - u_m^n + v_m^n + v_{m-1}^n - \frac{h^2}{4} F'$$
$$\cdot \left(\frac{v_m^n + v_{m-1}^n}{2} \right) + O\left(h^4 \right) \tag{7.61}$$

$$v_m^{k_1+1} = - v_m^h + u_{m+1}^{n+1} + u_m^{n+1} - \frac{h^2}{4} F'$$
$$\cdot \left(\frac{u_{m+1}^{n+1} + u_m^{n+1}}{2} \right) + O\left(h^4 \right) \tag{7.62}$$

and take the period condition $u^n_{2p+h} = u^n_h$, where $2p$ is the period.

The collision between two solitons of the DSG equation and the φ^4_- equation was calculated using the formats (7.60), (7.61), and (7.62). The calculation results are shown in Fig.7-7.

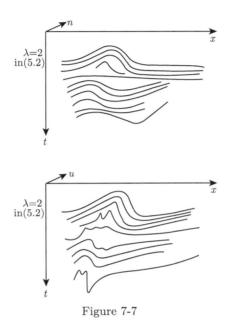

Figure 7-7

7.6 Numerical Computation of a Class of Nonlinear Wave Stability Problems

The plasma dynamics equation system is a complex set of equations. Yu Min provided a system of high and low frequency, dual flow plasma dynamics equations with no external magnetic field and uniform initial density. In [29], under certain assumptions, the planar one-dimensional form of this system of equations was also given

$$\frac{\partial n_i}{\partial t} + \frac{\partial (n_i v_i)}{\partial x} = 0 \tag{7.63}$$

$$\frac{\partial v_i}{\partial t} + v_i \frac{\partial v_i}{\partial x} + \frac{\partial \varphi}{\partial x} = 0 \tag{7.64}$$

$$\frac{\partial^2 \varphi}{\partial x^2} - e^{\varphi - \psi_1^2 - \psi_2^2} + n_i = 0 \tag{7.65}$$

$$\mu \frac{\partial \psi_1}{\partial t} + \frac{\partial^2 \psi_2}{\partial x^2} - \left(e^{\varphi - \psi_1^2 - \psi_2^2} - 1\right) \psi_2 = 0 \tag{7.66}$$

$$\mu\frac{\partial\psi_2}{\partial t} - \frac{\partial^2\psi_1}{\partial x^2} + \left(e^{\varphi-\psi_1^2-\psi_2^2} - 1\right)\psi_1 = 0 \tag{7.67}$$

where n_i is the ion density, v_i is the ion velocity, $n_e = e^{\varphi-\psi_1^2-\psi_2^2}$ is the low-frequency electron number density, φ is the potential function, ψ_1, ψ_2 are quantities describing the amplitude of high-frequency fields, and μ is a constant.

If $n_i(x,t), v_i(x,t), \varphi(x,t)$, and

$$\psi(x,t) = \sqrt{\psi_1^2(x,t) + \psi_2^2(x,t)}$$

all are functions of $\xi = x - ct$, the equation that the solitary wave solution of this set of equations should satisfy can be obtained as follows

$$\frac{d^2\varphi}{d\xi^2} = e^{\varphi-\psi^2} - \left(1 - \frac{2}{c^2}\right)\varphi^{-\frac{1}{2}} \tag{7.68}$$

$$\frac{d^2\psi}{d\xi^2} = \left(e^{\varphi-\psi^2} - 1 + a^2\right)\psi \tag{7.69}$$

$$n_i = \left(1 - \frac{2}{c^2}\right)\varphi^{-\frac{1}{2}} \tag{7.70}$$

$$v_i = c\left[1 - \left(1 - \frac{2}{c^2}\varphi\right)^{-\frac{1}{2}}\right] \tag{7.71}$$

where a, c are parameters, c represents the propagation speed of solitary waves, and a represents the deviation of high-frequency electric fields relative to a fixed frequency.

(7.68), (7.69) form a closed system, and its boundary condition is

$$\varphi = 0, x \to \pm\infty \tag{7.72}$$

$$\psi = 0, x \to \pm\infty \tag{7.73}$$

In [2], the solutions and properties of (7.68), (7.69), (7.72) and (7.75) were discussed, and solitary wave solutions for various parameters a, c were obtained through numerical calculations. Are these solitary waves stable? This is a question that people care about and must answer. Due to the complexity of equations (7.63) - (7.67), it is difficult to obtain answers through analytical qualitative analysis. Shen Longjun and others considered the stability of such waves through actual numerical calculations.

The initial conditions for (7.63) - (7.67) are

$$n_i(x,0)| = n_i^0(x), \ v_i(x,0) = v_i^0(x), \ \psi_1|_{t=0} = \psi_1^0(x)$$

$$\psi_2|_{t=0} = \psi_2^0(x) \tag{7.74}$$

and assuming that when

$$x \to \pm\infty, \text{ there are } n_i \to 1, v_i \to 0, \ \varphi \to 0, \psi_1 \to 0, \psi_2 \to 0 \tag{7.75}$$

It is obviously that the Cauchy problems of problems (7.63) - (7.67) are problems on infinite intervals, and for finite difference calculations, finite interval approximation always brings some errors. So we can implement a transformation of spatial variables

$$\xi = \text{ th } \lambda x \tag{7.76}$$

This transformation transforms the interval $(-\infty, \infty)$ of x into the interval $(-1, +1)$ of ξ, and equations (7.63) - (7.67) accordingly become

$$\frac{\partial n_i}{\partial t} + \lambda \left(1 - \xi^2\right) \frac{\partial n_i v_i}{\partial \xi} = 0 \tag{7.77}$$

$$\frac{\partial v_i}{\partial t} + \lambda \left(1 - \xi^2\right) v_i \frac{\partial v_i}{\partial \xi} + \lambda \left(1 - \xi^2\right) \frac{\partial \varphi}{\partial \xi} = 0 \tag{7.78}$$

$$\lambda^2 \left(1 - \xi^2\right) \frac{\partial}{\partial \xi} \left(1 - \xi^2\right) \frac{\partial \varphi}{\partial \xi} - e^{\varphi - \psi_1^2 - \psi_2^2} + n_i = 0 \tag{7.79}$$

$$\mu \frac{\partial \psi_1}{\partial t} + \lambda^2 \left(1 - \xi^2\right) \frac{\partial \psi_2}{\partial \xi} - \left(e^{\varphi - \psi_1^2 - \psi_2^2} - 1\right) \psi_2 = 0 \tag{7.80}$$

$$\begin{aligned} \mu \frac{\partial \psi_2}{\partial t} - \lambda^2 \left(1 - \xi^2\right) \frac{\partial}{\partial \xi} \left(1 - \xi^2\right) \frac{\partial \psi_1}{\partial \xi} \\ + \left(e^{\varphi - \psi_1^2 - \psi_2^2} - 1\right) \psi_1 = 0 \end{aligned} \tag{7.81}$$

It is obviously that $\xi \to \pm 1, n_i \to 1, v_i \to 0, \varphi \to 0, \psi_1 \to 0, \psi_2 \to 0$.

(7.77), (7.78) are fluid mechanics equations that can be solved using the Richtimyer method:

$$\begin{aligned} \frac{v_{j+\frac{1}{2}}^{k+1} - v_{j+\frac{1}{2}}^{k}}{\Delta t} + \lambda \left(1 - \xi_{j+\frac{1}{2}}^2\right) v_{j+\frac{1}{2}}^k \frac{v_{j+\frac{1}{2}}^k - v_{j-\frac{1}{2}}^k}{\Delta \xi_j} \\ + \lambda \left(1 - \xi_{j+\frac{1}{2}}^2\right) \frac{\varphi_{j+1}^k - \varphi_j^k}{\Delta \xi_{j+\frac{1}{2}}} = 0, \text{ when } v_{j+\frac{1}{2}}^k \geqslant 0 \end{aligned} \tag{7.82}$$

$$\begin{aligned} \frac{v_{j+\frac{1}{2}}^{k+1} - v_{j+\frac{1}{2}}^{k}}{\Delta t} + \lambda \left(1 - \xi_{j+\frac{1}{2}}^2\right) v_{j+\frac{1}{2}}^k \frac{v_{j+\frac{1}{2}}^k - v_{j+\frac{3}{2}}^k}{\Delta \xi_{j+1}} \\ + \lambda \left(1 - \xi_{j+\frac{1}{2}}^2\right) \frac{\varphi_{j+1}^k - \varphi_j^k}{\Delta \xi_{j+\frac{1}{2}}} = 0, \text{ when } v_{j+\frac{1}{2}}^k < 0 \end{aligned} \tag{7.83}$$

$$\frac{n_j^{k+1} - n_j^k}{\Delta t} + \lambda \left(1 - \xi_j^2\right) v_{j-\frac{1}{2}}^{k+1} \frac{n_j^k - n_{j-1}^k}{\Delta \xi_{j-\frac{1}{2}}}$$

$$+ \lambda \left(1 - \xi_j^2\right) n_j^k \frac{v_{j+\frac{1}{2}}^{k+1} - v_{j-\frac{1}{2}}^{k+1}}{\Delta \xi_j} = 0, \tag{7.84}$$

when $v_{j-\frac{1}{2}}^{k+1} \geqslant 0$

$$\frac{n_j^{k+1} - n_j^k}{\Delta t} + \lambda \left(1 - \xi_j^2\right) v_{j-\frac{1}{2}}^{k+1} \frac{n_{j+1}^k - n_j^k}{\Delta \xi_{j+\frac{1}{2}}}$$

$$+ \lambda \left(1 - \xi_j^2\right) n_j^k \frac{v_{j+\frac{1}{2}}^{k+1} - v_{j-\frac{1}{2}}^{k+1}}{\Delta \xi_j} = 0, \tag{7.85}$$

when $v_{j-\frac{1}{2}}^{k+1} < 0$

As we known, for (7.79) and (7.80), their explicit formats are absolutely unstable, while their implicit formats are absolutely stable. However, for the sake of computational convenience, the following semi explicit and semi implicit formats can be used:

$$\mu \frac{\psi_{1j}^{k+1} - \psi_{1j}^k}{\Delta t} = - \lambda^2 \left(1 - \xi_j^2\right) \frac{1}{\Delta \xi_j}$$

$$\cdot \left[\frac{1 - \xi_{j+\frac{1}{2}}^2}{\Delta \xi_{j+\frac{1}{2}}} \psi_{2j+1}^k - \left(\frac{1 - \xi_{j+\frac{1}{2}}^2}{\Delta \xi_{j+\frac{1}{2}}} + \frac{1 - \xi_{j-\frac{1}{2}}^2}{\Delta \xi_{j-\frac{1}{2}}} \right) \right. \tag{7.86}$$

$$\left. \cdot \psi_{2j}^k + \frac{1 - \xi_{j-\frac{1}{2}}^2}{\Delta \xi_{j-\frac{1}{2}}} \psi_{2j-1}^k \right] + \left(e^{\varphi_j^k - \left(\psi_{1j}^k\right)^2 - \left(\psi_{2j}^k\right)^2} - 1 \right) \psi_{2j}^k$$

$$\mu \frac{\psi_{2j}^{k+1} - \psi_{2j}^k}{\Delta t} = \lambda^2 \left(1 - \xi_j^2\right) \frac{1}{\Delta \xi_j}$$

$$\cdot \left[\frac{1 - \xi_{j+\frac{1}{2}}^2}{\Delta \xi_{j+\frac{1}{2}}} \psi_{1j+1}^{k+1} - \left(\frac{1 - \xi_{j+\frac{1}{2}}^2}{\Delta \xi_{j+\frac{1}{2}}} + \frac{1 - \xi_{j-\frac{1}{2}}^2}{\Delta \xi_{j-\frac{1}{2}}} \right) \right. \tag{7.87}$$

$$\left. \cdot \psi_{1j}^{k+1} + \frac{1 - \xi_{j-\frac{1}{2}}^2}{\Delta \xi_{j-\frac{1}{2}}} \psi_{1j+1}^{k+1} \right] - \left(e^{\varphi_j^k - \left(\psi_{1j}^k\right)^2 - \left(\psi_{2j}^k\right)^2} - 1 \right) \psi_{1j}^k$$

In this way, these two equations can be solved directly without iteration. It is easy to deduce that this format needs to meet stability requirements:

$$\Delta t \leqslant \frac{\mu}{2} \Delta x^2$$

After obtaining n_i, ψ_1, ψ_2, the iterative method can be used to solve (7.78):

$$
\begin{aligned}
\lambda^2 (1 - \xi_j^2) \frac{1}{\Delta \xi_j} \Bigg[& \frac{1 - \xi_{j+\frac{1}{2}}^2}{\Delta \xi_{j+\frac{1}{2}}} \varphi_{j+1}^{k+1,s+1} \\
& - \left(\frac{1 - \xi_{j+\frac{1}{2}}^2}{\Delta \xi_{j+\frac{1}{2}}} + \frac{1 - \xi_{j-\frac{1}{2}}^2}{\Delta \xi_{j-\frac{1}{2}}} \right) \varphi_i^{k+1,s+1} \\
& + \frac{1 - \xi_{j-\frac{1}{2}}^2}{\Delta \xi_{j-\frac{1}{2}}} \psi_{i-1}^{k+1,s+1} \Bigg] \\
& - e^{\varphi_j^{k+1,s}} - \left(\psi_{1j}^{k+1} \right)^2 - \left(\psi_{2j}^{k+1} \right)^2 + n_j^{(k+1)} = 0
\end{aligned}
\tag{7.88}
$$

Shen Longjun et al. numerically calculated a class of so-called nonlinear unimodal initial conditions for problems (7.77) - (7.79) using difference schemes (7.82) - (7.88). The numerical calculation results show that there is no change in this type of waveform, demonstrating the good stability of solitary waves in problems (7.63) and (7.67) as shown in Fig.7-8.

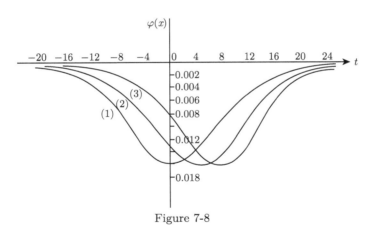

Figure 7-8

In addition, the conservation quantities of equations (7.63) - (7.67) can be obtained as follows

$$
\int_{-\infty}^{\infty} (n_1 - 1) \, dx = \text{const}
$$

$$
\int_{-\infty}^{\infty} v_i \, dx = \text{const}
$$

$$
\int_{-\infty}^{\infty} \left(e^{\varphi - \psi_1^2 - \psi_2^2} - 1 \right) dx = \text{const}
$$

$$\int_{-\infty}^{\infty} |\psi|^2 dx = \text{const}$$

$$\int_{-\infty}^{\infty} (\psi_{1x}^2 - \psi_{2x}^2 + \frac{1}{2}\varphi \left(n + e^{\varphi - \psi_1^2 - \psi_2^2} \right) + \frac{1}{2}nv^2)dx = \text{const}$$

Based on the above conservation quantities, we can use them to test the conservation of the difference scheme.

References

[1]   Zabusky N J, Kruskal M D. Interaction of" solitons" in a collisionless plasma and the recurrence of initial states[J]. Physical review letters, 1965, 15(6): 240.

[2]   Berezin Y A, Karpman V I. Nonlinear evolution of disturbances in plasmas and other dispersive media[J]. Soviet Physics JETP, 1967, 24(5): 1049-1056.

[3]   Vliegenthart A C. On finite-difference methods for the Korteweg-de Vries equation[J]. Journal of Engineering Mathematics, 1971, 5(2): 137-155.

[4]   Greig I S, Morris J L. A hopscotch method for the Korteweg-de-Vries equation[J]. Journal of Computational Physics, 1976, 20(1): 64-80.

[5]   Eilbeck J C, McGuire G R. Numerical study of the regularized long-wave equation I: numerical methods[J]. Journal of Computational Physics, 1975, 19(1): 43-57.

[6]   Eilbeck J C, McGuire G R. Numerical study of the regularized long-wave equation. II: Interaction of solitary waves[J]. Journal of Computational Physics, 1977, 23(1): 63-73.

[7]   郭柏灵. 一类非线性 Schrödinger 方程及其方程组的数值计算问题 [J]. 计算数学, 1981, 3(3): 211-223.

[8]   郭柏灵, 许林宝, 黄书科. 一类多维非线性 Schrödinger 方程及其方程组的数值计算问题 [J]. 中国科学数学: 中国科学, 1983, 26(6): 485-494.

[9]   郭柏灵, 常谦顺. 高阶非线性波动方程组的差分方法 [J]. 中山大学学报 (自然科学版)(中英文), 1985,24(3):56-63.

[10] 郭柏灵. 一类多维高阶广义 BBM-KdV 方程组的 Galerkin 有限元解法和误差估计 [J]. Chinese Science Bulletin, 1982, 27(16): 962-965.

[11] 常谦顺. 高阶 Schrödinger 方程的恒稳显式与半显式差分格式 [J]. 科学通报, 1981, 26(18): 1094-1097.

[12] 郭柏灵. 关于 TOR 法的敛散性 [J]. 计算数学, 1982, 4(4):365-372.

[13] 郭柏灵, 梁华湘. 具波动算子的一类非线性 Schrödinger 方程组的数值计算问题 [J]. 数值计算与计算机应用, 1983, 4(3):176-182.

[14] Bo-Ling G, Qian-Shun C. Difference Method for Multi-Dimensional Nonlinear Schrödinger Equations with Wave Operator[J]. Journal of Computational Mathematics, 1983, 1(4): 346-352.

[15] Benjamin T B, Bona J L, Mahony J J. Model equations for long waves in nonlinear dispersive systems[J]. Philosophical Transactions of the Royal Society of London. Series A, Mathematical and Physical Sciences, 1972, 272(1220): 47-78

[16] Ablowitz M J, Kruskal M D, Ladik J F. Solitary wave collisions[J]. SIAM Journal on Applied Mathematics, 1979, 36(3): 428-437.

[17] 郭本瑜. Soliton 的数值计算 [J]. 科学通报, 1978(10): 592-597.

[18] 邬华谟, 郭本瑜. KdV-Burgers-RLW 方程的高精度差分格式 [J]. 计算数学, 1983, 5(1): 90-98.

[19] 郭柏灵, 常谦顺. GALERKIN FINITE ELEMENT METHOD AND ERROR ESTIMATES FOR THE SYSTEM OF MULTI-DIMENSIONAL HIGHORDER GENERALIZED BBM-KDV EQUATIONS[J]. 科学通报: 英文版, 1983 (3): 310-315.

[20] Yu-Lin Z, Bo-Ling G. Finite difference solutions of the boundary problems for the systems of ferro-magnetic chain[J]. Journal of Computational Mathematics, 1983, 1(3): 294-302.

[21] 常谦顺, 郭柏灵. 一类非线性波动方程组差分格式的理论分析 [J]. 科学通报, 1984, 29(2):68-68

[22] Bo-Ling G, Guang-Nan C. THE CONVERGENCE OF GALERKIN-FOURIER METHOD FOR A SYSTEM OF EQUATIONS OF SCHRODINGER-BOUSSINESQ FIELD[J]. Journal of Computational Mathematics, 1984, 2(4): 344.

[23] Baker G A, Dougalis V A, Karakashian O A. Convergence of Galerkin approximations for the Korteweg-de Vries equation[J]. Mathematics of Computation, 1983, 40(162): 419-433.

[24] Abe K, Inoue O. Fourier expansion solution of the Korteweg-de Vries equation[J]. Journal of Computational Physics, 1980, 34(2): 202-210.

[25] Alexander M E, Morris J L. Galerkin methods applied to some model equations for non-linear dispersive waves[J]. Journal of Computational Physics, 1979, 30(3): 428-451.

[26] Winther R. A conservative finite element method for the Korteweg-de Vries equation[J]. Mathematics of computation, 1980, 34(149): 23-43.

[27] Qian-Shun C, Bo-Ling G. FINITE DIFFERENCE METHOD FOR A NONLINEAR WAVE EQUATION[J]. Journal of Computational Mathematics, 1984, 2(4): 297-304.

[28] 陈雅深, 高飞, 王大贤. 朗缪尔孤立波与光孤立波的耦合 [J]. 核聚变与等离子体物理, 1983(01): 12-17.

[29] 符鸿源. 扩散和色散方程单支格式的稳定性 [J]. 计算数学, 1983, 5(1): 36-50.

Chapter 8
The Geometric Theory of Solitons

8.1 Bäcklund Transform and Surface with Total Curvature $K = -1$

As we have known, the Bäcklund transform inspires that another soliton solution can be obtained from a soliton solution of the Sine-Gordon equation. Using the principle of nonlinear superposition, the special solution of the nonlinear equation can be easily obtained through algebraic operations, which is a very clever method. We will now introduce the geometric methods adopted by Chern and Terng. The study of the relationship between the sine-Gordon equation and the $K = -1$ surface, as well as the geometric characteristics of its solution, shows that the problem of solving the sine-Gordon equation comes down to the problem of finding another $K = -1$ surface from one $K = -1$ surface.

Let $M(u, v)$ be a planar region, R^3 be a three-dimensional Euclidean space, and $x : M \to R^3$ be a surface. Take a unit right-handed orthogonal frame at each point on the surface: $[x; e_1, e_2, e_3]$, $(e_\alpha, e_\beta) = \delta_{\alpha\beta}$, $(e_1, e_2, e_3) = 1$, $1 \leqslant \alpha, \beta \leqslant 3$. Assuming e_3 is the normal vector, there is an equation of motion

$$dx = \sum_\alpha w_\alpha e_\alpha, \quad w_3 = 0 \tag{8.1}$$

$$de_\alpha = \sum_\beta w_{\alpha\beta} e_\beta, \quad w_{\alpha\beta} + w_{\beta\alpha} = 0 \tag{8.2}$$

where $w_\alpha, w_{\alpha\beta}$ are both first-order differential equations.

We call

$$\mathrm{I} = w_1^2 + w_2^2 \quad (= dx \cdot dx)$$

$$\mathrm{II} = w_1 w_{13} + w_2 w_{23} \quad (= -de_3 dx)$$

are the first and second basic forms of the surface respectively.

By performing external differentiation on (8.1) and (8.2), separately, we obtain the structural equation:

$$\begin{cases} dw_1 = w_{12} \wedge w_2 \\ dw_2 = w_1 \wedge w_{12} \\ \mathrm{II} = -(dx, de_3) = a w_1^2 + 2b w_1 w_2 + c w_2^2 \end{cases} \tag{8.3}$$

here

$$\begin{cases} w_{13} = aw_1 + bw_2 \\ w_{23} = bw_1 + cw_2 \\ dw_{12} = -Kw_1 \wedge w_2 \quad (\text{ Gauss equation}) \end{cases} \qquad (8.4)$$

where, $K = ac - b^2$ is the Gauss curvature.

$$\begin{cases} dw_{13} = w_{12} \wedge w_{23} \\ dw_{23} = w_{12} \wedge w_{13} \end{cases} \qquad (8.5)$$

here w_{12} is only related to the first basic form, which is called the connected form of the surface. Equation (8.5) is called the Codazzi equation. Now consider the surface of $K = -1$. Take the curvature line coordinates

$$\begin{cases} w_1 = \sin\psi du, & w_2 = \cos\psi dv \\ w_{13} = \cos\psi du, & w_{23} = -\sin\psi dv \\ w_{12} = -\psi_v du - \psi_u dv \end{cases} \qquad (8.6)$$

By substituting the Gauss equation, we obtain

$$\psi_{uu} - \psi_{vv} = -\cos\psi\sin\psi \qquad (8.7)$$

From (8.7), we obtain

$$I = \sin^2\psi du^2 + \cos^2\psi dv^2$$
$$II = \sin\psi\cos\psi\left(dv^2 - du^2\right) \qquad (8.8)$$

From this, it can be seen that 2ψ is the angle between the asymptotes $\dfrac{d\psi}{du} = \pm 1$. From equation (8.8), it can be seen that 2ψ, as a function of (u, v), is the solution to the sine-Gordon equation. On the contrary, according to the fundamental theorem of surface theory, any solution of the sine-Gordon equation can be regarded as the angle between the asymptotes of a $K = -1$ surface. So the problem of solving sine-Gordon equation boils down to the problem of finding the surface with $K = -1$. We know that the surface with $K = -1$ is a pseudo sphere, which is a trumpet shaped surface with singularities. We can provide the correspondence between the $K = -1$ surfaces through a false ball cue.

The so-called tractrix refers to a family of lines in Euclidean space that depend on two parameters (u, v).

$$y = \chi(u, v) + \lambda n(u, v) \qquad (8.9)$$

$n^2 = 1$, $\chi(u,v)$ is generally represented as a surface. If (u,v) is fixed, then (8.9) represents a straight line passing through point $\chi(u,v)$ with direction $n(u,v)$. If $(u(t), v(t))$ is a curve on a surface, then:

$$y(t,\lambda) = \chi(u(t), v(t)) + \lambda n(u(t), v(t))$$

is a straight textured surface. The necessary and sufficient conditions for this straight textured surface to be a developable surface are:

$$|n, dx, dn| = 0 \qquad (8.10)$$

Here, $|n, dx, dn|$ represents the mixed product determinant of n, dx and dn, and (8.10) is a quadratic homogeneous equation about du, dv. If its ratio $\dfrac{du}{dv}$ has two different real solutions, then it corresponds to two families of curves with surfaces. The straight lines in the clues of each curve in the family form a developable surface, and then two families of developable surfaces are obtained. Each developable surface has a ridge line, and all the ridges of each family of developable surfaces form a surface, which is called a focal surface. Then we obtained two focal surfaces, denoted as s and s', where each line in tractrix (8.9) is a common tangent to s and s'. So through these common tangents, the tractrix (8.9) gives a transformation between the focal surface s and the focal surface s'

$$l : s \to s'$$

That is to say, if $p' = l(p)$, then $p(\in s)$ and $p'(\in s')$ have a common tangent l belonging to clue (8.9).

Definition 8.1 *A tractrix is pseudospherical, if 1) $|pp'| = \gamma$ (constant), which means the distance between corresponding points is a constant value 2) $\langle e3(p), e3'(p) \rangle = \tau$ (constant), which means the angle between the normal directions of the corresponding points is a constant value.*

Theorem 8.1 (Bäcklund) *The two focal surfaces of the pseudospherical tractrix have corresponding constant Gauss curvatures*

$$K = -\frac{\sin^2 \tau}{r^2}$$

Specifically, if $r = \sin\tau$, then $K = -1$, the problem of constructing another $K = -1$ surface from a given $K = -1$ surface comes down to the problem of constructing a false ball cue from a given $K = -1$ surface. Then, it is only necessary to determine the direction of each straight line in the desired tractrix. Then, it boils down to solving a fully integrable total differential equation

$$d\alpha + \sin\alpha w = \cos\tau w_{13} \qquad (8.11)$$

Here, α represents the angle between the main direction of the tangent point (i.e. the direction of the u curve) and the straight line in the desired tractrix. The following fully integrable system of first-order partial differential equations can be obtained from (8.6) and (8.11)

$$\begin{cases} \sin\tau\,(\alpha_u - \psi_v) = \cos\tau\cos\alpha\cos\psi + \sin\alpha\sin\psi \\ \sin\tau\,(\alpha_v - \psi_u) = -\cos\tau\sin\alpha\sin\psi - \cos\psi\cos\alpha \end{cases} \tag{8.12}$$

Then, the first and second basic forms of the required surface are respectively

$$I' = \cos^2\alpha du^2 + \sin^2\alpha dv^2$$

$$II' = \cos\alpha\sin\alpha\,(du^2 - dv^2)$$

It can be seen that $2\alpha(u,v)$ is a solution of the sine-Gordon equation. Therefore, from (8.12), if a solution 2ψ of the sine-Gordon equation is given, another solution 2α of the sine-Gordon equation can be obtained. Since (8.12) is completely integrable, essentially only one ordinary differential equation needs to be solved.

K. Tenenblat and C.L. Terng further discussed the Bäcklund theorem for n dimensional submanifolds in $2n-1$ dimensional Euclidean space, as well as the high-dimensional extensions of the sine-Gordon equation[2],[3].

8.2   Lie Group and Nonlinear Evolution Equations

In Chapter 2, we have pointed out that in detail the methods established by GGKM and AKNS to accurately solve nonlinear evolutionary equations using scattering inversion. Chern and Peng[1] pointed out that the algebraic basis of these equations lies in Lie groups and their structural equations. They started from the structure of the 2×2 real unitary Lie group SL(2) and naturally derived high-order KdV and MKdV equations, making the geometric meanings of these equations more clear. The use of group operations is also relatively convenient. Sasaki further established the connection between the AKNS equation and negative constant curvature surfaces in reference [4].

Let

$$SL(2,R) = \left\{ X = \begin{pmatrix} a & b \\ c & d \end{pmatrix} \middle| ad - bc = 1 \right\} \tag{8.13}$$

be the group of all 2×2 real unitary matrices. Its right invariant Mauer-Carter form is

$$w = dxX^{-1} = \begin{pmatrix} w_1^1, & w_1^2 \\ w_2^1, & w_2^2 \end{pmatrix} \tag{8.14}$$

where
$$w_1^1 + w_2^1 = 0$$

The structural equation of $SL(2; R)$ or the Manren-Cartan equation is

$$dw = w \wedge w \qquad (8.15)$$

or to be more specific,

$$\begin{cases} dw_1^1 = w_1^2 \wedge w_2^1 \\ dw_1^2 = 2w_1^1 \wedge w_1^2 \\ dw_2^1 = 2w_2^1 \wedge w_1^1 \end{cases} \qquad (8.16)$$

Let v be a neighborhood on the (x, t) plane, and consider a smooth mapping

$$f : v \to SL(2; R) \qquad (8.17)$$

After being mapped in the Lie group, these methods become functions of (x, t).

$$\begin{cases} w_1^1 = \eta dx + A dt \\ w_1^2 = q dx + B dt \\ w_2^1 = r dx + C dt \end{cases} \qquad (8.18)$$

here, the coefficients are functions of (x, t). Since

$$dw_1 = w_1^2 \wedge w_2^1 = (q dx + B dt) \wedge (r dx + C dt)$$

$$= qr dx \wedge dx + (qC - Br) dx \wedge dt + BC dt \wedge dt$$

on the other hand,

$$dw_1^1 = A_x dx \wedge dt + A_t dt \wedge dt + \eta d^2 x + A d^2 t$$

$$+ \eta_t dt \wedge dx + \eta_x dx \wedge dx,$$

$$dw_1^2 = 2w_1^1 \wedge w_1^2 = 2(\eta dx + A dt) \wedge (q dx + B dt)$$

$$= 2\eta q dx \wedge dx + 2(\eta B - Aq) dx \wedge dt$$

$$+ 2AB dt \wedge dt$$

Additionally,

$$dw_1^2 = q_x dx \wedge dx + q_t dt \wedge dx + q d^2 x + B_x dx dt$$

$$+ B_t dt \wedge dt + B d^2 t$$

Furthermore dw_2^1 can be calculated in the same way.

From this, it can be concluded that

$$\begin{cases} -\eta_t + A_x - qC + rB = 0 \\ -q_t + B_x - 2\eta B + 2qA = 0 \\ -r_t + C_x - 2rA + 2\eta A = 0 \end{cases} \tag{8.19}$$

Assuming $\eta = \text{const}$, which means η is a parameter independent of x, t. Now consider several special situations:

(1) $r = +1$, η is a constant, $q = u(x,t)$. Then A can be solved frcm the third equation in (8.19), and B can be solved from the first equation

$$\begin{cases} A = \eta C + \dfrac{1}{2}C_x \\ B = -\dfrac{1}{2}C_{xx} - \eta C_x + uC \end{cases} \tag{8.20}$$

Substituting (8.20) into the second equation of (8.19) yields:

$$u_t = K(u) \tag{8.21}$$

where

$$K(u) = u_x C + 2uC_x + 2\eta^2 C_x - \frac{1}{2}C_{x\pi x} \tag{8.22}$$

As an example, taking

$$C = \eta^2 - \frac{1}{2}u \tag{8.23}$$

from (8.21), it can be concluded that

$$u_t = \frac{1}{4}u_{xxx} - \frac{3}{2}uu_x \tag{8.24}$$

which is the well-known KdV equation.

We can naturally take any polynomial with C as η. Since in (8.22), only η^2 is included, C can be set as a polynomial of η^2. Let

$$C = \sum_{0 \leqslant j \leqslant n} C_j(x,t)\eta^{2(n-j)} \tag{8.25}$$

among them, $C_i(x,t)$ is a function of x,t. By substituting (8.25) into (8.22) and setting the coefficient of η^2 to 0, we can obtain

$$C_0 = \text{const} \tag{8.26}$$

$$C_{j+1,x} = -\frac{1}{2}u_x C_j - uC_{j,x} + \frac{1}{4}C_{j,xxx} \tag{8.27}$$

We noticed that the latter happens to be the cyclic formula for the conservation density of the KdV equation. The right end of (8.21) can be written as

$$K_n(x) = u_x C_n + 2uC_{n,x} - \frac{1}{2}C_{n,xxx} = -2C_{n+1,x} \tag{8.28}$$

The final equation is introduced as a definition, which further introduces an infinite sequence of C_i. Let (8.27) hold for all j, and $0 \leqslant j < \infty$. The equation

$$u_t = K_n(u) \tag{8.29}$$

is called the n-order KdV equation. It can be proved that C_j is a polynomial of u and its successive derivatives with respect to x. For example, it can be obtained

$$2C_1 = -u$$

$$2C_{j+1} = -\sum_{1\leqslant k\leqslant j} C_k C_{j+1-k} - u\sum_{0\leqslant k\leqslant j} C_k C_{j-k}$$

$$+\frac{1}{2}\sum_{0\leqslant k\leqslant j-1} C_k C_{j-k,xx} \tag{8.30}$$

$$-\frac{1}{4}\sum_{1\leqslant k\leqslant j} C_{kx}C_{j-k,x}, j = 1, 2, \cdots$$

Specifically, $2C_2 = \frac{3}{4}u^2 - \frac{1}{4}u_{xx}$

$$2C_3 = -\frac{5}{8}u^3 + \frac{5}{16}u_x^2 + \frac{5}{8}uu_{xx} - \frac{1}{16}u_{xxxx} \tag{8.31}$$

(2) In (8.19), $q = r = V(x,t)$, and η is a parameter independent of x, t, then (8.19) becomes

$$\begin{cases} A_x = V(C - B) \\ V_t = B_x - 2\eta B + 2VA \\ V_t = C_x + 2\eta C - 2VA \end{cases} \tag{8.32}$$

The last two equations in (8.32) can be written as

$$\begin{cases} (C - B)_x = 4VA - 2\eta(B + C) \\ V_t = \frac{1}{2}(B + C)_x + \eta(C - B) \end{cases} \tag{8.33}$$

Let

$$C - B = \eta P, \quad C + B = Q, \quad A = \eta R$$

the above equation becomes

$$\begin{cases} R_x = VP \\ P_x = 4VR - 2Q \\ V_t = \dfrac{1}{2}Q_x + \eta^2 P \end{cases} \tag{8.34}$$

By eliminating P and Q, we can obtain

$$V_t = M(V) \tag{8.35}$$

where

$$M(V) = \eta^2 \frac{R_x}{V} + (VR)_x - \frac{1}{4}\left(\frac{R_x}{V}\right)_{xx}$$

Taking $R = \eta^2 - \dfrac{1}{2}V^2$, equation (8.35) becomes

$$V_t = \frac{1}{4}V_{xxx} - \frac{3}{2}V^2 V_x \tag{8.36}$$

which is the famous MKdV equation.

8.3   The Prolongation Structure of Nonlinear Equations

For the differential manifold M and n form ideal I, the so-called extension refers to the $n - 1$ form P on M, with coefficients taking on differentiable functions on M, and meet

$$dP \subset F^*(M) \wedge P + I \tag{8.37}$$

where, $F^*(M)$ is the $1-$form on M. In 1975, Wohlquist and Estabrook first proposed the concept of extension and applied it to the KdV equation. They represented the KdV equation as a set of closed ideals in equivalent external differential form. And by extending this closed ideal, the scattering inversion problem and Bäcklund transform of the KdV equation were successfully found, and a similar discussion was made on the nonlinear Schrödinger equation in [6]. Morris, Coroues, Gibbon and others had discussed the prolongation structures of shallow water wave equations with gravitational effects, Hirota equations, nonlinear Schrödinger equations, high-order KdV equations, and self-duality Yang Mills equations, as shown in [7], [8], [9], [11], [12], [14], [15], [16], [18], [19]. It can be seen that the extended structure method is not only applicable to a large number of nonlinear evolutionary equations, but also can be naturally extended to high-dimensional space. Therefore, in this respect, it has greater advantages than the scattering inversion method, and this differential geometry method may become the theoretical basis of the inverse scattering method.

Now we consider the extension structure method of the KdV equation. We begin from the KdV equation

$$u_t + u_{xxx} + 12uu_x = 0 \tag{8.38}$$

Let $z = u_x$, $p = z_x = u_{xx}$, then (8.38) can be written as a first-order equation

$$u_t + p_x + 12uz = 0 \tag{8.39}$$

For the 5-dimensional manifold $M\{x, t, u, z, p\}$, the basis of the dual space $T^*(M)$ of the tangent space is $\{dx, dt, du, dz, dp\}$. The 2−form group is introduced on the 2-dimensional submanifolds $\{x, t, u(x,t), z(x,t), p(x,t)\}$ of M as

$$
\begin{cases}
\alpha_1 = du \wedge dt - z dx \wedge dt \\
\alpha_2 = dz \wedge dt - p dx \wedge dt \\
\alpha_3 = -du \wedge dx + dp \wedge dt + 12uz dx \wedge dt
\end{cases}
\tag{8.40}
$$

where d represents the outer derivative, and $\wedge$ represents the outer product. The first two terms of (8.40) correspond to the terms which are introduced new variables, while the latter term corresponds to the term of the original equation, which can be directly calculated

$$
\begin{cases}
d\alpha_1 = dx \wedge \alpha_2 \\
d\alpha_2 = dx \wedge \alpha_3 \\
d\alpha_3 = -12 dx \wedge (z\alpha_1 + u\alpha_2)
\end{cases}
\tag{8.41}
$$

Therefore, $\{\alpha_1, \alpha_2, \alpha_3\}$ form a closed ideal on manifold M, and when truncated on manifold $S_2 = \{u(x,t), z(x,t), p(x,t)\}$, the 2−form (8.40) is zero. This is the external derivative KdV equation. For the given 5-dimensional differentiable manifold M and the closed ideals generated by α_i and $d\alpha_i$, there exists an additional extension variable $y^i (i = 1, 2, \cdots, m)$, and $\{y^i\}$ extends an $m+5$ dimensional bundle at each point of the original manifold $M(x, t, u, z, p)$. Therefore, an expanded ideal I' can be generated in the fiber bundle. The generator of I' not only includes α_i, but also m 1−forms w_i introduced by the extension variable y^i. These W_i are called extension forms, and for the extension variable y^i, there is a Pfaff form w_K.

$$\omega_K = dy^K + F^K\left(x, t, u, z, p, y^i\right) dx$$
$$+ G^K\left(x, t, u, z, p, y^i\right) dt \tag{8.42}$$

They must meet the closed ideal conditions

$$dw_K = \sum_{i=1}^{3} f_k^i \alpha_i + \sum^{n} \eta_k^i \wedge w_i \tag{8.43}$$

where η_k^i is $1-$form. The first-order partial differential equations of F^K and G^K can be obtained from (8.41) and (8.42). This system of equations is generally nonlinear because it contains commutator sub items

$$\sum_i \left(G^i \frac{\partial F^k}{\partial y^i} - F^i \frac{\partial G^k}{\partial y^i}\right) dx \wedge dt \tag{8.44}$$

If F^k and G^k depend only on y^k, then y^k determines a usual conservation law. Then y^i is called potential. If F^K and G^K depend on the extended variable $y^i (i \neq k)$, this elbow is called y^k decay potential. The existence of commission (quasi) potential is the key to the Bäcklund transformation.

We define commutator

$$[F \cdot G]^k \equiv F^i G^k_{,y_i} - G^i F^k_{,y_i} \tag{8.45}$$

By eliminating f_k^i according to (8.43), the partial differential equations satisfied by $F^k(u, z, p, y^i), G^k(u, z, p, y^i)$ can be obtained

$$F_{,z}^k = 0, F_{,p}^k = 0, F_{,u}^k + G_{,p}^k = 0 \tag{8.46}$$

$$zG_{,u}^k + pG_{,z}^k - 12uzG_{,p}^k + G^i F_{,,y_i}^k - F^i G_{,y_i}^k = 0 \tag{8.47}$$

The integrability conditions of equations (8.46) and (8.47) make it easy to obtain the expressions for F^k, G^k

$$\begin{cases} F^k = 2X^k + 2uX_2^k + 3u^2 X_3^k \\ G^k = -2\left(p + 6u^2\right) X_2^k + 3\left(z^2 - 8u^3 - 2up\right) X_3^k \\ \quad + 8X_4^k + 8uX_5^k + 4u^2 X_6^k + 4zX_7^k \end{cases} \tag{8.48}$$

Substituting the forms of F^k, G^k given in equation (8.48) to (8.47) yields a series of commutator relationships

$$\begin{cases} [X_1, X_3] = [X_2, X_3] = [X_1, X_4] = [X_2, X_6] = 0 \\ [X_1, X_2] = -X_7, [X_1, X_7] = X_5, [X_2, X_7] = X_6 \\ [X_1, X_5] + [X_2, X_4] = 0, [X_3, X_4] + [X_1, X_6] + X_7 = 0 \end{cases} \tag{8.49}$$

Forcing this open algebraic structure to close into a finite dimensional Lie algebra, and using the Jocobi identity, we can obtain further relations. Introducing new generators X_8, X_9,

$$[X_3, X_4] = -X_8, \quad [X_1, X_5] = X_8$$

We demand

$$X_9 = \sum_{m=1}^{8} c_m X_m \tag{8.50}$$

where c_m are constants. Assuming that the 1 to 8 generators in (8.49) are linearly independent and using the Jocobi identity, we can obtain

$$c_m = 0 (m \neq 7, 8), \quad c_7 = -c_8 \equiv \lambda$$

where λ is an arbitrary constant. Finally, we obtain a closed Lie algebra consisting of $\{X_1, \cdots, X_8\}$.

$$\begin{cases} [X_1, X_2] = -X_7, \ [X_2, X_5] = -X_9/\lambda \\ [X_4, X_7] = -\lambda X_5 \\ [X_1, X_5] = X_9, \ [X_2, X_3] = X_6, \ [X_5, X_6] = X_9/\lambda \\ [X_1, X_6] = -X_9/\lambda, \ [X_3, X_4] = -X_6 \\ [X_5, X_7] = -X_5 - \lambda X_6 \\ [X_1, X_7] = X_5, \ [X_4, X_5] = -\lambda X_9, \ [X_6, X_7] = X_6 \\ [X_2, X_4] = -X_9, \ [X_4, X_6] = X_9, \ X_9 \equiv \lambda (X_7 - X_8) \end{cases} \tag{8.51}$$

It is not difficult to obtain the 8-dimensional relationship of this algebra. Take the base vector as

$$b_k = \frac{\partial}{\partial y^k} (k = 1, \cdots, 8) \tag{8.52}$$

where the coordinate set of the extension and transformation of y^k, and the non degenerate representation of the generator are

$$\begin{cases} X_1 = \dfrac{1}{2}\left[b_1 + \exp\left(2y_3\right)^{-b_2} + y_8 b_3 + y_7 b_3 + \left(y_8^2 - \lambda\right)b_8\right] \\[2mm] X_2 = \dfrac{1}{2}\left(b_7 + 2b_8\right) \\[2mm] X_3 = \dfrac{1}{3}b_6 \\[2mm] X_4 = -\dfrac{1}{2}\lambda\left[b_1 + \exp\left(2y_3\right)b_2 + y_8 b_3 - b_4\right. \\[2mm] \qquad\qquad \left. + \left(\dfrac{3}{2\lambda}\right)y_6 b_5 + \left(y_8^2 - \lambda\right)b_8\right] \\[2mm] X_5 = -\dfrac{1}{2}\left[\exp\left(2y_3\right)b_2 + y_8 b_3 + \left(y_8^2 + \lambda\right)b_8\right] \\[2mm] X_6 = b_8 \\[2mm] X_7 = \dfrac{1}{2}\left[b_3 + \dfrac{1}{2}b_5 + 2y_8 b_8\right] \\[2mm] X_8 = \dfrac{1}{4}b_5 \end{cases} \qquad (8.53)$$

From (8.48), 8 obvious expressions in Pfaff form are

$$\begin{cases} w_k = dy^k + F^k dx + G^k dt \\[2mm] w_1 = dy_1 + dx - 4\lambda dt \\[2mm] w_2 = dy_2 + \exp\left(2y_3\right)dx - 4\exp\left(2y_3\right)(u+\lambda)dt \\[2mm] w_3 = dy_3 + y_8 dx + \left[2z - 4y_8(u+\lambda)\right]dt \\[2mm] w_4 = dy_4 + 4\lambda dt \\[2mm] w_5 = dy_5 + y_7 dx + (z - 6y_6)\,dt \\[2mm] w_6 = dy_6 + u^2 dx + \left(z^2 - 8u^3 - 2up\right)dt \\[2mm] w_7 = dy_7 + u dx - \left(p + 6u^2\right)dt \\[2mm] w_8 = dy_8 + \left(2u + y_8^2 - \lambda\right)dx - 4 \\[2mm] \qquad\quad \cdot \left[(u+\lambda)\left(2u + y_8^2 - \lambda\right) - \dfrac{1}{2}p - zy_8\right]dt \end{cases} \qquad (8.54)$$

Using (8.54), we can obtain the soliton solutions of the KdV equation, the Bäcklund transform, and the corresponding scattering inversion problem.

In fact, starting from $w_3 = 0$ and recording $y_8 = y$, there is

$$\begin{cases} y_x = -\left(2u + y^2 - \lambda\right) \\ y_t = -4 \cdot \left[(u + \lambda)\left(2u + y^2 - \lambda\right) + \dfrac{1}{2}p - zy\right] \end{cases} \tag{8.55}$$

The first equation of (8.55) is the Riccati equation. Set

$$y = \psi_x/\psi \tag{8.56}$$

we have

$$\psi_{xx} + (2u - x)\psi = 0 \tag{8.57}$$

this corresponds to the one-dimensional Schrödinger equation of the KdV equation. From Pfaff form w_3 there is

$$y = -y_{3,x}$$

From (8.56) and w_2, there is

$$y_3 = -\ln\psi$$

Let $\varphi = \psi_x$, then (8.56) gives

$$y = \varphi/\psi$$

Let

$$w_9 \equiv \psi w_8 - \varphi w_3$$
$$w_{10} \equiv -\psi w_3$$

we have

$$\begin{cases} w_9 = d\varphi - (2u - \lambda)\psi dx + \{2z\varphi - [4(u + \lambda) \\ \qquad \times (2u - \lambda) + 2p]\psi\}dt \\ w_{10} = d\varphi - \varphi dx - [2z\psi - 4(u + \lambda)\varphi]dt \end{cases} \tag{8.58}$$

From (8.58), it can be seen that $\psi_x = \varphi$, $\varphi_x + (2u - \lambda)\rho = 0$, which is the first-order scattering equation system for the KdV equation.

On the other hand, let KdV equation have other solution $u' = u'(u, z, p, y^i)$, which satisfies

$$\begin{cases} \alpha_1' = du' \wedge dt - z' dx \wedge dt \\ \alpha_2' = dz' \wedge dt - p' dx \wedge dt \\ \alpha_3' = -du' \wedge dx + dp' \wedge dt + 12u'z' dx \wedge dt \end{cases} \tag{8.59}$$

By direct calculation, we get

$$u' = -u - y^2 + \lambda \tag{8.60}$$

Since $u = 0$ is the solution of the KdV equation, $u'_0 = -y^2 + \lambda$ must be a solution. From (8.55)

$$\begin{cases} y_x = -\left(y^2 - \lambda\right) \\ y_t = 4\lambda\left(y^2 - \lambda\right) = -4\lambda y_x \end{cases} \tag{8.61}$$

Its analytical integral is $y = \lambda^{1/2}\tanh[\lambda^{1/2}(x - x_0 - 4\lambda t)]$, and u'_0 is the analytical soliton solution. Using the Pfaff form w_7, we have

$$u = -y_{7,x} = -w_x$$

(8.60) can be written as

$$-w'_x = w_x - y^2 + \lambda = w_x + y_x - 2w_x \tag{8.62}$$

Integrating it and incorporating the integral constant into the potential, there is

$$y = w - w' \tag{8.63}$$

Therefore, equation (8.60) can ultimately be written as

$$-w'_x - w_x = u' + u = \lambda - \left(w' - w\right)^2 \tag{8.64}$$

Letting $\lambda = k^2$ and using (8.61), (8.63), the second equation of (8.55) can be written as

$$w'_t + w_t = 4\left(u'^2 u^2 + u'u + u^2\right) + 2\left(w' - w\right)\left(z' - z\right) \tag{8.65}$$

Combining (8.64) and (8.65) yields the Bäcklund transformation of the KdV equation.

References

[1] Chern S, Peng C. Lie groups and KdV equations[J]. manuscripta mathematica, 1979, 28(1): 207-217.

[2] Tenenblat K, Terng C L. Bäcklund's theorem for n-dimensional submanifolds of R 2n-1[J]. Annals of Mathematics, 1980, 111(3): 477-490.

[3] Terng C L. A higher dimension generalization of the sine-Gordon equation and its soliton theory[J]. Annals of Mathematics, 1980, 111(3): 491-510.

[4] Sasaki R. Soliton equations and pseudospherical surfaces[J]. Nuclear Physics B, 1979, 154(2): 343-357.

[5] Wahlquist H D, Estabrook F B. Prolongation structures of nonlinear evolution equations[J]. Journal of Mathematical Physics, 1975, 16(1): 1-7.

[6] Estabrook F B, Wahlquist H D. Prolongation structures of nonlinear evolution equations. II[J]. Journal of Mathematical Physics, 1976, 17(7): 1293-1297.

[7] Morris H C. Prolongation structures and a generalized inverse scattering problem[J]. Journal of Mathematical Physics, 1976, 17(10): 1867-1869.

[8]   Corones J. Solitons and simple pseudopotentials[J]. Journal of Mathematical Physics, 1976, 17(5): 756-759.

[9]   Corones J. Solitons, pseudopotentials, and certain Lie algebras[J]. Journal of Mathematical Physics, 1977, 18(1): 163-164.

[10] Morris H C. Soliton solutions and the higher order Korteweg–de Vries equations[J]. Journal of Mathematical Physics, 1977, 18(3): 530-532.

[11] Chowdhury A R, Roy T. Prolongation structure for a nonlinear equation with explicit space dependence[J]. Journal of Mathematical Physics, 1980, 21(6): 1416-1417.

[12] Morris H C. Inverse scattering problems in higher dimensions: Yang–Mills fields and the supersymmetric sine-Gordon equation[J]. J. Math. Phys.(NY);(United States), 1980, 21(2).

[13] Morris H C. Prolongation structures and nonlinear evolution equations in two spatial dimensions. II. A generalized nonlinear Schrödinger equation[J]. Journal of Mathematical Physics, 1977, 18(2): 285-288.

[14] Dodd R K, Gibbon J D. The prolongation structures of a class of nonlinear evolution equations[J]. Proceedings of the Royal Society of London. A. Mathematical and Physical Sciences, 1978, 359(1699): 411-433.

[15] Dodd R K, Gibbon J D. The prolongation structure of a higher order Korteweg-de Vries equation[J]. Proceedings of the Royal Society of London. A. Mathematical and Physical Sciences, 1978, 358(1694): 287-296.

[16] Morris H C. A prolongation structure for the AKNS system and its generalization[J]. Journal of Mathematical Physics, 1977, 18(3): 533-536.

[17] Li Y S. ONE SPECIAL INVERSE PROBLEM OF THE SECOND-ORDER DIFFER-ENTIAL-EQUATION ON THE WHOLE REAL AXIS[J]. CHINESE ANNALS OF MATHEMATICS SERIES B, 1981, 2(2): 147-155.

[18] Morris H C. Prolongation structures and nonlinear evolution equations in two spatial dimensions[J]. Journal of Mathematical Physics, 1976, 17(10): 1870-1872.

[19] Morris H C. Prolongation structures and a generalized inverse scattering problem[J]. Journal of Mathematical Physics, 1976, 17(10): 1867-1869.

[20] 卢文, 姚起元, 区智, 等. 解 KdV 方程的微分几何方法 [J]. 中山大学学报 (自然科学版)(中英文), 1979,18(3): 29-43.

[21] 郭汉英, 向延育, 吴可. 纤维丛联络论与非线性演化方程的延拓结构 [J]. 数学物理学报, 1983(02): 135-143.

[22] 郭汉英, 吴可, 侯伯宇, 等. Soliton 方程的紧致代数结构旋量 AKNS 系统及其球面实现 [J]. 数学物理学报, 1983(03): 241-247.

Chapter 9

The Global Solution and "Blow up" Problem of Nonlinear Evolution Equations

9.1 Nonlinear Evolutionary Equations and the Integral Estimation Method

Recently, with the development of soliton problems and their theories, a large number of nonlinear evolutionary equations with soliton solutions have attracted increasing attention, such as: KdV equations, nonlinear Schrödinger equations, RLW equations, nonlinear Klein-Gordon equations, etc. In addition to the important characteristic of solitons, these equations also have other obvious physical properties: the unity of dispersion and nonlinearity. It has a certain degree of volatility, but its solution also has a certain degree of smoothness, attenuation and scattering properties of solutions when $t \to \infty$ (or $x \to \infty$). Due to its close connection with physical problems, the theoretical research on its solutions and properties has exceeded traditional research methods. For example, the emergence of scattering inversion, a new, accurate and very important solution method, has opened up new avenues for the theoretical study of differential equations; the Bäcklund transformation; the extended structural method established using differential geometry and Lie groups. At the same time, in terms of theoretical research on such nonlinear partial differential equations, it is no longer possible to simply copy some traditional methods from the past. For example, for the KdV equation and nonlinear Schrödinger equation, although their solutions have good smoothness, there is no extremum principle, so they can only be estimated using energy integration. And what sets this integral estimation apart from the usual ones is that it must fully use its various conservation laws. As Lax pointed out:"For the KdV equation, having infinitely many conservation laws is its main characteristic". From the current proof of the existence and uniqueness of global solutions for such nonlinear equations, there are several methods: (1) First, make a prior estimate of the integral, and then use various approximation methods to establish a local solution at $[0, t_1]$, t_1 depends on the initial function, and then establish a local solution at $[t_1, t_2]$, $t_2 - t_1$ depends on the modulus $\|u(t_1)\|$. Since the prior estimate $\|u(t)\| \leqslant$ const, it can be extended from $t_1, t_2, \cdots$ to any finite interval $[0, T]$. (2) The viscous elimination method (also known as the parabolic regularization method) seeks the global solution u_ε of the viscous approximation equation, and then use the uniform boundedness of u_ε and some of its derivatives with respect to the small parameter ε, we obtains the solution we need by taking $\varepsilon \to 0$. (3) The method of functional analysis is to transform the original equation into the standard form of differential operators, and use the known theorems of differential operators to obtain the existence theorem of global solutions. Here, it is necessary to specifically verify whether the conditions for the existence of solutions to differential operator equations are met. (4) The Galerkin approximation method, which use the consistency estimation of the Galerkin approximation solution to directly obtain a large-scale global solution. Among the various methods for proving the existence of a global solution, regardless of which one is used, prior integral estimation plays a decisive role. In fact, not all nonlinear evolutionary equations can have global solutions through prior estimation. For example, some multidimensional nonlinear wave equations and nonlinear Schrödinger equations exhibit the phenomenon of "blow up" of solutions, where the L_2 modulus of a solution or its first-order derivative tends to infinity when $t \to t_1 (t_1$ is finite). But if the L_2 modulus of the initial condition is appropriately small, its global solution can be obtained. These phenomena have attracted great attention and interest. At present, most of the solutions for such nonlinear equations are limited to periodic initial value problems and initial value problems, with a few considering their initial and boundary value problems. Generally speaking, there are many difficulties in formulating boundary value problems for such equations (such as KdV equations) and studying the existence of their solutions. Up to now, there are few research results in this area. For initial value problems, we usually assume that its solution tends to zero when $|x| \to \infty$, and this requirement is acceptable. For example, for the KdV equation, it has been proven that as long as the initial condition $|x| \to \infty$ tends to zero at a certain decay rate, its solution also obtains the corresponding decay rate. Of course, this requirement is not necessary. If periodic boundary conditions are used to approximate the initial conditions, other conditions for the initial conditions can be used to replace this assumption. We still make this assumption for the initial value problem. In addition, many of the results obtained from the following definite solution problems are equally valid for periodic initial value problems or initial value problems, and we do not intend to provide further explanation.

9.2   The Periodic Initial Value Problem and Initial Value Problem of the KdV Equation

The existence and uniqueness of solutions to the KdV equation were first obtained by Sjöberg in [1], [2], who considered the following initial and boundary problem

$$\begin{cases} u_t = uu_x + \delta u_{xxx}, & \delta \neq 0 \\ u(x,0) = f(x), & \forall x \in R \\ u(x,t) = u(x+1,t), & \forall x, t \end{cases} \tag{9.1}$$

which has the following results:

Theorem 9.1   *If $\delta \neq 0$ and $f(x)$ is a function with a period of 1, and its derivative up to the third order belongs to L_2, then there exists a unique solution to problem (9.1).*

Using the differential difference format corresponding to (9.1) as follows:

$$\begin{cases} \dfrac{\partial}{\partial t} u_N(x,t) = \left[u_N\left(x_r,t\right) D_0 u_N\left(x_r,t\right) + D_0 u_N^2\left(x_r,t\right)\right]/3 \\ \qquad + \delta D_+ D_-^2 u_N\left(x_r,t\right) \quad (r = 1,2,\cdots,N) \\ u_N\left(x_r,0\right) = f\left(x_r\right) \quad (r = 1,2,\cdots,N) \\ u_N\left(x_r,t\right) = u_N\left(x_{r+N},t\right), \forall x, t \end{cases} \tag{9.2}$$

therefore $h = \dfrac{1}{N}$, $x_r = rh$, D_+, D_-, D_0 represent differential operators and are defined as

$$hD_+g\left(x_r\right) = g\left(x_{r+1}\right) - g\left(x_r\right)$$
$$hD_-g\left(x_r\right) = g\left(x_r\right) - g\left(x_{r-1}\right)$$
$$2hD_0g\left(x_r\right) = g\left(x_{r+1}\right) - g\left(x_{r-1}\right)$$

It can be proven the existence of local solutions of the problem (9.1). Reuse the three conservation laws of (9.1)

$$\int_0^1 u^2(x,t)dx = \int_0^1 f^2(x)dx = \alpha_1 = \text{ const} \tag{9.3}$$

$$\int_0^1 \left(\frac{u^2}{3} - \delta u_x^2\right) dx$$
$$= \int_0^1 \left(\frac{f^3}{3} - \delta f'^2(x)\right) dx = \alpha_2 = \text{ const} \tag{9.4}$$

$$\int_0^1 \left(u^4 - 12\delta uu_x^2 + 36\delta^2 u_{xx}^2/5\right) dx$$

$$= \int_0^1 \left(f^4 - 12\delta f f'^2 + 36\delta^2 {f''}^2 /5 \right) dx = \alpha_3 = \text{const.} \tag{9.5}$$

By performing an integral prior estimation, the existence and uniqueness of the global solution to problem (9.1) have been proven, which can be easily obtained by the energy inequality. Lax first proved the uniqueness of the solution to the Cauchy problem

$$u|_{t=0} = u_0(x) \quad (-\infty < x < +\infty) \tag{9.6}$$

of the KdV equation

$$u_t + uu_x + u_{xxx} = 0 \quad (-\infty < x < +\infty, t > 0) \tag{9.7}$$

on the interval $(-\infty, +\infty)$ in [3], where the solution refers to $u(\cdot, t) \in C^\infty(-\infty < x < +\infty)$, u and all its derivatives with respect to x tend to zero $(|x| \to \infty)$. Let v be other solutions to problems (9.6) and (9.7):

$$\begin{cases} v_t + vv_x + v_{xxx} = 0 \\ v|_{t=0} = u_0(x) \end{cases}$$

then set $w = u - v$, we can obtain the linear equation for w

$$w_t + uw_x + wv_x + w_{xxx} = 0$$

Multiply the above equation by w, integrate $x \in (-\infty, +\infty)$, and integrate by part to obtain the relationship equation:

$$\frac{d}{dt} \frac{1}{2} \int_{-\infty}^{+\infty} w^2 dx + \int_{-\infty}^{+\infty} \left(v_x - \frac{1}{2} u_x \right) w^2 dx = 0 \tag{9.8}$$

Let

$$E(t) = \frac{1}{2} \int_{-\infty}^{\infty} w^2 dx, \quad \max |2v_x - u_x| = m$$

From (9.8), it can be concluded that

$$\frac{d}{dt} E(t) \leqslant mE(t)$$
$$\therefore \quad E(t) \leqslant E(0)e^{mt}$$

$$E(0) = 0 \Longrightarrow E(t) = 0 \ (t > 0) \Longrightarrow w \equiv 0$$

Using the fourth-order small parameter method to the following periodic initial value problem of the KdV equation in [4]

$$
\begin{cases}
u_t + uu_x + \mu u_{xxx} = 0 & (0 \leqslant t \leqslant T, 0 < x < 1) \\
u(x,0) = u_0(x) & (0 \leqslant x \leqslant 1) \\
u(0,t) = u(1,t) & (0 \leqslant t \leqslant T) \\
u_x(0,t) = u_x(1,t) & (0 \leqslant t \leqslant T) \\
u_{xx}(0,t) = u_{xx}(1,t) & (0 \leqslant t \leqslant T)
\end{cases}
\tag{9.9}
$$

which means considering the solution $u_\varepsilon(x,t)$ of the following initial and boundary problem corresponding to problem (9.9)

$$
u_{\varepsilon t} + u_\varepsilon u_{\varepsilon x} + \mu u_{\varepsilon xxx} + \varepsilon u_{\varepsilon xxxx} = 0 \ (\varepsilon > 0)
$$
$$
(0 < t \leqslant T, 0 < x < 1)
\tag{9.10}
$$

$$
u_\varepsilon(x,0) = u_{0\varepsilon}(x) \quad (0 \leqslant x \leqslant 1)
\tag{9.11}
$$

$$
\frac{\partial^j u_\varepsilon}{\partial x^j}(0,t) = \frac{\partial^j u_\varepsilon}{\partial x^j}(1,t)
$$

$$
(0 \leqslant t \leqslant T) \quad (j = 0,1,2,3)
\tag{9.12}
$$

When $\varepsilon \to 0$, it tends towards the solution of (9.9), where $u_{0\varepsilon} \in C^\infty([0,1])$, and such that

$$
\frac{du_0^j\varepsilon(0)}{dx^j} = \frac{du_0^j(1)}{dx^j}, \forall j \geqslant 0
$$

and let $u_{0\varepsilon}$ weakly converge to u_0 according to $H^1(\Omega)(\varepsilon \to 0)$. Here we use Ω to represent the interval $(0,1)$, and $H^s(\Omega)(s \geqslant 0$, which is an integer) is the Sobolev space

$$
\{v \mid v(x) \in L^2(\Omega), D^j v(x) \in L^2(\Omega), 0 \leqslant j \leqslant s\}
$$

$$
\|v\|_{H^s(\Omega)} = \left\{ \sum_{j=0}^s \left\| \frac{\partial^j v}{\partial x^z} \right\|_{L^2(\Omega)}^2 \right\}^{1/2}
$$

From Chapter 1 [5], it is easy to know the solution u_ε to the problems (9.10) - (9.12) exists and satisfies

$$
u_s \in L^\infty\left(0,T; L^2(\Omega)\right) \cap L^2\left(0,T; H^2(\Omega)\right)
\tag{9.13}
$$

where $L^\infty(0,T; H^s)$ represents the function space defined on $[0,T]$ and values in H^s, and $u(x,t)$ belongs to H^s as a function of x. For

$$
t \in [0,T], \ \sup_{0 \leqslant t \leqslant T} \|u(\cdot,t)\|_s < \infty
$$

$L^2(0, T, H^s)$ represents the function space, $u(x, t)$ as a function of x belongs to H^s for each $t \in [0, T]$ and $\int_0^T \|u(x, t)\|_s^2 dt < \infty$. From (9.13) we have

$$\frac{\partial u_\varepsilon}{\partial x} \in L^2(0, T; H^1(\Omega)) \subset L^2(0, T; L^\infty(\Omega))$$

$$u_\varepsilon \frac{\partial u_\varepsilon}{\partial x} \in L^2(0, T; L^2(\Omega)) \tag{9.14}$$

From equation (9.10), it can be derived that

$$\frac{\partial u_\varepsilon}{\partial t} + \mu \frac{\partial^3 u_\varepsilon}{\partial x^3} + \varepsilon \frac{\partial^4 u_\varepsilon}{\partial x^4} = -u_\varepsilon \frac{\partial u_\varepsilon}{\partial x} \in L^2 \tag{9.15}$$

Based on the smoothness theorem of linear equation solutions and boundary conditions, it can be inferred that

$$\frac{\partial u_\varepsilon}{\partial t} \in L^2(Q), u_\varepsilon \in L^2\left(0, T; H^4(Q)\right)$$

$$Q = \Omega \times [0, T] \tag{9.16}$$

We are now making prior estimates for the solutions to problems (9.10), (9.11) and (9.12).

Lemma 9.1   *If $u_0 \in L^2(\Omega)$, then there is*

$$\|u_\varepsilon\|_{L^\infty(0.T, L^2(\Omega))} \leqslant c \tag{9.17}$$

$$\sqrt{\varepsilon} \left\| \frac{\partial^2 u_\varepsilon}{\partial x^2} \right\|_{L^2(\Omega)} \leqslant c \tag{9.18}$$

where the constant c is independent of ε.

Proof   By multiplying equation (2.10) by u_ε and integrating x, under the periodic conditions (9.11), (9.12) , we obtain

$$\frac{1}{2} \frac{d}{dt} \|u_\varepsilon(t)\|_{L^2}^2 + \varepsilon \int_\Omega \left(\frac{\partial^2 u_\varepsilon}{\partial x^2} \right)^2 dx = 0$$

which leads to (9.17) and (9.18).   □

Lemma 9.2   *For all functions $v(x) \in H^3(\Omega)$, there is*

$$\|v\|_{L^4(\Omega)} \leqslant c \|v\|_{L^2(\Omega)}^{11/12} \left(\|v\|_{L^2(\Omega)} + \left\| \frac{d^3 v}{dx^3} \right\|_{L^2(\Omega)} \right)^{\frac{1}{12}} \tag{9.19}$$

$$\left\|\frac{dv}{dx}\right\|_{L^4(\Omega)} \leqslant c\|v\|_{L^2(\Omega)}^{7/12} \left(\|v\|_{L^2(\Omega)} + \left\|\frac{d^3v}{dx^3}\right\|_{L^2(\Omega)}\right)^{\frac{5}{12}} \tag{9.20}$$

Proof   According to the interpolation representation in Chapter 1[6], there is

$$\left[H^3(\Omega), H^0(\Omega)\right]_{\frac{11}{12}} = H^{\frac{1}{4}}(\Omega), H^0(\Omega) = L^2(\Omega)$$

And there is: $H^{\frac{1}{4}}(\Omega) \subset L^4(\Omega)$, which leads to

$$\|\nu\|_{L^4(\Omega)} \leqslant c\|\nu\|_{H^{\frac{1}{4}}(\Omega)} \leqslant c\|\nu\|_{H^3(\Omega)}^{\frac{1}{12}} \|\nu\|_{H^0(\Omega)}^{\frac{11}{12}}$$

which leads to (9.19). Similarly, $[H^3(\Omega), H^0(\Omega)]_{7/12} = H^{\frac{5}{4}}(\Omega)$

$$\left\|\frac{dV}{dx}\right\|_{L^4(\Omega)} \leqslant c\|v\|_{H^{\frac{5}{4}}(\Omega)}$$

which leads to (9.20). $\square$

Lemma 9.3   *If $u_0(x) \in H^1(\Omega)$, then there is*

$$\|\frac{\partial u_\varepsilon}{\partial x}\|_{L^\infty(0,T;L^2(\Omega))} \leqslant c \tag{9.21}$$

$$\sqrt{\varepsilon}\|\frac{\partial^3 u_\varepsilon}{\partial x^3}\|_{L^2(\Omega)} \leqslant c \tag{9.22}$$

Here, the constant c is independent of ε.

Proof   Multiplying equation (9.10) by

$$\psi_1(u_\varepsilon) = u_\varepsilon^2 + 2\alpha \frac{\partial^2 u_\varepsilon}{\partial x^2}$$

and integrating x, using periodic boundary conditions yields

$$\frac{d}{dt}\int_\Omega \left[\frac{1}{3}u_\varepsilon^3 - \alpha\left(\frac{\partial u_\varepsilon}{\partial x}\right)^2\right] dx$$

$$+\varepsilon\int_\Omega \frac{\partial^4 u_\varepsilon}{\partial x^4}\left(u_\varepsilon^2 + 2\alpha\frac{\partial^2 u_\varepsilon}{\partial x^2}\right) dx = 0$$

or

$$\frac{d}{dt}\int_\Omega \left[\frac{1}{3}u_\varepsilon^3 - \alpha\left(\frac{\partial u_\varepsilon}{\partial x}\right)^2\right] dx$$

$$-2\varepsilon\int_\Omega \frac{\partial^3 u_\varepsilon}{\partial x^3}u_\varepsilon\frac{\partial u_\varepsilon}{\partial x}dx \tag{9.23}$$

$$-2\alpha\varepsilon\int_\Omega \left(\frac{\partial^3 u_\varepsilon}{\partial x^3}\right)^2 dx = 0$$

From this, we obtain

$$
\alpha \frac{d}{dt}\left\|\frac{\partial u_\varepsilon}{\partial x}\right\|_{L^2(\Omega)}^2 + 2\alpha\varepsilon\left\|\frac{\partial^3 u_\varepsilon}{\partial x^3}\right\|_{L^2(\Omega)}^2
$$
$$
=\frac{1}{3}\frac{d}{dt}\int_\Omega u_\varepsilon^3 dx - 2\varepsilon\int_\Omega u_8\frac{\partial u_s}{\partial x}\cdot\frac{\partial^3 u_\varepsilon}{\partial x^3}dx
$$

(9.24)

Integrating t with (9.24) and dividing by α yields

$$
\left\|\frac{\partial u_\varepsilon}{\partial x}\right\|_{L^2(\Omega)}^2 + 2\varepsilon\int_0^t\left\|\frac{\partial^3 u_\varepsilon}{\partial x^3}(\sigma)\right\|_{L^2}^2 d\sigma
$$
$$
=\left\|\frac{du_0}{dx}\right\|_{L^2}^2 + \frac{1}{3\alpha}\int_\Omega u_\varepsilon^3(x,t)dx
$$
$$
-\frac{1}{3\alpha}\int_\Omega u_0^3(x)dx - \frac{2\varepsilon}{\alpha}\int_0^t u_\varepsilon\frac{\partial u_\varepsilon}{\partial x}
$$
$$
\cdot\frac{\partial^3 u_\varepsilon}{\partial x^3}dxd\sigma
$$

(9.25)

Since

$$
|\int_\Omega u_\varepsilon^3(t)dt| \leqslant |u_\varepsilon(t)\|_{L^\infty(\Omega)}\|u_\varepsilon(t)\|_{L^2(\Omega)}^2
$$
$$
\leqslant c_1\|u_\varepsilon(t)\|_{L^\infty(\Omega)} \quad (\text{ According to (9.17)})
$$
$$
\leqslant c_2\|u_\varepsilon(t)\|_{L^2(\Omega)}^{\frac{1}{2}}\left(\|u_\varepsilon(t)\|_{L^2(\Omega)} + \left\|\frac{\partial u_\varepsilon}{\partial x}(t)\right\|_{L^2(\Omega)}\right)^{1/2}
$$

(Note that for $v \in H^1(\Omega)$, we have$\|v\|_{L^\infty(\Omega)} \leqslant c\|v\|_{L^2}^{1/2}\cdot\left(\|v\|_{L^2} + \|\frac{\partial u}{\partial x}t\|_{L^2}\right)^{1/2}$)

$$
\leqslant c_3\left(1 + \left\|\frac{\partial u_\varepsilon}{\partial x}(t)\right\|_{L^2}\right)^{1/2}
$$
$$
\leqslant c_4 + \frac{3|\alpha|}{2}\left\|\frac{\partial u_\varepsilon}{\partial x}(t)\right\|_{L^2}^2
$$
$$
|\int_\Omega u_\varepsilon\frac{\partial u_\varepsilon}{\partial x}\cdot\frac{\partial^3 u_\varepsilon}{\partial x^3}dx|
$$

$$\leqslant \|u_\varepsilon(t)\|_{L^4(\Omega)} \left\|\frac{\partial u_\varepsilon}{\partial x}\right\|_{L^4(\Omega)} \left\|\frac{\partial^3 u_\varepsilon}{\partial x^3}\right\|_{L^2(\Omega)}$$

$$\leqslant c_5 \|u_\varepsilon(t)\|_{L^2}^{3/2} \left(\|u_\varepsilon(t)\|_{L^2} + \left\|\frac{\partial^3 u_\varepsilon}{\partial x^3}(t)\right\|_{L^2}\right)^{1/2}$$

$$\cdot \left\|\frac{\partial^3 u_\varepsilon}{\partial x^3}(t)\right\|_{L^2(\Omega)} \qquad \text{(According to (9.19), (9.20))}$$

$$\leqslant c_6 \left(1 + \left\|\frac{\partial^3 u_\varepsilon}{\partial x^3}(t)\right\|_{L^2}\right)^{1/2} \left\|\frac{\partial^3 u_\varepsilon(t)}{\partial x^3}\right\|_{L^2(\Omega)}$$

$$\leqslant c_7 + \frac{|\alpha|}{2} \left\|\frac{\partial^3 u_\varepsilon(t)}{\partial x^3}\right\|_{L^2(\Omega)}^2 \qquad \text{(According to (9.17))}$$

Considering the final inequality, from (9.25), we can obtain

$$\left\|\frac{\partial u_\varepsilon}{\partial x}(t)\right\|_{L^2(\Omega)}^2 + 2\varepsilon \int_0^t \left\|\frac{\partial^3 u_\varepsilon(\sigma)}{\partial x^3}\right\|_{L^2(\Omega)}^2 d\sigma$$

$$\leqslant \left\|\frac{du_0}{dx}\right\|_{L^2(\Omega)}^2 + \frac{1}{3|\alpha|} \int_\Omega |u_0|^3 dx \qquad (9.26)$$

$$+ \frac{1}{2}\left\|\frac{\partial u_\varepsilon(t)}{\partial x}\right\|_{L^2(\Omega)}^2 + \varepsilon \int_0^t \left\|\frac{\partial^3 u_\varepsilon(\sigma)}{\partial x^3}\right\|_{L^2(\Omega)}^2 d\sigma + c$$

from this, we get

$$\left\|\frac{\partial u_\varepsilon}{\partial x}(t)\right\|_{L^2(\Omega)}^2 + \varepsilon \int_0^t \left\|\frac{\partial^3 u_\varepsilon}{\partial x^3}(\sigma)\right\|_{L^2(\Omega)}^2 d\sigma \leqslant c$$

Thus, (9.21) and (9.22) has been confirmed.

Using (9.10)

$$\frac{\partial u_\varepsilon}{\partial t} = -u_\varepsilon \frac{\partial u_\varepsilon}{\partial x} - \alpha \frac{\partial^3 u_\varepsilon}{\partial x^3} - \varepsilon \frac{\partial^4 u_\varepsilon}{\partial x^4}$$

and (9.21), (9.22), we get

$$\frac{\partial u_\varepsilon}{\partial t} \text{ is uniformly bounded on } L^2\left(0, T; H^{-2}(\Omega)\right) \qquad (9.27)$$

From (9.1), (9.18), (9.21), (9.22), (9.27), it is possible to select the subsequence u_ε such that $u_\varepsilon \to u$, and weakly converges in $L^\infty(0, T; L^2(\Omega))$.

$$\frac{\partial u_\varepsilon}{\partial x} \to \frac{\partial u}{\partial x}, \text{ weakly converges in } L^\infty(0, T; L^2(\Omega))$$

$$\frac{\partial u_\varepsilon}{\partial t} \to \frac{\partial u}{\partial t}, \text{ weakly converges in } L^2(0, T; H^{-2}(\Omega))$$

From the first and second results, it can be inferred that $u_\varepsilon \to u$ weakly converges in $L^\infty(0, T; H^1(\Omega))$. Derive from the second result that $u_\varepsilon \to u$ strong convergence in $L^\infty(0, T; L^2(\Omega))$. It is not difficult to take the limit from (9.13), (9.10), (9.11), (9.12) and

$$u_\varepsilon \frac{\partial u_\varepsilon}{\partial x} \to u \frac{\partial u}{\partial x} \tag{9.28}$$

weakly converges in $L^\infty(0, T; L^1(\Omega))$. Therefore, equation $(9.10) \to u_t + uu_x + \mu u_{xxx} = 0$, which is the solution we require. We have the following theorem.

Theorem 9.2 *Let $\mu \in R, \mu \neq 0, u_0(x) \in H^1(\Omega), u_0(0) = u_0(1)$, then there exists a function $u(x, t), u \in L^\infty(0, T; H^1(\Omega))$ and satisfies*

$$u, u_x \in L^\infty\left(0, T; L^2(\Omega)\right) \tag{9.29}$$

$$u_t + uu_x + \mu u_{xxx} = 0 \tag{9.30}$$

$$u(x, 0) = u_0(x) \tag{9.31}$$

$$u(0, t) = u(1, t) \tag{9.32}$$

Theorem 9.3 *Let $\mu \in R, \mu \neq 0, u_0 \in H^2(\Omega), \dfrac{d^j u(0)}{dx^j} = \dfrac{d^j u(1)}{dx^j} (j = 0, 1)$. Then there exists a unique solution to the initial and boundary problem (9.9).*

Note If $u_0 \in L^\infty(\bar{\Omega})$, and

$$\frac{d^j u_0(0)}{dx^j} = \frac{du_0^j(1)}{dx^j}, \forall j \geqslant 0$$

the solution of (9.9) is $u \in L^\infty(\bar{\Omega})$.

For the initial value problem in [7]

$$\begin{cases} u_t + uu_x + u_{xxx} - \varepsilon u_{xxt} = 0 \\ \quad (t > 0, -\infty < x < +\infty) \\ u(x, 0) = g(x) \quad (-\infty < x < +\infty) \end{cases} \tag{9.33}$$

establishing a consistent prior estimate of its solution u_ε with respect to ε, thus the existence and uniqueness of solutions for the initial value problem of the KdV equation

$$\begin{cases} u_t + uu_x + u_{xxx} = 0 \quad (-\infty < x < +\infty, t > 0) \\ u(x, 0) = g(x) \quad\quad\quad (-\infty < x < +\infty) \end{cases} \tag{9.34}$$

are proved. For a wider range of KdV equations, complex KdV equations and higher-order KdV equation systems, the existence and uniqueness theorems of solutions to various initial and boundary problems can be found in [8] − [15].

9.3   Periodic Initial Value Problem for a Class of Nonlinear Schrödinger Equations

We consider the periodic initial value problem for a class of nonlinear Schrödinger equations

$$iu_{jt} - u_{jx} + \beta(x)q(\sigma_{21}|u_1|^2 + \sigma_{31}|u_2|^2)u_j$$
$$+ k_j(x)u_j = 0 \ (j = 1, 2, 0 < x < 2\pi, t > 0) \tag{9.35}$$

$$u_j|_{t=0} = u_0^j(x), 0 \leqslant x \leqslant 2\pi, j = 1, 2 \tag{9.36}$$

$$u_j(x, t) = u_j(x + 2\pi, t), \forall x, t \geqslant 0, j = 1, 2 \tag{9.37}$$

where $i = \sqrt{-1}$, σ_{21}, σ_{31} are normal numbers; $\beta(x)$ is a bounded real function with a period of 2π, $q(s) \geqslant 0$, $s \in [0, +\infty)$, $k_j(x)(j = 1, 2)$ is a bounded real function with a period of 2π. $u_j(x, t)$ is a complex valued unknown function $u_j^0(x)(j = 1, 2)$ is a known complex valued function with a period of 2π. Define inner product

$$(f, g) = \int_0^{2\pi} f\bar{g}dx, \quad a(u, v) = \int_0^{2\pi} \frac{\partial u}{\partial x}\frac{\partial \bar{v}}{\partial x}dx$$

We first make a prior estimate of the solutions to problems (9.35) - (9.37).

Lemma 9.4   *If (i) $\beta(x)$, $q(s)$, $k_j(x)$ are real function, (ii) $u_0^j(x) \in L_2$, then the solution $u^j(x, t)$ to problems (9.35) - (9.37) has an equation*

$$\left\|u^j(t)\right\|_{L_2}^2 = \left\|u_0^j\right\|_{L_2}^2 \ (j = 1, 2) \tag{9.38}$$

Proof   Multiply (9.35) by $\bar{u}_j$ and integrate x to obtain

$$i\left(u_{jt}, u_j\right) + a\left(u_j, u_j\right) + \left(\beta(x)q\left(\sigma_{21}|u_1|^2\right.\right.$$
$$\left.\left. +\sigma_{31}|u_2|^2\right)u_j, u_j\right) + (k_j(x)u_j, u_j) = 0 \tag{9.39}$$

Since $a(u_j, u_j) \geqslant 0$, $\beta(x)$ is a real function,

$$(\beta q u_j, u_j) = \int_0^{2\pi} \beta(x)q|u_j|^2 dx$$

$$(k_j u_j, u_j) = \int_0^{2\pi} k_j(x)|u_j|^2 dx$$

Taking the imaginary part from (9.39) yields (9.38).

Lemma 9.5 *If (i) σ_{21}, σ_{31} are real numbers, $\beta(x)$, $k_j(x)$ and $q(s)$ are all real functions; (ii) $u_0^j(x) \in L_2, \beta(x)$, $Q(\sigma_{21}|u_0^1|^2+\sigma_{31}|u_0^2|^2) \in L_1$, where $Q(s) = \int_0^s q(z)dz$, the solutions to problems (9.35) - (9.37) satisfy*

$$\sigma_{21} \|u_{1x}\|_{L_2}^2 + \sigma_{31} \|u_{2x}\| L_{L_2}^2 + \int_0^{2\pi} \beta(x)Q\left(\sigma_{21}|u_1|^2 + \sigma_{31}|u_2|^2\right) dx$$

$$+ \int_0^{2\pi} \left[k_1(x)\sigma_{21}|u_1|^2 + k_2(x)\sigma_{31}|u_2|^2 \, dx \right.$$

$$= \sigma_{21} \|u_{0x}^1\| \|_{L_2}^2 + \sigma_{31} \|u_{0x}^2\| L_{L_2}^2 \tag{9.40}$$

$$+ \int_0^{2\pi} \beta(x)Q\left(\sigma_{21}|u_0^1|^2 + \sigma_{31}|u_0^2|^2\right) dx$$

$$+ \int_0^{2\pi} \left[k_1(x)\sigma_{21}|u_0^1|^2 + k_2(x)\sigma_{31}|u_0^2|^2 \right] dx$$

Proof Multiply (9.35) by $\bar{u}_{jt}$ and integrate x to obtain

$$i\left(u_{jt}, u_{jt}\right) + \left(u_{jx}, u_{jxt}\right) + \left(\beta(x)q\left(\sigma_{21}|u_1|^2 + \sigma_{31}|u_2|^2\right)u_j, u_{jt}\right)$$

$$+ (k_j u_j, u_{jt}) = 0 \tag{9.41}$$

Since

$$\mathrm{Re}\left(u_{jx}, u_{jxt}\right) = \frac{1}{2}\frac{d}{dt} \|u_{jx}\| L_2^2$$

$$\mathrm{Re}\left(k_j u_j, u_{jt}\right) = \frac{d}{dt}\frac{1}{2} \int_0^{2\pi} k_j(x) |u_j|^2 \, dx$$

$$\mathrm{Re}\left(\beta(x)q\left(\sigma_{21}|u_1|^2 + \sigma_{31}|u_2|^2\right)u_j, \sigma_{21}u_{1t}\right)$$

$$+\mathrm{Re}\left(\beta(x)q\left(\sigma_{21}|u_1|^2 + \sigma_{31}|u_2|^2\right)u_2, \sigma_{31}u_{2t}\right)$$

$$= \frac{1}{2} \int_0^{2\pi} \beta(x)q\left(\sigma_{21}|u_1|^2 + \sigma_{31}|u_2|^2\right)$$

$$\cdot \frac{\partial}{\partial t}\left(\sigma_{21}|u_1|^2 + \sigma_{31}|u_2|^2\right) dx$$

$$= \frac{d}{dt}\frac{1}{2} \int_0^{2\pi} \beta(x)Q\left(\sigma_{21}|u_1|^2 + \sigma_{31}|u_2|^2\right) dx$$

taking the real part from (9.41) and multiply σ_{21} for $j = 1$; Multiply σ_{31} for $j = 2$. Adding the two equations together we obtain

$$\frac{1}{2}\frac{d}{dt}\left(\sigma_{21}\|u_1\|_{L_2}^2 + \sigma_{31}\|u_2\|_{L_2}^2\right)$$

$$+ \frac{1}{2}\frac{d}{dt}\int_0^{2\pi} \beta(x)Q\left(\sigma_{21}|u_1|^2 + \sigma_{31}|u_2|^2\right)dx$$

$$+ \frac{1}{2}\frac{d}{dt}\int_0^{2\pi} \left[k_1(x)\sigma_{21}|u_1|^2 + k_2(x)\sigma_{31}|u_2|^2\right]dx$$

$$= 0$$

Then we get (9.40).

Lemma 9.6   *If the conditions of Lemma 9.5 are satisfied, and $k_j(x)(j = 1, 2)$ is a bounded real function, $q(s) \geqslant 0$, $\beta(x) \geqslant 0$ is bounded, and $\sigma_{21} > 0$, $\sigma_{31} > 0$, then the solutions to problems (9.35)-(9.37) are estimated:*

$$\|u_{1x}\|_{L_2}^2 \leqslant c_1, \quad \|u_{2x}\|_{L_2}^2 \leqslant c_2$$

$$\int_0^{2\pi} \beta(x)Q\left(\sigma_{21}|u_1|^2 + \sigma_{31}|u_2|^2\right)dx \leqslant c_3 \tag{9.42}$$

where the constants c_1, c_2 only depend on the initial function and its derivatives.

Proof   It can be obtained from (9.40) and the conditions of lemma.

Corollary 9.1

$$\sum_{j=1}^2 \|u_j\|_{L^\infty} \leqslant c_4 \tag{9.43}$$

where the constant c_4 only depends on the initial function and its derivatives.

Proof   It can be obtained from the conclusion of lemma and the Sobolev inequality.

Lemma 9.7   *If the condition of Lemma 9.6 is satisfied and $u_0^j(x) \in H^2$ $(j = 1, 2)$ is assumed, then the solutions to problems (9.35)-(9.37) have estimates:*

$$\sup_{0 \leqslant t \leqslant T} \|u_{1t}\|_{L_2}^2 \leqslant c_5, \quad \sup_{0 \leqslant t \leqslant T} \|u_{2t}\|_{L_2}^2 \leqslant c_6 \tag{9.44}$$

where the constants c_5, c_6 depend on the derivative of the initial function up to the second order.

Proof Differentiate (9.35) with t, and multiply by $\bar{u}_{jt}$, then integrate to obtain

$$i\left(u_{jtt}, u_{jt}\right) + \left(u_{jx_t}, u_{jxt}\right)$$

$$+ \left(\beta(x)\frac{d}{dt}u_j q\left(\sigma_{21}|u_1|^2 + \sigma_{31}|u_2|^2\right), u_{jt}\right) \tag{9.45}$$

$$+ \left(k_j u_{jt}, u_{jt}\right) = 0$$

since

$$\left(\beta(x)\frac{d}{dt}u_j q\left(\sigma_{21}|u_1|^2 + \sigma_{31}|u_2|^2\right), u_{jt}\right)$$

$$= \int_0^{2\pi} \beta(x)q\left(\sigma_{21}|u_1|^2 + \sigma_{31}|u_2|^2\right)|u_{jt}|^2\,dx$$

$$+ \int_0^{2\pi} \beta(x)q'\frac{\partial}{\partial t}\left(\sigma_{21}|u_1|^2 + \sigma_{31}|u_2|^2\right)$$

$$\cdot u_j\bar{u}_{jt}dx, \quad \frac{\partial}{\partial t}|u_j(t)|^2 = u_{jt}\bar{u}_j + u_j\bar{u}_{jt}$$

$$\therefore$$

$$\left|\int_0^{2\pi} \beta q^1 \frac{\partial}{\partial t}(\sigma_{21}|u_1|^2 + \sigma_{31}|u_2|^2)u_j\bar{u}_{jt}dx\right|$$

$$\leqslant c_1\left\|q'\left(\sigma_{21}|u_1|^2 + \sigma_{31}|u_2|^2\right)\right\|_{L^\infty}$$

$$\cdot \int_0^{2\pi} dx\left[|u_{1t}|\cdot|\bar{u}_1| + |u_2|\cdot|u_{1t}|\right]|u_j||u_{jt}|$$

$$\leqslant c_2\left[\|u_{1t}\|_{L_2}^2 + \|u_{2t}\|_{L_2}^2\right]$$

By taking the imaginary part from (9.45), we can obtain:

$$\frac{d}{dt}\left[\|u_{1t}(t)\|_{L_2}^2 + \|u_{2t}\|_{L_2}^2\right] \leqslant c_3\left[\|u_{1t}(t)\|_{L_2}^2 + \|u_{2t}(t)\|_{L_2}^2\right]$$

(9.44) can be obtained from the Gronwall inequality and conditions of lemma.

We now define the generalized solutions of initial and boundary problem (9.35) - (9.37). The function $u_j(x,t) \in L^\infty(0,T;H^1), u_{jt} \in L^\infty(0,T;L_2)\,(j=1,2)$ with a period of 2π in space, which satisfies $Q\left(\sigma_{21}|u_0^1|^2 + \sigma_{31}|u_0^2|^2\right) \in L_1$ $\left(Q(s) = \int_0^s q(z)dz\right)$, is called the generalized solution of the initial and boundary problem (9.35)-(9.37). If it satisfies the integral equation:

$$i\left(u_{jt}, v_{j}\right) + \left(u_{jx}, v_{jx}\right) + \left(\beta u^{j} q \left(\sigma_{21} |u_{1}|^{2} \right.\right.$$

$$\left.\left. + \sigma_{31} |u_{2}|^{2}\right), v_{j}\right) + (k_{j} u_{j}, v_{j}) = 0, \tag{9.46}$$

$$\forall v_{j}(x) \in H^{1}, \ t \geqslant 0, j = 1, 2$$

$$\left(u^{j}\big|_{t=0}, v_{j}\right) = \left(u_{0}^{j}(x), v_{j}\right) (j = 1, 2) \tag{9.47}$$

By the Galerkin approximation method or rewriting (9.35) as an integral equation

$$u^{j}(x,t) = s(t)u_{0}^{j}(x) + \int_{0}^{t} s(t-\tau) \left[q\left(\sigma_{21} |u_{1}(x,\tau)|^{2}\right.\right.$$

$$\left.\left. + \sigma_{1} |u_{2}(x,\tau)|^{2}\right) u_{j} + k_{j}(x) \cdot u_{j}(x,\tau)\right] d\tau$$

where $s(t) = \dfrac{1}{\sqrt{2\pi i t}} e^{-\frac{x^{2}}{4\pi i}t}$. Using the principle of compressive mapping, we can easily obtain the existence of local solutions for problems (9.35) - (9.37). Based on the above prior estimates, it can be concluded that

Theorem 9.4   *If the following conditions are satisfied: (i) $\sigma_{21} \geqslant 0, \sigma_{31} \geqslant 0$, a real function $\beta(x) \geqslant 0$ with a period of 2π, (ii) $q(s)$ is a real function, $q(s) \in c^{1}$, $q(s) \geqslant 0$, $s \in [0, \infty)$; $k_{j}(x)$ is a bounded real periodic function with a period of 2π, (iii) $u_{0}^{j}(x)$ is a periodic complex valued function and $u_{0}^{j}(x) \in H^{2}$, then the generalized solution to the initial and boundary problem (9.35)-(9.37) exists.*

Theorem 9.5   *If $q(s) \in c^{1}, s \in [0, +\infty), k_{j}(x)$ is a bounded function, then the generalized solution to the initial and boundary problem (9.35)-(9.37) is unique.*

Proof   Supposing that there are two sets of generalized solutions $u^{j}, z^{j} (j = 1, 2)$ for (9.35)-(9.37). Let $w_{j} = u_{j} - z_{j}, j = 1, 2$. From (3.12), we can obtain:

$$i(w_{jt}, v_{j}) + (w_{jx}, v_{jx}) + (\beta(x)u_{j}q(\sigma_{21}|u_{1}|^{2}$$

$$+ \sigma_{31}|u_{2}|^{2}) - \beta(x)z_{j}q(\sigma_{21}|z_{1}|^{2}$$

$$+ \sigma_{31}|z_{2}|^{2}), v_{j}) + (k_{j} \cdot w_{j}, v_{j}) = 0, \tag{9.48}$$

$$v_{j} \in H^{1}, \ t > 0$$

$$w_{j}\big|_{t=0} = 0 \tag{9.49}$$

Specifically, let $v_j = w_j$, and due to

$$q\left(\sigma_{21}|u_1|^2 + \sigma_{31}|u_2|^2\right)u_j - q\left(\sigma_{21}|z_1|^2 + \sigma_{31}|z_2|^2\right)z_j$$

$$=q'(\tilde{z})\left[\sigma_{21}\left(|u_1|^2 - |z_1|^2\right) + \sigma_{31}\left(|u_2|^2 - |z_2|^2\right)\right]u_j$$

$$+ q\left(\sigma_{21}|z_1|^2 + \sigma_{31}|z_2|^2\right)(u_j - z_j)$$

where $\tilde{z}$ is between $|u_1|^2 + |u_2|^2$ and $|z_1|^2 + |z_2|^2$, therefore

$$(\beta(x)q(\sigma_{21}|u_1|^2 + \sigma_{31}|u_2|^2)u_j - \beta(x)q(\sigma_{21}|z_1|^2$$

$$+ \sigma_{31}|z_2|^2)z_j, w_j)$$

$$\leqslant \max|\beta(x)|\bigg[(|\sigma_{21}| + |\sigma_{31}|)\left\|q^1(\tilde{z})\right\|_{L^\infty}$$

$$\cdot \|u_j\|_{L^\infty} \cdot \sum_{j=1}^{2}\left(\|u_j\|_{L^\infty} + \|z_j\|_{L^\infty}\right)$$

$$+ \left\|q\left(\sigma_{21}|z_1|^2 + \sigma_{31}|z_2|^2\right)\right\|_{L^\infty}\bigg]$$

$$\cdot\left(\sum_{k=1}^{2}|w_k|, |w_j|\right)$$

By taking the real part from (3.14) and summing by j, we can obtain

$$\frac{1}{2}\frac{d}{dt}\sum_{j=1}^{2}\|w_j\|_{L_2}^2 \leqslant c\sum_{j=1}^{2}\|w_j\|_{L_2}^2$$

From the Gronwall inequality and $w_j(0) = 0$, we obtain $u_j = z_j$.

Note 1   For the initial and boundary value problems of the nonlinear Schrödinger equation system (9.35), its solution satisfies not only (9.35) but the initial condition (9.36)

$$u_j\big|_{x=0} = u_j\big|_{x=1} = 0 \tag{9.50}$$

The conclusions of Theorem 9.4 and Theorem 9.5 mentioned above still stand.

Note 2   If the initial function and coefficients of equation have higher smoothness, the classical global solutions to problems (9.35)-(9.37) can be obtained using common methods for calculating equation derivatives.

The well-posedness of global solution of a broader class of nonlinear Schrödinger equation systems, multidimensional nonlinear Schrödinger equations, and integral type nonlinear Schrödinger equation systems, can be found in [16] − [25].

9.4   Initial Value Problem of Nonlinear Klein-Gordon Equation

We now consider the initial value problem of the nonlinear Klein-Gordon equation:

$$\begin{cases} \dfrac{\partial^2 u}{\partial t^2} - \Delta u + m^2 u = -\lambda |u|^2 u, & x \in R^3,\ t > 0 \\[2mm] u(x,0) = f(x), & x \in R^3 \\[2mm] \dfrac{\partial u}{\partial t}(x,0) = g(x), & x \in R^3 \end{cases} \tag{9.51}$$

where $m, \lambda > 0$, $\Delta \equiv \dfrac{\partial^2}{\partial x_1^2} + \dfrac{\partial^2}{\partial x_2^2} + \dfrac{\partial^2}{\partial x_3^2}$. We will use the functional method of abstract differential operators to prove the existence of a global solution to the initial value problem (9.50).

We first transform problem (9.51) into a first-order system of equations for the variable t:

$$\begin{cases} \dfrac{\partial v}{\partial t} - \Delta u + m^2 u = -\lambda |u|^2 u \\[2mm] \dfrac{\partial u}{\partial t} = v \\[2mm] u(x,0) = f(x),\, v(x,0) = g(x) \end{cases}$$

or

$$\begin{cases} \dfrac{d\varphi(t)}{dt} - i \begin{pmatrix} 0 & I \\ \Delta - m^2 & 0 \end{pmatrix} \varphi(t) = J(\varphi(t)) \\[4mm] \varphi(x,0) = \begin{pmatrix} f(x) \\ g(x) \end{pmatrix} \end{cases} \tag{9.52}$$

where $\varphi = \begin{pmatrix} u \\ v \end{pmatrix}$, $J(\varphi) = \begin{pmatrix} 0 \\ -\lambda |u|^2 u \end{pmatrix}$, and I is the identity operator.

We will apply the general theorem of Hilbert space to prove the existence and uniqueness of the solution to problem (9.51). We first take the Hilbert space $\mathscr{H} = L^2(R^3)$. Set $B^2 = -\Delta + m^2$, and it is easy to know that B^2 is closed. Using $\mathscr{H}_B$ represents direct sum $\mathscr{H}_B = D(B) \oplus D(\mathscr{H})$, which has an inner product

$$(\langle u, v \rangle, \langle u, v \rangle)_B \equiv (Bu, Bu) + (v, v)$$

Set

$$A = i \begin{pmatrix} 0 & I \\ -B^2 & 0 \end{pmatrix}, \quad i = \sqrt{-1} \tag{9.53}$$

It is easy to verify that A is a symmetric operator in $\mathcal{H}_B$ with a domain $D \equiv D(B^2) \oplus D(B)$. A is also closed. We can further rewrite (9.52) in the form of an operator equation:

$$\begin{cases} \dfrac{d\varphi}{dt} = -iA\varphi + J(\varphi) \\ \varphi(0) = \varphi_0 = \langle f(x), g(x) \rangle \end{cases} \tag{9.54}$$

Now we estimate the solution of (9.51) firstly.

Lemma 9.8 *Let $u \in c_0^\infty \left(R^3\right)$, then*

$$\|u\|_6 \leqslant k\|Bu\|_2 \tag{9.55}$$

Proof Note $\dfrac{\partial u}{\partial x_i} \triangleq u_{xi}$, then there is

$$|u(x)|^4 \leqslant 4 \int \left| u_{xi} u^3 \right| dx_i$$

Integrating x_j here, and for a fixed $j \neq i$, there is

$$|u(x)|^6 \leqslant K \left(\int \left| u_{x_1} u^3 \right| dx_1 \right)^{1/2} \left(\int \left| u_{x_2} u^3 \right| dx_2 \right)^{1/2}$$
$$\cdot \left(\int \left| u_{x_3} u^3 \right| dx_3 \right)^{1/2}$$

By integrating both sides of the above inequality and applying the Schwartz inequality, we obtain

$$\int_{R^3} |u|^6 dx \leqslant K \left(\int_{R^3} \left| u_{x_1} u^3 \right| dx \right)^{1/2}$$
$$\cdot \left(\int_{R^3} \left| u_{x_2} u^3 \right| dx \right)^{1/2} \left(\int_{R^3} \left| u_{x_3} u^3 \right| dx \right)^{1/2}$$
$$\leqslant K \left(\int_{R^3} |u|^6 dx \right)^{3/4} \left(\int_{R^3} \left| u_{x_1} \right|^2 dx \right)^{1/4}$$
$$\cdot \left(\int_{R^3} \left| u_{x_2} \right|^2 dx \right)^{1/4} \left(\int_{R^3} \left| u_{x_3} \right|^2 dx \right)^{1/4}$$

It is easy to obtain

$$\left(\int_{R^2} |u|^6 dx \right)^{1/6} \leqslant K \left(\|u_{x_1}\|_2 + \|u_{x_2}\|_2 + \|u_{x_3}\|_2 \right)$$

$$= K \left(\|k_1 \hat{u}\|_2 + \|k_2 \hat{u}\|_2 + \|k_3 \hat{u}\|_2 \right)$$

$$\leqslant K \left\| \left(\Sigma k_i^2 + m^2 \right)^{1/2} \hat{u} \right\|_2 = K \|Bu\|_2$$

where

$$\hat{u}(t, k) = \frac{1}{2\pi^{3/2}} \int_{R^3} e^{-ix \cdot k} u(x, t) dx$$

$$\left(x \cdot k = \sum_{i=1}^{3} x_i \cdot k_i \right)$$

Lemma 9.9   *Let* $u_1, u_2, u_3 \in D(B)$, *then*

$$\|u_1 u_2 u_3\|_2 \leqslant K \|Bu_1\|_2 \|Bu_2\|_2 \|Bu_3\|_2 \tag{9.56}$$

Proof   Set $u \in D(B)$, since B is essentially self conjugate at $c_0^\infty (R^3)$, we can choose the function sequence $u_n \in c_0^\infty (R^3)$, making $u_n \xrightarrow{L_2} u, Bu_n \xrightarrow{L_2} B u$. Selecting a subsequence, still denoted it as u_n, then u_n point state converges to u, since

$$\left\| u_n^3 - u_m^3 \right\|_2 = \left\| (u_n - u_m) \left(u_n^2 + u_n u_m + u_m^2 \right) \right\|_2$$

$$\leqslant K \left\| u_n - u_m \right\|_6 \left\| \left(u_n^2 + u_n u_m + u_m^2 \right) \right\|_3$$

$$\leqslant K \left\| u_n - u_m \right\|_6 \left(\|u_n\|_6^2 + \|u_n\|_6 \cdot \|u_m\|_6 + \|u_m\|_6^2 \right)$$

$$\leqslant K \left\| Bu_n - Bu_m \right\|_2 \left(\|Bu_m\|_2^2 \right.$$

$$\left. + \|Bu_n\|_2 \|Bu_m\|_2 + \|Bu_m\|_2^2 \right)$$

Therefore, $\{u_n^3\}$ is the Cauchy sequence in L_2 and it point state converges to u^3, so $u^3 \in L_2$. The inequality above has a limit

$$\|u\|_6^3 = \|u^3\|_2 \leqslant K \|Bu\|_2^3$$

The proof of lemma is obtained by applying the Hölder inequality twice.

Lemma 9.10   *For all* $\varphi_1, \varphi_2 \in \mathscr{H}$, *J satisfies*

$$\|J(\varphi_1)\| \leqslant K \|\varphi_1\|^3$$

$$\|J(\varphi_1) - J(\varphi_2)\| \leqslant C (\|\varphi_1\|, \|\varphi_2\|) \|\varphi_1 - \varphi_2\|$$

Proof   If $\varphi_i = \langle u_i, v_i \rangle$, then from Lemma 9.9,

$$\|J(\varphi_1)\| = \|\lambda u_1^2 \bar{u}_1\|_2 \leqslant K \|Bu_1\|_2^3 \leqslant K \|\varphi_1\|^3$$

$$\|J(\psi_1) - J(\varphi_2)\| = \|\lambda(u_1^2 \bar{u}_1 - u_2^2 \bar{u}_2)\|_2$$

$$\leqslant K \|B(u_1 - u_2)\|_2 \left(\|Bu_1\|_2^2 + \|Bu_1\|_2 \|Bu_2\|_2 + \|Bu_2\|_2^2 \right)$$

$$\leqslant K \|\varphi_1 - \varphi_2\| \left(\|\varphi_1\|^2 + \|\varphi_1\| \|\varphi_2\| + \|\varphi_2\|^2 \right)$$

The lemma is obtained.

Lemma 9.11   *If $\varphi_1, \varphi_2 \in D(A)$, then*

$$\|AJ(\varphi_1)\| \leqslant K \|\varphi_1\|^2 \|A\varphi_1\|$$

$$\|A(J(\varphi_1) - J(\varphi_2))\|$$

$$\leqslant C(\|\varphi_1\|, \|\varphi_2\|, \|A\varphi_1\|, \|A\varphi_2\|)(\|A\varphi_1 - A\varphi_2\|)$$

Proof   Let $\varphi_i = \langle u_i, v_i \rangle$, where $u_i \in D(B^2)$, $v_i \in D(B)$, we calculate

$$\|Bu_{xi}\|^2 = \left\| \left(\Sigma k_i^2 + m^2 \right)^{1/2} k_i \hat{u} \right\|_2^2 \leqslant \left\| \left(\Sigma k_i^2 + m^2 \right) \hat{u} \right\|_2^2$$

$$= \left\| B^2 u \right\|_2^2$$

Therefore, according to Lemma 9.9, there is

$$\left\| \left(u^2 \bar{u} \right)_{xi} \right\|_2 = \left\| 2u u_{xi} \bar{u} + u^2 \bar{u}_{xi} \right\|_2 \leqslant K \|Bu\|_2^2 \cdot \|Bu_{xi}\|_2$$

$$\leqslant K \|Bu\|_2^2 \left\| B^2 u \right\|_2$$

Therefore

$$\|AJ(\varphi_1)\|^2 = \lambda^2 \left\| Bu_1^2 \bar{u}_1 \right\|_2^2$$

$$= \lambda^2 \sum_{i=1}^{3} \left\| \left(u_1^2 \bar{u}_1 \right)_{xi} \right\|_2^2 + \lambda^2 m^2 \left\| u_1^2 \bar{u}_1 \right\|_2^2$$

$$\leqslant K \left(\|Bu_1\|_2^4 \left\| B^2 u_1 \right\|_2^2 + m^2 \|Bu_1\|_2^6 \right)$$

$$\leqslant K \|Bu_1\|_2^4 \left\| B^2 u_1 \right\|_2^2 \leqslant K \|\varphi_1\|^4 \|A\varphi_1\|^2$$

We have proven the first equation. To prove the second equation, from Lemma 9.9 and the above equation, we have

$$\frac{1}{4} \left\| \left(u^2 \bar{u}_1 - u_1^2 \bar{u}_2 \right)_{x_i} \right\|_2^2$$

$$\leqslant \left\| u_1^2 \left(\bar{u}_1 - \bar{u}_2 \right)_{x_i} \right\|_2^2 + \left\| \left(u_1^2 - u_2^2 \right) \left(\bar{u}_2 \right)_{x_i} \right\|_2^2$$

$$+ \left\| 2 \left(u_1 \right)_{x_i} \left(|u_1|^2 - |u_2|^2 \right) \right\|_2^2$$

$$+ \left\| 2 \left(u_1 - u_2 \right)_{x_i} |u_2|^2 \right\|_2^2$$

$$\leqslant K \left(\| Bu_1 \|_2^4 \left\| B^2 \left(u_1 - u_2 \right) \right\|_2^2 + \left\| B^2 u_2 \right\|_2^2 \| B \left(u_1 \right.$$

$$+ u_2) \|_2^2 \| B^2 \left(u_1 - u_2 \right) \|_2^2)$$

$$\leqslant K \left(\| \varphi_1 \|^4 \| A \left(\varphi_1 - \varphi_2 \right) \|^2 + \| A \varphi_2 \|^2 \left(\| \varphi_1 \| \right.$$

$$+ \| \varphi_2 \|) \| A \left(\varphi_1 - \varphi_2 \right) \|^2)$$

Therefore,

$$\| A \left(J \left(\varphi_1 \right) - J \left(\varphi_2 \right) \right) \|^2 = \lambda^2 \left\| B \left(u_1^2 \bar{u}_1 - u_2^2 \bar{u}_2 \right) \right\|_2^2$$

$$= \lambda^2 \sum_{i=1}^3 \left\| \left(u_1^2 \bar{u}_1 - \bar{u}_2 u_2^2 \right)_{x_i} \right\|_2^2 + m^2 \lambda^2 \left\| u_1^2 \bar{u}_2 - u_2^2 \bar{u}_2 \right\|_2^2$$

$$\leqslant C \left(\| \varphi_1 \|, \| \varphi_2 \|, \| A \varphi_2 \| \right) \| A \left(\varphi_1 - \varphi_2 \right) \|^2$$

$$+ C \left(\| \varphi_1 \|, \| \varphi_2 \| \right) \| A \left(\varphi_1 - \varphi_2 \right) \|^2$$

where we have repeatedly used the inequality $\| Bu \|_2 \leqslant K \| B^2 u \|_2$, and the lemma has been proven.

Lemma 9.12   *Let $u(x, t)$ be the solution of* (9.51) *on* $[0, T]$, *where* $u(x, 0) = f(x) \in D(B^2), u_t(x, 0) = g(x) \in D(B)$, *then*

$$E(t) = \frac{1}{2} \int \left[|Bu(x, t)|^2 + |u_t(x, t)|^2 \right.$$

$$\left. + \frac{\lambda}{2} |u(x, t)|^4 \right] d^3 x$$

is independent of t.

Proof   Let $\varphi(t) = \langle u(x, t), u_t(x, t) \rangle$. Since for each $t \in [0, T], \varphi(t) \in D(A)$, we have $u(\cdot, t) \in D \left(B^2 \right), u_t(\cdot, t) \in D(B), \forall t \in [0, T]$. Since $\varphi(t)$ is strongly differentiable, the function u, u_t on $L^2(R^3)$ are strongly differentiable and

$$\begin{cases} \left\| B \left(\dfrac{u(t+h) - u(t)}{h} - u_t(t) \right) \right\|_2 \to 0, \ h \to 0 \\[4mm] \left\| \dfrac{u_t(t+h) - u_t(t)}{h} - u_{tt}(t) \right\|_2 \to 0, \ h \to 0 \end{cases} \tag{9.57}$$

so the first two terms of $E(t)$ are differentiable. To prove that its third term is also differentiable, we derive from Lemma 9.8 and Hölder inequality that

$$\left\| u \left(\frac{u(t+h) - u(t)}{h} - u_t(t) \right) \right\|_2$$

$$\leqslant \|u\|_2^{1/2} \|Bu\|_2^{1/2} \|B \left(\frac{u(t+h) - u(t)}{h} - u_t(t) \right) \|_2$$

It can be inferred from (9.57) that $u^2(x, t)$ is strongly differentiable, thus

$$\int |u(t, x)|^4 dx = \left(u^2(t), u^2(t) \right)_2$$

is differentiable, therefore $E(t)$ is differentiable, and

$$E'(t) = \frac{1}{2} (Bu_t, Bu) + \frac{1}{2} (u_{tt}, u_t)$$

$$+ \frac{\lambda}{2} (uu_t, u^2) + \frac{1}{2} (Bu, Bu_t)$$

$$+ \frac{1}{2} (u_t, u_{tt}) + \frac{\lambda}{2} (u^2, uu_t)$$

$$= \frac{1}{2} (u_t, B^2 u + u_{tt} + \lambda |u|^2 u)$$

$$+ \frac{1}{2} (B^2 u + u_{tt} + \lambda |u|^2 u, u_t) = 0$$

where we used the differential equation satisfied by u.

With the basic estimation of the solution to problem (9.51), and by applying the existence theorem of the solution of the abstract differential operator as follows, we can obtain the existence, uniqueness, and smoothness of the solution to problem (9.51).

We now consider the operator equation (9.54), where A is the self conjugate operator on a Hilbert space $\mathscr{H}$. Let J be a nonlinear mapping from $D(A)$ to $\mathscr{H}$. Our problem is to find what conditions J should satisfy to ensure that for any $\varphi_0 \in D(A)$ there exists a unique function $\varphi(t)$, $t \in [0, \infty)$ satisfying

$$\begin{cases} \dfrac{d\varphi}{dt} = -iA\varphi + J(\varphi) \\ \varphi(0) = \varphi_0 \end{cases} \tag{9.58}$$

We have the following theorem.

Theorem 9.6 (Local Existence)   *Let A be a self conjugate operator on Hilbert space $\mathscr{H}$, and J be a mapping of $D(A) \to D(A)$, satisfying*

$$(H_0) \quad \|J(\varphi)\| \leqslant C(\|\varphi\|)\|\varphi\|$$

$$(H_1) \quad \|AJ(\varphi)\| \leqslant C(\|\varphi\|, \|A\varphi\|)\|A\varphi\|$$

$$(H_0^2) \quad \|J(\varphi) - J(\psi)\| \leqslant C(\|\varphi\|, \|\psi\|)\|\varphi - \psi\|$$

$$(H_1^2) \quad \|A(J(\varphi) - J(\psi))\| \leqslant C(\|\varphi\|, \|A\varphi\|, \|\psi\|,$$

$$\|A\psi\|)\|A\varphi - A\psi\|, \ \forall \varphi \cdot \psi \in D(A)$$

here, each constant C is a single increasing function of the module it refers to. Then for all $\varphi_0 \in D(A)$, there exists $T > 0$, so that (9.58) has a unique continuous and differentiable solution on $[0, T)$, and for all $\varphi_0 \in \{\varphi \mid \|\varphi\| \leqslant a, \|A\varphi\| \leqslant b\}$, T can be chosen to hold uniformly.

Theorem 9.7 (Local Smoothness)   (a) *Let A be a self conjugate operator on Hilbert $\mathscr{H}$, and J be the reflection of*

$$D\left(A^j\right) \to D\left(A^j\right), \ (1 \leqslant j \leqslant n)$$

which satisfies (for $j = 0, 1, \cdots, n$):

$$(H_j) \ \left\|A^j J(\varphi)\right\| \leqslant C\left(\|\varphi\|, \cdots \|A^j \varphi\|\right) \left\|A^j \varphi\right\|$$

$$(H_j^2) \ \left\|A^j(J(\varphi) - J(\psi))\right\| \leqslant C(\|\varphi\|, \|\psi\|, \cdots$$

$$\left\|A^j \varphi\right\|, \left\|A^j \psi\right\|) \cdot \left\|A^j \varphi - A^j \psi\right\|$$

$$\forall \varphi, \psi \in D\left(A^j\right)$$

where each constant C is a single increasing function of its variable, so for all $\varphi_0 \in D(A^n)$, $n \geqslant 1$, there exists $T_n > 0$, such that (9.58) has a unique solution $\varphi(t) \in D\left(A^n\right)$, $t \in [0, T_n)$. For example, the $\varphi_0 \in$ set

$$\{\varphi \mid \|A^j \varphi\| \leqslant a_j, j = 0, 1, \cdots, n\}$$

then T can be selected to hold consistently.

  (b) *If the assumption is increased compared to (a), for $j < n$, J has the following properties: if φ is j times strongly continuous differentiable, $\varphi^{(k)}(t) \in D(A^{n-k})$, $A^{n-k}\varphi^{(k)}(t)$ is continuous (for all $k \leqslant i$), then $J(\varphi(t))$ is j times differentiable.*

$$\frac{d^j J(\varphi(t))}{dt^j} \in D\left(A^{n-j-1}\right), \ A^{n-j-1}d^j J(\varphi(t))/dt^j$$

is continuous, then the solution obtained from (a) is n times strongly differentiable for t, and

$$\frac{d^j \varphi(t)}{dt^j} \in D\left(A^{n-j}\right)$$

Theorem 9.8 (Global Existence and Smoothness)   *Let A be a self-adjoint operators on Hilbert space $\mathscr{H}$, and n be a positive integer. J is a mapping $(1 \leqslant j \leqslant n)$ of $D\left(A^j\right) \to D\left(A^j\right)$, and satisfies (for $0 \leqslant j \leqslant n$)*

$$(H_0) \quad \|J(\varphi)\| \leqslant C(\|\varphi\|)\|\varphi\|$$

$$(H_j') \quad \left\|A^j J(\varphi)\right\| \leqslant C\left(\|\varphi\|, \cdots, \left\|A^{j-1}\varphi\right\|\right) \left\|A^j \varphi\right\|,$$

$$j = 1, 2, \cdots, n$$

$$(H_j^L) \quad \left\|A^j(J(\varphi) - J(\psi))\right\| \leqslant C\left(\|\varphi\|, \cdots, \left\|A^{j-1}\varphi\right\|\right)$$

$$\left\|A^j \varphi - A^j \psi\right\|, \forall \varphi, \psi \in D\left(A^j\right)$$

$$j = 1, 2, \cdots, n$$

where C is the single increasing function of all its variables. If $\varphi_0 \in D\left(A^n\right)$ and $\|\varphi(t)\|$ is bounded on any finite interval guaranteed by, then there exists a strongly differentiable function $\varphi(t)$ on $D(A^n)$, which on $[0, \infty)$ satisfies

$$\begin{cases} \varphi'(t) = -iA\varphi(t) + J(\varphi(t)) \\ \varphi(0) = \varphi_0 \end{cases} \tag{9.59}$$

Furthermore, if J satisfies the assumption of part (b) in Theorem 2, then $\varphi(t)$ is n times strongly differentiable, and $\dfrac{d^j \varphi(t)}{dt^j} \in D\left(A^{n-j}\right)$.

By using the prior estimates of Lemma 9.8-Lemma 9.12 and Theorem 9.6-Theorem 9.8, we can obtain the existence theorem of the solution to problem (9.51).

Theorem 9.9   *Let $\lambda > 0$, $m > 0$, and*

$$f \in D\left(-\Delta + m^2\right), \ g \in D\left(\left(-\Delta + m^2\right)^{1/2}\right)$$

then there exists a unique function $u(x, t)$, $t \in R$, $x \in R^3$, such that $t \to u(\cdot, t)$ is a quadratic strongly differentiable function of t on $L_2\left(R^3\right)$, and for all t, $u(\cdot, t) \in D\left(-\Delta + m^2\right)$, $u(x, 0) = f(x)$, $u_t(x, 0) = g(x)$, and satisfies

$$u_{tt} - \Delta u + m^2 u = -\lambda |u|^2 u \tag{9.60}$$

For all t, mapping $\langle f, g \rangle \mapsto \langle u(\cdot, t), u_t(\cdot, t) \rangle$ is continuous.

Proof   According to Lemma 9.10 and Lemma 9.11, it can be inferred that J satisfies the conditions (H_0^2), (H_1^2), and $(H_0)(H_1')$ of Theorem 9.6. Therefore, the only local solution $\varphi(t)$ exists on $[0, T)$, and by Lemma 9.12, $E(t)$ is constant for all $t \in [0, T)$, since

$$\frac{1}{2}\|\varphi(t)\|^2 \leqslant \frac{1}{2}\|\varphi(t)\|^2$$
$$+ \frac{\lambda}{4} \int_{R^3} |u(x, t)|^4 dx^3 = E(t) = E(0)$$

$\|\varphi(t)\|$ is bounded on $[0, T)$, and the solution exists according to Theorem 9.8 for all $t \geqslant 0$. The theorem has been proven.

For other nonlinear evolution equations and their systems of equations, the use of abstract differential operators to prove their existence and uniqueness is referenced in $[26] - [29]$.

9.5   The RLW Equation and the Galerkin Method

We use the Galerkin method to prove the existence of solutions to the RLW equation and discuss the smoothness of the solutions.

Consider the initial and boundary value problems of the general RLW equation as follows

$$u_t + f(u)_x - u_{xxt} = g(x, t) \tag{9.61}$$

$$u|_{t=0} = u_0(x) \tag{9.62}$$

$$u|_{x=0} = u|_{x=1} = 0 \tag{9.63}$$

In order to prove the existence of solutions to problems (9.61) - (9.63), the following two lemmas in Sobolev space are required.

Lemma 9.13   *If $u \in H^1(0, 1)$, then there exists a constant $C > 0$, which is independent of u, such that,*

$$\sup_{0 \leqslant x \leqslant 1} |u(x)| \leqslant C\|u\|^{1/2} (\|u\| + \|u_x\|)^{1/2} \tag{9.64}$$

Lemma 9.14   *Let $f \in C^k(R)$, $k \geqslant 1$, and $f(0) = 0$, if $u(x, t) \in L^\infty (0, T; H^k(0, 1))$, then $f(u(x, t)) \in L^\infty (0, T; H^k(0, 1))$, and there are inequalities:*

$$\|f(u(t))\|_{H^1(0,1)} \leqslant M\|u(t)\|_{H^1(0,1)}$$

and

$$\|f(u(t))\|_{H^k(0,1)} \leqslant C_k \left(1 + \|u(t)\|_{H^{k-1}}^{k-1}\right) \cdot \|u(t)\|_{H^k} \quad (k \geqslant 2) \tag{9.65}$$

where M and C_k are constants.

Setting $(0,1) = Q$, $\mathcal{Q} = Q \times [0,T]$, $T > 0$, we have the following theorem.

Theorem 9.10   Let $T > 0$ be a real number,

$$g(x,t) \in L^\infty\left(0,T;L^2(\Omega)\right), u_0(x) \in H_0^1(\Omega),$$

and $f(s) \in C^1(R)$, then there exists a unique function $u(x,t)$, $(x,t) \in \mathcal{Q}$, which satisfies the conditions

$$u \in L^\infty\left(0,T;H_0^1(\Omega)\right) \tag{9.66}$$

$$u_t \in L^\infty\left(0,T;H_0^1(\Omega)\right) \tag{9.67}$$

$$u_t + (f(u))_x - u_{xxt} = g(x,t), \text{ on } Q \tag{9.68}$$

$$u(x,0) = u_0(x) \tag{9.69}$$

Note   Without loss of generality, we always assume that $f(0) = 0$. In fact, if $f(0) \neq 0$, there is an equation

$$u_t + (h(u))_x - u_{xxt} = g(x,t)$$

where $h(s) = f(s) - f(0)$, which is equivalent to equation (9.68).

Proof   Firstly, we see that (9.69) is meaningful. In fact, according to Lemma 6.1 in Chapter 6 and (9.66), (9.67), it can be concluded that $u(x,t)$ can be determined at $t = 0$. This theorem is proved in three steps using the Galerkin method: (i) Construct an approximate solution to equation (9.68); (ii) Make prior estimates of approximate solutions; (iii) Limit the approximate solution.

The first step is to construct an approximate solution. Let $\{\omega_\nu\}$ be the basis function of the space $H_0^1(\Omega)$, and for any $m \in N$, $w_1, w_2, \cdots, w_n$ are linearly independent. Taking an approximate solution of (9.68), $u^m = u^m(x,t) = \sum_{\nu=1}^m g_{\nu m}(t) w_\nu(x)$, where the coefficients $g_{\nu m}(t)$ can be determined by the equation

$$(u_t^m, w_\nu) + a(u^m, w_\nu) + ((f(u^m))_x, w_\nu) = (g, w_\nu)$$
$$\nu = 1, 2, \cdots, m \tag{9.70}$$

Here $a(u,\nu) = \int_0^1 \frac{\partial u}{\partial x} \cdot \frac{\partial \nu}{\partial x} dx$. Since $u_1 \in H_0^1(\Omega)$, there exists a constant $C_{\nu m}(\nu = 1, 2, \cdots, m)$, such that if $u_{0m} = \sum_{\nu=1}^m C_{\nu m} w_\nu$, we have

$$u_{0m} \xrightarrow{\text{strong}} u_0, \text{ in } H_0^1(\Omega), m \to \infty \tag{9.71}$$

If we add equation (9.70) with initial conditions, $u^m(0) = u_{0m}$, then we obtain a system of ordinary differential equations for the unknown function $g_{\nu m}$, with initial conditions $g_{\nu m}(0) = C_{\nu m}$. Due to the fact that the basis function $\{w_\nu\}_{\nu \in N}$ is linearly independent, the coefficient matrix of $g_{\nu m}$ is invertible, and its local solutions exist. Therefore, on $[0, t_m)$, there exists a solution $g_{\nu m}(t)_{1 \leqslant \nu \leqslant m}$ to the system of equations (9.70). In the next step, due to the establishment of a prior estimate, it is known that its solution can be extended from $[0, t_m)$ to $[0, T]$, where T is any finite positive number.

The second step is to make a prior estimate, multiply the equation (9.70) satisfied by the approximate solution with $g_{\nu m}(t)$ on both sides, and sum ν from 1 to m. We obtain

$$\frac{1}{2}\frac{d}{dt}\left(\|u^m\|^2 + a\left(u^m, u^m\right)\right)$$
$$-\left(f\left(u^m\right), \frac{\partial u^m}{\partial x}\right) = (g, u^m) \tag{9.72}$$

If let

$$h(x, t) = \int_0^{u^m(x,t)} f(s)ds$$

then

$$\frac{\partial h}{\partial x} = f\left(u^m\right)\frac{\partial u^m}{\partial x}$$

Since $w_\nu \in H_0^1(\Omega)$, $h(0, t) = h(1, t) = 0$, we have

$$\left(f\left(u^m\right), \frac{\partial u^m}{\partial x}\right) = h(1, t) - h(0, t), \quad \forall\, t$$

According to (9.72), there are

$$\frac{d}{dt}\left(\|u^m\|^2 + a\left(u^m, u^m\right)\right) \leqslant \|g(t)\|^2 + \|u^m\|^2$$

From the Gronwall lemma and $u_{0m} \xrightarrow{H_0^1(\Omega)} u_0$, we obtain

$$\|u^m\|_{H_0^1(\Omega)} \leqslant c, \ \forall t \in [0, T] \tag{9.73}$$

where the constant c is independent of m. From this, it can be inferred that when $m \to \infty$, u^m belongs to the bounded set of $L^\infty\left(0, T; H_0^1(\Omega)\right)$. Therefore, $u^m \xrightarrow{\text{weak} *} u$ is in $L^\infty(0, T; H_0^1(\Omega))$.

Next we estimate u_t^m. Multiply (9.70) by $g'_{\nu m}(t)$ and sum ν from 1 to m. We obtain

$$\|u_t^m\|^2 + a\left(u_t^m, u_t^m\right) \leqslant |(f\left(u^m\right)_x, u_t^m)| + |(g, u_t^m)|$$

$$\leqslant \|f\left(u^m\right)_x\|\,\|u_t^m\| + \|g(t)\|\,\|u_t^m\|$$

From Lemma 9.14 and the estimation (9.73), it can be concluded that

$$\|u_t^m\|_{H_0^1(0,1)} \leqslant c,\ t \in [0, T] \tag{9.74}$$

where the constant c is independent of m. Thus u^m belongs to the bounded set of $L^\infty\left(0, T; H_0^1(\Omega)\right)$. Thus, there exists a subsequence $[u_t^m]$ such that

$$u_t^m \xrightarrow{\text{weak*}} u_t \text{ in } L^\infty\left(0, T; H_0^1(\Omega)\right) \tag{9.75}$$

Therefore, for all m, u^m belongs to the bounded set of $H^1(Q)$, according to the Rellich embedding theorem, $u^m \to u$ exists in $L_2(Q)$, and there exists a subsequence u^m that converges almost everywhere to u.

The third step is the limit of approximate solution. Firstly, we consider the nonlinear terms. According to Lemma 9.14 and (9.73), we have

$$\|f\left(u^m\right)\|_{H^1(\Omega)} \leqslant c,\ t \in [0, T] \tag{9.76}$$

where the constant c is independent of m. Therefore,

$$F(u^m)_x \xrightarrow{\text{weak *}} \chi, \text{ in } L^\infty(0, T; L_2(\Omega)) \tag{9.77}$$

Due to the bounded set of u^m belonging to $H^1(\Omega)$ and the continuity of f on R, we have

$$f\left(u^m\right) \to f(u), \text{ almost everywhere in } Q \tag{9.78}$$

Due to (9.73), Lemma 9.13 and Lemma 9.14, there is

$$\|f\left(u^m\right)\|_{L_2(\Omega)} \leqslant c,\ \forall m, t \in [0, T] \tag{9.79}$$

From (9.78) and (9.79)

$$f(u^m) \xrightarrow{\text{week *}} f(u), \text{in } L_2(Q) \tag{9.80}$$

which conclude that

$$f(u^m)_x \to f(u)_x, \text{ holds in a distributed sense over } Q \tag{9.81}$$

We have $f(u)_x = \chi$ from (9.67) and (9.81), therefore,

$$f(u^m)_x \xrightarrow{\text{weak *}} f(u)_x, \text{ in } L^\infty(0, T; L_2(\Omega)) \tag{9.82}$$

If $m > v$, then (9.66) becomes

$$(u_t^m, w_\nu) + a(u_t^m, w_\nu) + (f(u^m)_x, w_\nu) = (g, w_\nu) \tag{9.83}$$

From (9.75) and (9.82), we obtain:

$$(u_t^m, w_\nu) \xrightarrow{\text{weak} *} (u_t, w_\nu), \text{in } L^\infty(0, T)$$

$$a(u_t^m, w_\nu) \xrightarrow{\text{weak} *} a(u_t, w_\nu), \text{in } L^\infty(0, T) \tag{9.84}$$

$$(f(u^m)_x, w_\nu) \xrightarrow{\text{weak} *} (f(u)_x, w_\nu), \text{in } L^\infty(0, T)$$

Therefore, from (9.83), let $m \to \infty$, we obtain

$$(u_t, w_\nu) + a(u_t, w_\nu) + (f(u)_x, w_\nu) = (g, w_\nu),$$

$$\text{for all } \nu. \tag{9.85}$$

Due to the density of $\{w_\nu\}$ in $H_0^1(\Omega)$, there is

$$(u_t, v) + a(u_t, v) + (f(u)_x, v) = (g, v), \ \forall v \in H_0^1(\Omega)$$

and u satisfies conditions (i) - (iii) of Theorem 9.10.

Now let's verify that u meets the initial conditions. In fact, since in $L^\infty(0, T; L_2(\Omega))$,
$u^m \xrightarrow{\text{weak} *} u$, then

$$\int_0^T (u^m, v) \, dt \to \int_0^T (u, v) dt, \ \forall v \in L^1(0, T; L_2(\Omega)) \tag{9.86}$$

From (9.75), there are:

$$\int_0^T (u_t^m, v) \, dt \to \int_0^T (u_t, v) \, dt, \ \forall v \in L^1(0, T; L_2(\Omega)) \tag{9.87}$$

Now consider $v(x, t) = \theta(t)w(x)$, $w(x) \in L_2(\Omega)$, $\theta \in C^1(0, T)$ such that $\theta(0) = 1$, $\theta(T) = 0$. If we take $v = \theta' w$ in (9.86) and $v = \theta w$ in (9.87), we can obtain

$$\lim_{m \to \infty} (u^m(0), w) = (u(0), w). \quad \forall w \in L_2(\Omega).$$

That is, $u^m(0) \xrightarrow{\text{weak}} u(0)$ in $L_2(\Omega)$, so $u(0) = u_0$.

To prove the uniqueness of the solution, assuming that there are two solutions u, v corresponding to the same initial condition. If let $w = u - v$, there are

$$\begin{cases} w_t - w_{xxt} + f(u)_x - f(v)_x = 0 \\ w(x, 0) = 0, \quad w(0, t) = w(1, t) = 0 \end{cases}$$

So w satisfies the equation

$$\frac{1}{2}\left(\frac{d}{dt}\right)\left(\|w\|^2 + \|w_x\|^2\right) = (f(u) - f(v), w_x) \tag{9.88}$$

From Lemma 9.13, Lemma 9.14, and (9.68), we obtain

$$\frac{d}{dt}\left(\|w\|^2 + \|w_x\|^2\right) \leqslant c\|w_x\|^2$$

where $c > 0$ is independent of t. Therefore, $w \equiv 0$.

Next, we will discuss the regularity of weak solutions. We select the basis function of the Galerkin method as the characteristic function of the one-dimensional Laplace operator. Then a higher smoothness of the solution can be obtained. Let $\{\psi_\nu\}_{\nu \in N}$ be the characteristic function of the one-dimensional Laplace operator in $L_2(\Omega)$, as shown in [6]. $\{\psi_\nu\}$ is a complete orthogonal system in $L_2(\Omega)$ and $H_0^1(\Omega)$. It is known that $\psi_\nu \in H^k(\Omega)$, where k is any positive integer and V^k represents the closure of a finite linear combination of ψ_ν in $H^k(\Omega)$.

Theorem 9.11 *Let $g(x,t) \in L^\infty(0,T;H^k(\Omega))$, such that*

$$D^{2\nu}g \in L^\infty\left(0,T;H_0^1(\Omega)\right), \nu = 0,1,2,\cdots,j$$

and $k - 2j \geqslant 1$; $f(s) = \dfrac{s^2}{2}$, $u_0 \in V^{k+1}$, then for every non negative integer k, there exists only one function $u(x,t)$ defined on Q and satisfying the conditions

$$u \in L^\infty\left(0,T;H^{k+1}(\Omega)\right) \tag{9.89}$$

$$D^{2\nu}u \in L^\infty\left(0,T;H_0^1(\Omega)\right), v = 0,1,2,\cdots,j$$

$$k + 1 - 2j \geqslant 1 \tag{9.90}$$

$$u_t \in L^\infty\left(0,T;H^{k+2}(\Omega)\right) \tag{9.91}$$

$$D^{2\nu}u_t \in L^\infty\left(0,T;H_0^1(\Omega)\right), v = 0,1,2,\cdots,k$$

$$k + 2 - 2j \geqslant 1 \tag{9.92}$$

$$u_t + uu_x - u_{xxt} = g, \quad \text{in } L^\infty\left(0,T;L_2(\Omega)\right) \tag{9.93}$$

$$u(x,0) = u_0(x) \tag{9.94}$$

Proof Assuming $u^m(x,t)$ is an approximate solution defined by Theorem 9.10 and replacing w_ν by the characteristic function ψ_ν, we obtain

$$\begin{aligned}(u_t^m, \psi_\nu) + (f\left(u^m\right)_x, \psi_\nu) - (u_{xxt}^m, \psi_\nu) \\ = (g, \psi_\nu), \ \nu = 1,\cdots,m \end{aligned} \tag{9.95}$$

$$u^m(0) = u_{0m} \qquad (9.96)$$

where $u_{0m} \xrightarrow{\text{strong}} u_0$ in $H^{k+1}(\Omega)$. We notice that

$$\Delta^p \psi_\nu = (-\lambda_\nu)^p \, \psi_\nu$$

holds for all p, where p is a non negative integer.

We will use induction to prove (under the assumption of the theorem)

$$\|u^m(t)\|_{H^{k+1}} \leqslant C, \quad \forall t \in [0, T] \qquad (9.97)$$

$$\|u_t^m(t)\|_{H^{k+2}} \leqslant C, \quad \forall t \in [0, T] \qquad (9.98)$$

where the constant $C > 0$ is independent of m, t.

Firstly, as proven in Theorem 9.10, we take the basis function as the characteristic function of the one-dimensional Laplace operator. According to the assumptions (9.73) and (9.74) of Theorem 9.10, there is an estimate that:

$$\|u_{xxt}^m(t)\| \leqslant C \qquad (9.99)$$

where the constant $C > 0$ is independent of m, t. In fact, multiplying both sides of (9.70) by $(-\lambda_\nu)g'_{\nu m}$ and summing by v yields

$$\|u_{xt}^m\|^2 + \|u_{xxt}^m\|^2 = -(g, u_{xxt}^m) + (f(u^m)_x, u_{xxt}^m)$$

$$\leqslant \|g\| \|u_{xxt}^m\| + \|f(u^m)_x\| \|u_{xxt}^m\|$$

From Lemma 9.14 and (9.73), we obtain (9.99). Therefore, for $k = 0$, (9.97) and (9.98) hold.

Note If replacing any base $\{w_\nu\}$, considering the special base $\{\psi_\nu\}$, using the assumptions of Theorem 9.10, we can replace (9.97) and (9.98) respectively for better results

$$u_t \in L^\infty(0, T; H^2(\Omega) \cap H_0^1(\Omega))$$
$$u_t + (f(u))_x - u_{xxt} = g(x, t), \text{in the sense of } L^\infty(0, T; L_2(\Omega)) \text{ weak*}$$

Assuming (9.97) and (9.98) hold for $k \geqslant 0$, we prove that they also hold for $k + 1$. Note that when q is an odd positive integer, we have

$$D^q f(u^m) = c_0 u^m D^q u^m + c_1 D u^m D^{q-1} u^m + \cdots + c_{\frac{q-1}{2}} \cdot D^{\frac{q-1}{2}} u^m$$

therefore

$$(Df(u^m), D^{2k+2}u^m) = \pm(D^{k+1}f(u^m), D^{k+2}u^m) \qquad (9.100)$$

$$(Df(u^m), D^{2k+4}u^m) = \pm(D^{k+2}f(u^m), D^{k+3}u^m) \qquad (9.101)$$

Multiplying both sides by $(-\lambda_\nu)^{k+1} g_{\nu m}$ in (9.95), and sum up ν, we have

$$\left(u_t^m, D^{2k+2} u^m\right) + \left(f\left(u^m\right)_x, D^{2k+2} u^m\right) - \left(D^2 u_t^m, D^{2k+2} u^m\right) = \left(g, D^{2k+2} u^m\right)$$

From (9.100), there are

$$\left(D^{k+1} u_t^m, D^{k+1} u^m\right) + \left(D^{k+2} u_t^m, D^{k+2} u^m\right)$$
$$= -\left(D^k g, D^{k+2} u^m\right) \pm \left(D^{k+1} f\left(u^m\right), D^{k+2} u^m\right)$$

or

$$\frac{1}{2}\left(\frac{d}{dt}\right)\left(\left\|D^{k+1} u^m\right\|^2 + \left\|D^{k+2} u^m\right\|^2\right)$$
$$\leqslant \frac{1}{2}\left\|D^k g\right\|^2 + \frac{1}{2}\left\|D^{k+1} f\left(u^m\right)\right\| + \left\|D^{k+2} u^m\right\|^2$$

Based on the inductive assumption and Lemma 9.14 , there is

$$\left\|D^{k+2} u^m(t)\right\| \leqslant C, \quad \forall t \in [0, T] \tag{9.102}$$

where the constant $C > 0$ is independent of m, t. Similarly, multiply both sides by

$$(-\lambda_\nu)^{k+2} \cdot g'_{\nu m}$$

and sum up ν, there is

$$\left\|D^{k+2} u_t^m\right\|^2 + \left\|D^{k+3} u_t^m\right\|$$
$$= -\left(D^{k+1} g, D^{k+3} u_t^m\right) \pm \left(D^{k+2} f\left(u^m\right), D^{k+3} u_t^m\right)$$
$$\leqslant 2\left\|D^{k+1} g\right\|^2 + 2\left\|D^{k+2} f\left(u^m\right)\right\|^2 + \frac{1}{4}\left\|D^{k+3} u_t^m\right\|^2$$

Based on the inductive assumption, Lemma 9.14 and (9.102), there is

$$\left\|D^{k+2} u_t^m(t)\right\| \leqslant C, \ \forall t \in [0, T] \tag{9.103}$$

where the constant $C > 0$ is independent of m, t. From (9.102) and (9.103), we obtain (9.97) and (9.98).

Note Theorem 9.11 also holds for $f(s) = C s^n$, where C is a constant, n is an even positive integer.

The study of other nonlinear evolution equations using the Galerkin method and viscous elimination method are mentioned in [30] − [35].

9.6   The Asymptotic Behavior of Solutions and "Blow up" Problem for $t \to \infty$

For a class of nonlinear evolutionary equations, it has the asymptotic property that the local L^2 modulus of its smooth solution tends to zero when $t \to \infty$. We will use a simpler proof method to illustrate it. In addition, for some nonlinear evolutionary equations, although their local solutions exist, their global solutions do not exist. In fact, when $t \to t_1$ (t_1 is finite), the L_2 modulus of its solution will tend to infinity, this phenomenon known as the "Blow up" of the solution. It has been discovered that many nonlinear evolutionary equations possess this property. In this section, we will give two examples to illustrate it.

Now we consider the initial value problem of the generalized KdV equation as follows.

$$u_t + (u_{xx} - f(u) - u)_x = 0, \ (x \in R, t > 0) \tag{9.104}$$

$$u|_{t=0} = \varphi(x), \ x \in R \tag{9.105}$$

Lemma 9.15   Let $u(x,t)$ be the classical solution to problems (9.104) and (9.105) and satisfy:

(i) $\lim\limits_{|x| \to \infty} (|u| + |u_x| + |u_{xx}|) (x,t) = 0, \ \forall t \geqslant 0$;

(ii) $f(s)$ is a real valued continuous function, and $f(s)s \geqslant 0$.

If $F(u) = \displaystyle\int_0^u f(s)ds$, then $F \geqslant 0$. We have

$$\|u(t)\| = \|\varphi\|, \ \forall t \geqslant 0$$

Proof   Multiplying (9.104) by u yields

$$\left(u^2\right)_t + \left(u^2\right)_{xxx} - \left(3\left(u_x\right)^2\right)_x - (2f(u)u)_x$$
$$+ (2F(u))_x - \left(u^2\right)_x = 0$$

Integrating x on interval $(-\infty, \infty)$ yields the lemma.

Theorem 9.12   Let $u(x,t)$ be the classical solution to problems (9.104) and (9.105), and satisfy (i) and (ii). We also assume:

$$f(u)u \geqslant F(u) \tag{9.106}$$

then there is

$$\int_0^\infty \int_{-r}^r \left(|u|^2 + |u_x|^2\right) dxdt < \infty, \ \forall r > 0 \tag{9.107}$$

If we further assume that there exists a normal number α such that

$$(1 - \alpha)f(u)u \geqslant F(u), \ 0 < \alpha < 1 \tag{9.108}$$

then there is

$$\int_{-r}^{r} |u|^2 dx \to 0, \ t \to \infty, \ \forall r > 0 \tag{9.109}$$

Proof   Set A is a function of x, and $A \in C^3$, then multiplying (9.105) by A to obtain

$$
\begin{aligned}
&\left(Au^2\right)_t + \left\{ A\left(u^2\right)_{xx} - A_x\left(u^2\right)_x + A_{xx}u^2 - 3A\left(u_x\right)^2 \right. \\
&\left. -2Af(u)^2 + 2AF(u) - Au^2 \right\}_x \\
&+ \left(-A_{xx} + A_x\right)u^2 + 3A_x\left(u_x\right)^2 \\
&+ 2A_x(f(u)u - F(u)) = 0
\end{aligned}
\tag{9.110}
$$

Further assuming that A satisfies $A_x > 0, -A_{xx} + A_x > 0$ and $|A|, |A_x|$ and $|A_{xx}|$ are bounded, obviously, this type of A is easy to find. Integrating (9.110) on $(-\infty, \infty) \times [0, T]$, from (9.106) and Lemma 9.15, we have

$$\int_0^\infty \int_{-r}^{r} \left(u^2 + u_x^2\right) dxdt < \infty \tag{9.111}$$

If it still satisfies (9.108) , then there is

$$\int_0^\infty \int_{-r}^{r} f(u)u(x,t)dxdt < \infty \tag{9.112}$$

We now use the concept of Morawetz[37] to prove

$$\int_{-r}^{r} u^2 dx \to 0, \ t \to \infty, \ \forall r \geqslant 0$$

Let $B(x) \in C_0^\infty(R), \ 0 \leqslant B(x) \leqslant 1$, and

$$B(x) = \left\{ \begin{array}{ll} 1, & |x| \leqslant r \\ 0, & |x| \geqslant 2r \end{array} \right.$$

then from (9.110), we have

$$\left| \int_{-2r}^{2r} Bu \cdot u_t dx \right| \leqslant C \int_{-2r}^{2r} \left[u^2 + u_x^2 + f(u)u\right] dx$$

If $0 < t_1 < t$, then

$$(t - t_1) \int_{-r}^{r} u^2(x, t)dx \leqslant (t - t_1) \int_{-2r}^{2r} Bu^2(x, t)dx$$

$$\leqslant \int_{0}^{t} \int_{-2r}^{2r} Bu^2(x, \tau)dxd\tau$$

$$+ 2 \int_{t_1}^{t} (\tau - t_1) \left| \int_{-2r}^{2r} Bu \cdot u_t(x, \tau)dx \right| d\tau$$

If $t_1 = t - 1$, then

$$\int_{-r}^{r} u^2(x, t)dx \leqslant C \int_{t-1}^{t} \int_{-2r}^{2r} \left[u^2 + u_x^2 + f(u)u \right] dxd\tau$$

From (9.111) and (9.112) , we can obtain

$$\int_{-r}^{r} u^2(x, t)dx \to 0, \ t \to \infty, \ \forall r > 0$$

Note   If $f(u) = u^p$, $p \geqslant 3$, p is an odd integer, it is easy to verify that condition (ii) of Theorem 1 and (9.106), (9.108) are all satisfied.

We are now considering two examples of "Blow up".

Example 1   Consider the initial value problem:

$$u_{tt} - u_{xx} = u^n \quad (n > 1) \tag{9.113}$$

$$u(x, 0) = u_0(x) \quad (x \in R) \tag{9.114}$$

$$u_t(x, 0) = v_0(x) \quad (x \in R) \tag{9.115}$$

It is easy to prove that there exists a local solution u for problems (9.113)-(9.115) if $u_0, v_0 \in C_0^\infty(R)$. We will prove that if u_0, v_0 is appropriately selected, then $F(t) = \int_R u^2(x, t)dx$ will tend towards infinity in finite time. Now it is possible to find $\alpha > 0$ and the initial values u_0, v_0, such that

$$(A) : \left(F(t)^{-\alpha} \right)'' \leqslant 0, \ t \geqslant 0$$

$$(B) : \left(F(t)^{-\alpha} \right)' < 0, \ t < 0$$

It is evident that $F(t)^{-\alpha}$ tends to zero within a finite time. Therefore, $F(t) \to \infty$, as shown in Fig. 9-1.

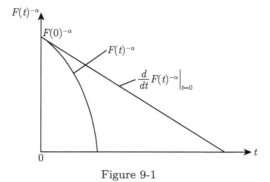

Figure 9-1

For condition (B), as long as u_0 and v_0 are selected with the same sign on $(-\infty, +\infty)$, they will automatically be satisfied since

$$\left(F(0)^{-\alpha}\right)' = -\alpha F(0)^{-1-\alpha} F'(0)$$

$$= -2\alpha F(0)^{-1-\alpha} \int u_0 v_0 dx < 0$$

Therefore, only condition (A) needs to be verified. In order to prove (A), it only needs to prove $Q(t) \geqslant 0$ since $F(t) \geqslant 0$. Here,

$$Q(t) = (-\alpha)^{-1} F^{a+2} \left(F^{-\alpha}\right)'' = F'' F - (\alpha + 1)\left(F'\right)^2$$

Since

$$F'(t) = 2 \int u u_t dx$$

$$F''(t) = 2 \int (u u_{tt} + u_t^2) dx$$

$$= 4(\alpha + 1) \int u_t^2 dx + 2 \int (u u_{tt} - (2\alpha + 1) u_t^2) dx$$

$$Q(t) = 4(\alpha + 1)t\{(\int u^2 dx)(\int u_t^2 dx) - (\int u u_t dx)^2\}$$

$$+ 2F(t)\left\{\int u u_{tt} dx - \int (2\alpha + 1) u_t^2 dx\right\}$$

The first term on the right-hand side of the above equation is positive by the Schwartz inequality, so it is only necessary to let $H(t) \geqslant 0$, here

$$H(t) = \int u u_{tt} dx - (2\alpha + 1) \int u_t^2 dx$$

$$-\int u^{n+1}dx + \int uu_{xx}dx - (2\alpha + 1)\int u_t^2 dx$$

$$= \int u^{n+1}dx - \int u_x^2 dx - (2\alpha + 1)\int u_t^2 dx$$

The energy conservation of (9.113) - (9.115) is

$$E(t) = \frac{1}{2}\int (u_t^2 + u_x^2)\, dx - \frac{1}{n+1}\int u^{n+1}dx$$

which means $E(t)$ is independent of t, therefore, if α is selected as

$$2(2\alpha + 1) = n + 1$$

we have

$$H(t) = -(n+1)E(t) + 2\alpha\int u_x^2 dx$$

$$= -(n+1)E(0) + 2\alpha\int u_x^2 dx \tag{9.116}$$

As a result, if $E(0) < 0$ and $\alpha = \dfrac{1}{4}(n-1) > 0$, then H is always strictly positive. Now selecting $u_0 \geqslant 0$, $v_0 \geqslant 0$, then condition (B) is satisfied. We multiply u_0 by a normal number such that $E(0) < 0$ (it is possible when $n + 1 > 2$), and for any such initial value, $F(t)$ tends to infinity in finite time. If we consider the equation

$$u_{tt} - u_{xx} = -u^n \tag{9.117}$$

it is easy to know that $H(t)$ still satisfies (9.116). If n is even and $u_0(x) \leqslant 0$, $v_0(x) \leqslant 0$, u_0 is sufficiently large such that $E(0) \leqslant 0$, then condition (B) is satisfied. Its solution also "Blow up" in a finite interval. On the other hand, n is odd because $E(t) \geqslant 0$. The conclusion is unclear, which is not surprising. For example, in the case of $-u^3$, we know that it also has a global solution.

Example 2   Considering the initial value problem of high-dimensional nonlinear Schrödinger equations

$$\begin{cases} iu_t = \Delta u + |u|^{p-1}u, & x \in R^n, t > 0 \\ u|_{t=0} = \varphi(x), & x \in R^n \end{cases} \tag{9.118}$$

We have the following results.

Theorem 9.13 *If the following conditions are satisfied*

$$\text{(i) } E(0) = \int_{R^n} \left(|\nabla\varphi|^2 - \frac{2}{p+1}|\varphi|^{p+1} \right) dx \leqslant 0$$

$$\text{(ii) Im} \int_{R^n} r\bar\varphi\varphi_r dx > 0, \text{ where } r^2 = |x|^2$$

$$\text{(iii) } P > 1 + \frac{4}{n}$$

then $\|\nabla u(t)\|_{L_2}$ and $\|u(t)\|_{L^\infty}$ tend to infinity in finite time.

   Note   Condition (ii) is also easy to verify, such as taking

$$\varphi(x) = e^{i|x|^2}\psi(x)$$

where $\psi(x)$ is any real valued function, then it can be directly calculated that

$$\text{Im} \int_{R^n} r\bar\varphi\varphi_r dx = 2\int r^2|\psi|^2 dx > 0$$

For the "Blow up" phenomenon of solutions to other nonlinear evolutionary equations and the asymptotic properties of $t \to \infty$, please refer to [36] − [44].

9.7 Well-Posedness Problems for the Zakharov System and Other Coupled Nonlinear Evolutionary Systems

In the study of solitons in plasma physics, Zakharov gives an important system of equations for the interaction between laser and plasma, commonly known as the Zakharov equation system. That is

$$\frac{\partial^2 n}{\partial t^2} - \frac{\partial^2 n}{\partial x^2} - \frac{\partial |\varepsilon|^2}{\partial x^2} = 0 \tag{9.119}$$

$$i\varepsilon_t + \varepsilon_{xx} - n\varepsilon = 0 \tag{9.120}$$

where n represents the disturbance (fluctuation) of ion density. It is a real valued function of variables x,t, where ε represents the electric field which is a complex valued function of variables x,t. Zakharov found soliton solutions for (9.119), (9.120) and studied the characteristics of these solitons. We are now studying it based on differential equations. For this reason, the potential function φ is introduced to transform (9.119) into a system of equations:

$$\frac{\partial n}{\partial t} - \frac{\partial^2\varphi}{\partial x^2} = 0 \tag{9.121}$$

$$\frac{\partial \varphi}{\partial t} - (n + |\varepsilon|^2) = 0 \tag{9.122}$$

We discuss the periodic initial value problem of equations (9.120) - (9.122), which is to find the solution $\varepsilon(x,t)$, $\varphi(x,t)$ of the system of equations (9.120)-(9.122) with a period of 2π for x to satisfy the initial conditions

$$n(x,0) = n_0(x), \ \varphi(x,0) = \varphi_0(x)$$
$$\varepsilon(x,0) = \varepsilon_0(x) \ (-\infty < x < \infty) \tag{9.123}$$

here $n_0(x)$, $\varphi_0(x)$, $\varepsilon_0(x)$ are all functions with a period of 2π.

We use the Galerkin method to construct approximate solutions for problems (9.120)-(9.122) and make prior estimates of the approximate solutions, thus the following theorems are given.

Theorem 9.14   *If $\varepsilon_0(x) \in H^6, \varphi_0(x) \in H^4, n_0(x) \in H^4$ are functions with a period of 2π, then local classical solutions to problems (9.120)-(9.122) exist.*

Based on prior estimation, we can extend the local solution to a large range, thereby the global solution is obtained. We have

Theorem 9.15   *If the conditions of Theorem 9.14 are satisfied, then the global classical solution to problems (9.120) - (9.122) exists.*

If we further improve the smoothness of the initial value function, we can obtain the following smooth solution.

Theorem 9.16   *If $\varepsilon_0(x) \in H^8, \varphi_0(x) \in H^6, n_0(x) \in H^6$ are functions with a period of 2π, the global smooth solution (the second-order derivative solution of t) of problems (9.120) - (9.122) exists and is unique.*

With the above results, it is easy to study the periodic initial value problem of the Zakharov equation: to find $n(x,t), \varepsilon(x,t)$ for x with periodic of 2π for (9.119), (9.120), so that they satisfy the initial conditions

$$n(x,0) = n_0(x), \ \frac{\partial n}{\partial t}(x,0) = n_1(x)$$

$$\varepsilon(x,0) = \varepsilon_0(x) \tag{9.124}$$

where $n_0(x)$, $n_1(x)$, $\varepsilon_0(x)$ are functions with a period of 2π.

Theorem 9.17   *If $n_0 \in H^6$, $n_1 \in H^4$, $\varepsilon_0 \in H^8$ are functions with a period of 2π, then the global classical solution for problems (9.119), (9.120) and (9.124) exists uniquely.*

For the Cauchy problem of a class of KdV-nonlinear Schrödinger coupled equations

$$i\varepsilon_t + a\varepsilon_{xx} - bn\varepsilon = 0 \tag{9.125}$$

$$n_t + \frac{1}{2} \left[\beta n_{xx} + n^2 + |\varepsilon|^2 \right]_x = 0 \tag{9.126}$$

$$\varepsilon|_{t=0} = \varepsilon_0(x), \quad n|_{t=0} = n_0(x) \ (-\infty < x < \infty) \tag{9.127}$$

we have the following results:

Theorem 9.18 *If (i) $\varepsilon_0(x)$, $\eta_0(x) \in H^s(s \geqslant 3)$ (ii) constant coefficients a and $b\beta$ have opposite signs, then the global solution to the Cauchy problem and periodic problem of the system of equations (9.125), (9.126) exists uniquely, and its solution $n(x,t)$, $\varepsilon(x,t) \in L^\infty(0,T;H^s)$.*

As for the global solution of the coupled equation systems of other nonlinear evolutionary equations, please refer to [43 − 45].

References

[1] A. Sjöberg. Numerical solution of the Korteweg de Vries equation[M], Rep., NR 25, Dept. of Computer Sci., Uppsala Univ., Sweden, 1969.

[2] Sjöberg A. On the Korteweg-de Vries equation: existence and uniqueness[J]. Journal of Mathematical Analysis and Applications, 1970, 29(3): 569-579.

[3] Lax P D. Integrals of nonlinear equations of evolution and solitary waves[J]. Communications on Pure and Applied Mathematics, 1968, 21(5): 467-490.

[4] Temam R. Sur un probleme non linéaire[J]. J. Math. Pures Appl, 1969, 48(2): 159-172.

[5] Rühs F. JL Lions, Équations Différentielles Opérationnelles et Problèmes aux Limites. IX+ 292 S. Berlin/Göttingen/Heidelberg 1961. Springer-Verlag. Preis geb. 64,—[J]. Zeitschrift Angewandte Mathematik und Mechanik, 1962, 42(7-8): 363-364.

[6] Lions J L. Quelques Méthodes de Résolution des Problèmes aux Limites Non-Linéaires[J]. Dunod, 1969.

[7] Bona J L, Smith R. The initial-value problem for the Korteweg-de Vries equation[J]. Philosophical Transactions of the Royal Society of London. Series A, Mathematical and Physical Sciences, 1975, 278(1287): 555-601.

[8] Dushane T E. On existence and uniqueness for a new class of nonlinear partial differential equations using compactness methods and differential difference schemes[J]. Transactions of the American Mathematical Society, 1974, 188: 77-96.

[9] 郭柏灵. 一类更广泛的 KdV 方程的整体解 [J]. 数学学报, 1982, 25(6): 641-656.

[10] 管志成. 一类更广泛的 KdV 方程解的存在性与唯一性 [J]. 浙江大学学报, 1979(04): 83-95.

[11] Николенко Н В. Инвариантные, асимптотически устойчивые торы возмущенного уравнения Кортевега-де Фриза[J]. Успехи математических наук, 1980, 35(5 (215): 121-180.

[12] 郭柏灵. 一类 KdV 非线性 Schrödinger 组合微分方程组周期初值问题和柯西问题整体解的存在性唯一性 [J]. 数学学报 1983, 26(05): 513-532

[13] Ton B A. Initial boundary-value problems for the Korteweg-de Vries equation[J]. Journal of Differential Equations, 1977, 25(3): 288-309.

[14]   周毓麟, 郭柏灵. 高阶广义 Korteweg-de Vries 型方程组的周期边界问题与初值问题 [J].
        数学学报 1984, 27(02): 154-176

[15]   Lin J E. Asymptotic behavior in time of the solutions of three nonlinear partial
        differential equations[J]. Journal of Differential Equations, 1978, 29(3): 467-473.

[16]   Strauss W A. Dispersion of low-energy waves for two conservative equations[J].
        Archive for Rational Mechanics and Analysis, 1974, 55: 86-92.

[17]   B. L. Guo, Proceedings of the 1980 Beijing Symposium on Differential Geometry and
        Differential Equations[M], Science Press, 1982, 3: 1227-1246

[18]   Guo Bailing. The Global Solution and Blow up Phenomenon for a Class of System of
        Nonlinear Schrodinger Equations with the Magnetic Field Effect[J]. Chinese Annals
        of Mathematics B,1985,6(3): 281288

[19]   郭柏灵. 一类具积分型非线性 Schrödinger 方程组的初值问题 [J]. Acta Mathematica
        Scientia, 1981, 1(Z1): 261-274.

[20]   Y. L. Zhou, B. L. Guo. On the solvability of the Inifial Value Problem for the
        Quasilinear Degenerate Parabolie System, In Proceedings of the 1982 Changchun
        Symposium on Differential Geometry and Differential Equations, Science Publishers,
        Inc.

[21]   郭柏灵. 某些非线性 Schrödinger 方程组柯西问题当 $t \to \infty$ 时解的渐近性质 [J]. 浙江
        大学学报 (工学版), 1983, 000(002): 23.

[22]   Ginibre J, Velo G. On a class of nonlinear Schrödinger equations. II. Scattering theory,
        general case[J]. Journal of Functional Analysis, 1979, 32(1): 33-71.

[23]   Y. L. Zhou, G. L. Guo, The Existence of weak solution of the boundary problem for the
        Systems of FerroMagnetic Chain, In Proceedings of the 1982 Changchun Symposium
        on Differential Germetry and Differential Equations, Science Publishers, Inc.

[24]   Pecher H, von Wahl W. Time dependent nonlinear Schrödinger equations[J]. manuscri-
        pta mathematica, 1979, 27(2): 125-157

[25]   Tsutsumi M, Hayashi N. Classical solutions of nonlinear Schrödinger equations in
        higher dimensions[J]. Mathematische Zeitschrift, 1981, 177(2): 217-234.

[26]   Reed M, Simon B. Methods of modern mathematical physics, 2. Fourier Analysis,
        Self-Adjointness[M]. New York, London: Academic Press, 1972.

[27]   Segal I. Non-linear semi-groups[J]. Annals of Mathematics, 1963, 78(2): 339-364.

[28]   Gross L. The Cauchy problem for the coupled Maxwell and Dirac equations[J].
        Communications on Pure and Applied Mathematics, 1966, 19(1): 1-15.

[29]   Chadam J M. On the Cauchy Problem for the Coupled Maxwell-Dirac Equations[J].
        Journal of Mathematical Physics, 1972, 13(5): 597-604.

[30]   郭柏灵. INITIAL VALUE PROBLEM AND PERIODIC BOUNDARY PROBLEM
        FOR SOME SYSTEMS OF MULTI-DIMENSIONAL AND NONLINEAR SCHR
        DINGER EQUATION OF HIGH ORDER[J]. 科学通报: 英文版, 1982 (9): 915-920

[31]   Guo Boling. The global solution for one class of generalized KdV equation[J]. Acta
        Math. Sinica, 1982, 25(6): 641-656

[32]   Guo Boling, Shen Long-jun. THE EXISTENCE AND UNIQUENESS OF THE
        CLASSICAL SOLUTION ON THE PERIODIC INITIAL VALUE PROBLEM FOR

3AXAPOB EQUATION. Acta Mathematicae Applicatae Sinica, 1982, 5(3): 310-324

[33] Guo B. The global solutions of some problems for a system of equations of Schrödinger-Klein-Gordon field[J]. Sci. Sin., Ser. A 1982, 25 (9): 897-910

[34] Guo B. Initial boundary value problem for one class of systems of multidimensional nonlinear Schrödinger equations with wave operator[J]. Sci. Sin., Ser. A 1983, 26 (6): 561-575.

[35] 郭柏灵. 一类 KdV 非线性 Schrödinger 组合微分方程组周期初值问题和柯西问题整体解的存在性唯一性 [J]. 数学学报, 1983(05): 513-532

[36] Medeiros L A, Miranda M M. Weak solutions for a nonlinear dispersive equation[J]. Journal of Mathematical Analysis and Applications, 1977, 59(3): 432-441.

[37] Morawetz C S. Time decay for the nonlinear Klein-Gordon equation[J]. Proceedings of the Royal Society of London. Series A. Mathematical and physical sciences, 1968, 306(1486): 291-296.

[38] Keller J B. On solutions of nonlinear wave equations[J]. Communications on Pure and Applied Mathematics, 1957, 10(4): 523-530.

[39] Levine H A. Some nonexistence and instability theorems for solutions of formally parabolic equations of the form Put=- Au+ $\mathcal{F}$ (u)[J]. Archive for Rational Mechanics and Analysis, 1973, 51(5): 371-386.

[40] John F. Blow-up of solutions of nonlinear wave equations in three space dimensions[J]. Manuscripta mathematica, 1979, 28(1): 235-268.

[41] Fujita H. On the blowing up of solutions of the Cauchy problem for $u_t = \Delta u + u <$ $1 + \sigma$[J]. J. Fac. Sci. Univ. Tokyo, 1966, 13: 109-124.

[42] Y. L. Zhou, G. L. Guo, The Periodic Boundary Problems and the Initial Value Problems for the systems of the Generalized korteweg-de Vries Type of High order, In Proceedings of the 1982 Changchum Symposium on Differential Geometry and Differential Equations, Science Press.

[43] Y. L. Zhou, Guo B Existence of weak solution for boundary problems of systems of ferro-magnetic chain[J]. Science in China, Ser. A, 1984.

[44] G. L. Guo, Some Problems for a Wide System of Zakharov Equations. In Proceedings of the 1982 ChangchumSymposiumonDifferentia Geometry andDifferential Equations, Science Press.

[45] 周毓麟, 符鸿源. 广义 Sine-Gordon 型非线性高阶双曲方程组 [J]. 数学学报, 1983(02): 234-249.

Chapter 10
Topological Solitons and Non-topological Solitons

10.1   Solitons and Elementary Particles

In this chapter, we study the motion of solitons in elementary particle physics. Although the discovery of solitary waves dates back more than three hundred years, this problem has only been brought to attention in recent years. The relationship between solitons and some observed elementary particles is worth in-depth research. Can solitons be equated to certain elementary particles, or do they play different roles? Since solitons originate from wave phenomena, it is generally believed that they are different from electrons and protons. Therefore, can they describe particles? There is still some controversy over this. To gain a preliminary understanding of this issue, we must trace the process of understanding elementary particles. About half a century ago, it was discovered that microscopic particles exhibit wave-particle duality. This was first observed in the phenomenon of light. Originally, light was considered to be an electromagnetic wave. According to this theory, the energy of light should be continuously distributed over the light wave. However, experiments such as the "photoelectric effect" and "Compton scattering" showed that the energy of light is discontinuous. For monochromatic light, each minimum energy portion is $h\nu$ (where ν is the frequency of the monochromatic light and h is Planck's constant), thus the concept of "photon" was born. Building on Planck's and Einstein's quantum theory of light, L. De Broglie later proposed that microscopic particles like electrons and protons also have wave properties, similar to photons. This wave nature can be clearly demonstrated in their propagation and other behaviors and is linked to particle properties via the relations $P = h/\lambda$ and $E = \hbar\nu$ (where P and E are the momentum and energy of the particle, and λ and ν are the wavelength and frequency of the wave). De Broglie's ideas were quickly confirmed by experimental physicists through diffraction patterns observed when electron beams passed through a crystal foil(as X-rays through the crystal). Later, similar experiments with molecular and atomic beams yielded the same results, demonstrating the wave-particle duality of microscopic particles. Quantum field theory is a scientific discipline that comprehensively describes these dual properties of microscopic particles.

Due to this duality of microscopic particles, when solitons were discovered and recognized to have both wave properties and the ability to maintain their shape and size, resembling particles, it was natural to introduce them into the study of microscopic particles. In fact, as early as the early 1930s, Born attempted to describe electrons by adding nonlinear correction terms to Maxwell's equations, resulting in what are known as Born-Infeld equations. Treating elementary particles as localized singularities of nonlinear field equations was a fascinating idea, but this theory could not incorporate some of the successful conclusions from wave mechanics and thus did not see much development for many years. Later, L. De Broglie developed this theory based on his "double solution theory" to attempt a localized description of microscopic particles, but the theory had some fatal weaknesses and did not progress much.

On the other hand, many experimental facts show that elementary particles have internal structures, such as the widely accepted conclusion that hadrons are composed of quarks or partons. Despite strong efforts, direct evidence of free quarks has not been found. In particular, the analysis of deep inelastic scattering experiments indicates that the mass of quarks inside hadrons is very small (about tens to hundreds of MeV). Considering such mass, quarks should have been detected in accelerator experiments long ago. Coordinating these two contradictory experimental facts leads to the belief that the interactions between the partons that make up hadrons make it difficult or impossible for individual quarks to escape from hadrons, a phenomenon known as "quark confinement."

In the study of "quark confinement," one possibility is that the mass of free quarks is very large, and be much greater than that of hadrons. Therefore, forming hadrons requires releasing a huge amount of binding energy, and separating quarks in hadrons would also require immense energy, referred to as "partial confinement." Another possibility is that quarks can never escape from hadrons individually, known as "permanent confinement." In 1974-1975, string models and bag models were proposed to phenomenologically describe quark "permanent confinement." The bag model, including the MIT bag and the SLAC bag, The two models have different starting points, but they both consider quarks to be confined in something resembling a bag. These so-called bags can be regarded as a certain form of solitons. The authors of the bag model compare quarks to small insects that like to burrow into sand and cover themselves. Solitons provide these insects (quarks) with an appropriate hiding place, or the insects can only move freely near the bag wall, requiring enormous energy (over 10 MeV) to move away from it, thus preventing the observation of free quarks.

In the "string model," hadrons are considered to be strings with quarks always "stuck" at the ends. The model uses a superconducting analogy: considering the vacuum outside hadrons as a kind of "superconducting phase," while the superconducting string inside hadrons corresponds to "magnetic flux lines," similar to how a magnetic field cannot penetrate a type-II superconductor but is confined in flux tubes. Similarly, the gluon field between quarks is confined in a string by the vacuum superconducting phase. Some speculate that the vacuum might be filled with many magnetic monopole pairs, transforming the vacuum phase into a special "confinement phase" that provides some vacuum pressure to confine the gluon field inside hadrons, like liquid squeezing a gas bubble.

As for magnetic monopoles, they were first proposed by Dirac in 1931. In 1974, G. 't Hooft discovered that these magnetic monopoles are also solutions to nonlinear equations, fitting the definition of solitons, making them another type of soliton.

From this perspective, the introduction of solitons is very helpful for solving some fundamental issues in elementary particle physics. Therefore, in recent years, people have been treating certain nonlinear equations as classical approximations of localized quantum field equations, deriving their soliton solutions, and studying their properties. Since elementary particle theory should be relativistically covariant, Lorentz-invariant equations like the sine-Gordon equation and the φ^4 field equation (i.e., the Higgs field equation) are most suitable for elementary particle theory. Researchers such as T. D. Lee have conducted in-depth studies on soliton motion in elementary particle physics. Their research suggests that quark fields are fundamental fields but not the lowest energy state, and soliton states are the lowest energy states. Thus, the hadrons observed in nature are actually a kind of soliton. Using this theory, they have explained the "quark confinement" problem. Recently, they proposed that as long as the vacuum is an ideal color dielectric, hadrons can be considered as solitons, explaining the "quark confinement" problem without necessarily introducing Higgs bosons. They have also classified existing solitons into topological solitons and non-topological solitons.

So, what are topological solitons and non-topological solitons? Their research indicates that in renormalizable relativistic localized field theory, all solitons, especially stable ones, must satisfy the Euler-Lagrange field equations derived from the Hamiltonian principle of least action $\delta S = 0$, and must also meet the stability requirement $\delta^2 S > 0$. There are two ways to achieve this. The difference between these methods divides solitons into topological and non-topological solitons. A necessary condition for the stable existence of topological solitons is the presence of degenerate vacuum (ground state) states. Therefore, different degenerate vacuum states (ground states) can exist at infinity in space, with different boundary conditions. If a soliton solution exists, the boundary conditions at infinity are different from those without a soliton solution. These inconsistent boundary conditions can be represented by different topologies. Introducing the quantum number of topological charge allows for the determination of its stability based on the conservation of topological charge.

Non-topological solitons, on the other hand, do not require degenerate vacuum states (ground states). All solutions, whether solitons exist or not, have the same boundary conditions at infinity. Bell-shaped solitons belong to this category. However, systems with non-topological soliton solutions must have an additive conservation law and a scalar field. Therefore, non-topological soliton solutions are universal and can exist in any dimensional space.

The most important application of topological solitons is in Quantum Chromo-dynamics (QCD). QCD is a theory of strong interactions developed in recent years. Its foundation is non-Abelian gauge theory. The field equations of non-Abelian gauge theory are all nonlinear and difficult to solve. Currently, topological soliton solutions of non-Abelian gauge fields have been found only in some special cases. For example, there are knot solutions in one-dimensional space (1D), vortex solutions in two-dimensional space (2D), G. 't Hooft's magnetic monopole solutions in three-dimensional space (3D), and instanton solutions in the four-dimensional Euclidean space pointed out by Polyakov. For both physical and mathematical considerations, this chapter mainly studies topological and non-topological solitons. Because topological solitons can help solve the annoying $UA(1)$ symmetry problem of local gauge invariance in QCD, possibly form the topological confinement picture of quarks and the so-called "mechanical color blindness" (i.e., all observed hadron states are colorless) topological scheme in QCD. As a static source-free solution of a non-Abelian gauge field, it provides an example of a singularity-free, self-sustaining stable organic unity. It might also offer dynamic explanations for quantum numbers such as baryon number, isospin, strangeness, and charm, among others. Moreover, there might exist real physical magnetic monopole solitons, among others. Lastly, since solitons are non-trivial solutions of the corresponding classical fields, we have reason to believe that such solutions dominate the contributions in the functional integral. Therefore, solitons may play an important role in calculating functional integrals.

From a mathematical perspective, solitons also play a significant role in the mathematical development of quantum field theory. Since the behavior of soliton amplitudes is approximately inversely proportional to the coupling constant, they might provide theoretical clues beyond perturbative content. The topological stability of solitons arises from the dynamic mechanisms and topological properties of field configurations. For non-topological solitons, stability is dynamically achieved through amplitude variations over time, essentially reflecting the conserved quantities given by Noether's theorem—Noether charges. Topological solitons are associated with topological conservation laws unrelated to the invariance of the Lagrangian function in non-Abelian theories. These laws stem from the intrinsic degrees of freedom of the field, and their topological stability encompasses some topological content, i.e., the manifold of the intrinsic space can be non-trivially mapped to the manifold of the real D-dimensional space of the theory. Therefore, researching these issues will promote the development of homotopy theory and fiber bundle theory in topology. Conversely, to understand topological and non-topological solitons deeply, one should also grasp some fundamental knowledge of related topology or homotopy theory. To this end, we will briefly introduce relevant knowledge in this area in the next section.

(Note: For convenience, natural units are used throughout this chapter.)

10.2   Preliminary Topological and Homotopy Theory

Firstly, let's understand the concept of manifolds. When studying nature, it is essential to extend calculus from a local theory to a broader one. Poincaré, through his research in celestial mechanics, first noted this problem and initially formed the concept of manifolds. He defined it in n-dimensional space using p equations, where the set of points satisfying these equations is a manifold. Thus, the movement of an $(n-1)$-dimensional manifold results in an n-dimensional manifold. In 1944, Chern gave an intrinsic proof of the Gauss-Bonnet theorem for differentiable manifolds, which significantly increased interest in differentiable manifold theory. The concept of a manifold is an extension of the surface concept to a global and abstract level. It can be seen as a generalization of the concept of curves or surfaces, and the imagery of surfaces is the key to understanding manifold theory.

When introducing parameter curves on a surface, one can intuitively imagine pasting a coordinate grid onto the surface. In studying the properties of surfaces, we generally cannot expect to cover the entire surface with a single coordinate system but instead use several coordinate neighborhoods to cover different parts of the surface. In regions where two or more coordinate neighborhoods overlap, coordinate transformations allow different coordinates of the same point to be interchanged. If one wishes to exclude certain troublesome points or regions from consideration, coordinate neighborhoods can be arranged to bypass these points or regions. Extending this idea, by replacing the surface with an abstract topological space with some restrictions and the coordinates from two real numbers to n real or complex numbers, and adding appropriate differentiability requirements, we obtain the concept of an n-dimensional differentiable manifold. That is, in our method of overlapping, if it is smooth, it is called a differentiable manifold. If it is angular, it is called a piecewise-linear manifold; if it is not very smooth but still maintains continuity, it is a topological manifold. In this way, manifolds look similar locally, differing only in the way they overlap. For example, a one-dimensional closed manifold is a circle. Additionally, differentiable manifolds can be equipped with various geometric structures under different conditions, and each differentiable manifold can globally be equipped with a positive definite Riemannian metric. A manifold equipped with a positive definite Riemannian metric is called a Riemannian manifold, and the global properties of a manifold often manifest as topological properties or are related to topological properties to varying degrees. Thus, the study of manifolds is the primary objective of differential topology, aiming to determine whether two given manifolds are the same or different, leading to classification, which can be topological or more refined.

So, what is topology? Here we define it by referring to the properties of open sets in metric spaces[1-3]. Let S be a set, and τ be a collection of subsets of S, whose members are called open sets of S (an open set is defined as follows: let A be a subset of a metric space X; if every point of A has a spherical neighborhood within A, then A is called an open set of X). If τ satisfies the following conditions: (1) S and the empty set are open sets; (2) the intersection of two open sets is an open set; (3) the union of any number of open sets is an open set, then τ is called a topology of the set S, and the pair$(S, t$ is called a topological space. Topology is the study of properties of shapes that remain invariant under continuous deformations.

An important concept in topology is mapping. If X and Y are both metric spaces, $f : X \to Y$ is a correspondence from X to Y, and for a point $x_0 \in X$, if for any neighborhood $U(f(x_0), \epsilon)$, there exists a neighborhood $x \in U(x_0, \delta)$ such that $f(x) \in U(f(x_0), \epsilon)$, then f is said to be continuous at the point x_0. If f is continuous at every point in X, then f is continuous on X, and $f : X \to Y$ is called a continuous mapping, or simply a mapping. X and Y are called the domain and codomain of f, respectively. If $f(X) = Y$, then f is a surjective mapping from X to Y.

If $f : X \to Y$ is a continuous surjective mapping from X to Y, and the inverse mapping $f^{-1} : Y \to X$ is also continuous, then f is called a topological mapping or homeomorphism. If there exists a topological mapping f, then spaces X and Y are called homeomorphic, denoted as $X \approx Y$. Properties of metric spaces that are invariant under any topological mapping are called topological properties. Quantities that remain invariant under continuous deformation (topological mapping) are called topological invariants. For example, for any polyhedron, regardless of its deformation, the number of vertices (V), edges (E), and faces (F) satisfy the invariant $V - E + F = 2$. For general cases, dividing a shape M into simple polyhedra gives the Euler-Poincar characteristic K, which is also a topological invariant and is very important. When a shape is deformed into another under a topological mapping, this topological invariant or characteristic number, if it is an integer, is often called a topological quantum number or topological charge. According to de. Rham's theory, it is defined by characteristic classes and characteristic forms. Chern believed that the $2K$-th characteristic form determines a cohomology class of dimension $2K$, hence it is called a characteristic class. From the perspective of differential geometry, characteristic forms depend on connections, but characteristic classes depend only on bundles. They are the simplest global invariants on bundles, emphasizing local properties, and characteristic forms contain more information than characteristic classes. When N is an oriented compact manifold, the integral of the highest-dimensional characteristic class (equal to the dimension of N) gives the characteristic number. When it is an integer, it is called a topological quantum number or topological charge, which is very important in topological solitons. For a physically existing topological soliton, there must be a topological charge, i.e., an internal degree of freedom of the field. It can be discrete, like inversion, or continuous, like isotropy, to ensure the stability of the soliton. The main idea behind topological stability is that the manifold of the internal field space may have a nontrivial mapping on the manifold of the real D-dimensional space of the theory, maintaining the soliton's stable shape (when the topological charge is nonzero). A nontrivial mapping is one that cannot be continuously deformed into a trivial mapping, where all points of one manifold are mapped to a single point of another (when the topological charge is zero). This mapping is directly related to homotopy.

However, homotopy is another extremely important concept in topology. It has an intuitive meaning and describes a fundamental property of topological spaces. For example, when X is the circle S^1, naturally, $f_0, f_1 : S^1 \rightarrow Y$ are called two closed paths in Y. The intuitive way to say that f_0 is homotopic to f_1 is that the closed path f_0 can be continuously deformed into the closed path f_1 in Y. The correct definition is:

Let $f_0, f_1 : X \rightarrow Y$ be maps between the topological spaces X and Y. If there exists a map $F : X \times I \rightarrow Y$ from the cylinder $X \times I$ (where I is the unit closed interval $0 \leqslant I \leqslant 1$ on the real number axis) to Y, such that

$$F(x, 0) = f_0(x), \quad F(x, 1) = f_1(x) \tag{10.1}$$

For any $x \in X$, we say that f_0 and f_1 are homotopic, i.e., f_0 and f_1 can be transformed into each other through a continuous deformation F. This is denoted as:

$$f_0 \simeq f_1 : X \rightarrow Y. \tag{10.2}$$

Such an F is called a homotopy or a homotopic transformation from f_0 to f_1. It can be proven that, in the set Y^X of all mappings from X to Y, the homotopy relation is an equivalence relation, i.e., if $F : f_0 \simeq f_1$ and $G : f_1 \simeq f_2$, then there exists an $H : f_0 \simeq f_2$.

$$H(x, t) = \begin{cases} F(x, 2t), & 0 \leqslant t \leqslant \dfrac{1}{2} \\ G(x, 2t - 1), & \dfrac{1}{2} < t \leqslant 1 \end{cases} \tag{10.3}$$

Thus, the set of homotopy equivalent mappings forms a class, denoted as $\{f\}$. Therefore, the set of mappings from X to Y, denoted as Y^X, can be divided into several equivalence classes under the homotopy relation. If the equivalence classes satisfy the addition law $\{f + g\} = \{g + f\}$, then it can be proven that the homotopy classes $\{f\}$ can form a homotopy group (which is essentially an invariant of the homotopy classes). To this end, take the space X to be a closed line segment $I = \{0, 1\}$ with endpoints identified. This manifold is topologically equivalent to a circle S^1. A reference point X on the boundary of the circle is identified with 0 and 1. If we only consider continuous mappings f that satisfy $f(0) = f(1) = g_0$, where y_0 is a fixed point in Y, then the equivalence classes of mappings $\{f\}, \{g\}$ from S^1 to Y form a group. The identity element $\{e\}$ is the homotopy class of the constant mapping C, as

$$C(x) = y_0, \forall(x) \tag{10.4}$$

and the inverse of $\{f\}$ is $\{f^{-1}\}$, where

$$f^{-1}(x) = f(1 - x) \tag{10.5}$$

The group multiplication is defined by the composability of the mappings: $\{f\} * \{g\} = \{f \cdot g\}$.

$$t \cdot g(x) = \begin{cases} f(2x), & 0 \leqslant x \leqslant \dfrac{1}{2} \\ g(2x - 1), & \dfrac{1}{2} < x \leqslant 1 \end{cases} \tag{10.6}$$

And this multiplication procedure is independent of the choice of representative mappings from $\{f\}$ and $\{g\}$ because

$$f_1 \sim f_2 \text{ and } g_1 \sim g_2 \rightarrow f_1 \cdot g_1 \sim f_2 \cdot g_2 \tag{10.7}$$

The above type of homotopy group is called the first homotopy group of Y, or the fundamental group, often denoted as $\pi_1(Y)$. It is also a topological invariant. The π_1 of a sphere contains only the zero element, which means all mappings from S^1 to S^2 are homotopic to a constant map. Shanker [4] provided an example of this group for $Y = \mathbb{R}^2 - (0, 0)$ (the Euclidean plane with the origin removed) and for y_0 being any point on the plane. Mappings from $[0, 1]$ or S^1 to Y can be represented by loops starting and ending at y_0 on the plane, as shown in Figure 10-1. Here, loops that do not encircle $(0, 0)$ can be shrunk to y_0, hence they belong to the identity element $\{e\} = \{0\}$. A loop encircling the origin once in a clockwise direction belongs to a class denoted by $\{1\}$. Encircling twice belongs to $\{2\}$, and so on, with encircling n times belonging to $\{n\}$ (counterclockwise encircling belongs to classes $\{-1\}, \{-2\}, \ldots$). Thus, this homotopy group can be denoted as

$$\pi_1(R^2 - (0, 0)) = Z.(The\ set\ of\ integers) \tag{10.8}$$

The group operation becomes ordinary addition, and the number n associated with $\{n\}$ is called the winding number (as shown in Figure 10-1).

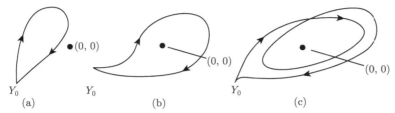

Figure 10-1   For a surface with a small part removed, different winding numbers 0, 1, and 2 represent topologically inequivalent mappings

The first homotopy group is an important tool for studying the global properties of compact Lie groups. The invariant $\pi(G)$ (which classifies mappings from S^1 to the set of elements in the group G) measures the connectivity properties of the group. After compactification, Lie groups are classified according to their local properties (Lie algebras). $\pi(G)$ is the only invariant used to classify groups. For example: $\pi(SU(2)) = 0$ (i.e., it contains only the identity element).

$$\pi_1(O(3)) = \pi_1(SU(2)/Z_2) = Z_2(The\ set\ of\ integers\ congruent\ to\ 2) \qquad (10.9)$$

That is, $SU(2)$ is simply connected, whereas $O(3)$ is doubly connected, reflecting the 2-to-1 homomorphism between $SU(2)$ and $O(3)$.

The first homotopy group can be used to determine whether Nielsen-Olesen vortices [5] exist in a specific gauge model, and if they exist, what kind of quantized fluxes are allowed. This will be discussed in detail below.

If we generalize from $X = S^1$ to $X = S^n$ (an n-dimensional sphere), or to something topologically equivalent to S^n like $I^n = \{(x_1, x_2, \ldots, x_n) \in \mathbb{R}^n \mid_{0 \leqslant x_i \leqslant 1}\}$ (an n-dimensional cube), whose boundary (faces) are identified and equivalent to the north pole of S^n, then the class of mappings with a fixed point $f(x_0) = y_0$ forms a group called the n-th homotopy group, denoted by $\pi_n(Y)$.

Table 10.1

Y	$U(1)$	$SU(2)$	$N > 3$ $SU(N)$	$SO(3)$	$SO(4)$	$SO(5)$	$SO(6)$	$N > 7$ $SO(N)$	$SP(N)$
$\pi_1(Y)$	Z	0	0	Z_2	Z_2	Z_2	Z_2	Z_2	0
$\pi_2(Y)$	0	0	0	0	0	0	0	0	0
$\pi_3(Y)$	0	Z	Z	Z	ZZ	Z	Z	Z	Z
$\pi_4(Y)$	0	Z_2	0	Z_2	$Z_2 Z_2$	Z_2	0	0	Z_1
$\pi_5(Y)$	0	Z_2	Z	Z_2	$Z_2 Z_2$	Z_2	Z	0	Z_2

For any space Y, determining the various $\pi_n(Y)$ is much more complex, and will not be discussed further here. Table 10-1 lists some results that may be related to physics [6]. Some are straightforward, while others are not. For example, several results come from $\pi_n(S^n) = Z$. This indicates that the class of mappings $f : S^n \to S^n$ is characterized by how many times one n-dimensional sphere covers another n-dimensional sphere. This number is called the wrapping number. For $SU(2)$ or higher-rank Yang-Mills theory, this is an extremely important result. Because ordinary space $\mathbb{R}^3$, when all points at infinity are identified, is equivalent to S^3. At the same time, all maps of the form:

$$\begin{pmatrix} a + ib & c + id \\ -c + id & a - ib \end{pmatrix}$$

The group $SU(2)$, consisting of 2×2 special unitary matrices such that $a^2 + b^2 + c^2 + d^2 = 1$, is also equivalent to S^3. Therefore, $\pi_3(SU(2)) = \pi_3(S^3) = Z$. This implies that the Yang-Mills theory has infinitely many topologically distinct vacua.

Additionally, if G is a simply connected compact Lie group, and H is a subgroup of G, then

$$\pi_2(G/H) = \pi_1(H) \tag{10.10}$$

where G/H is the left coset space. In the original model by G.'t Hooft and Polyakov [7],

$$\begin{cases} G = SU(2), H = U(1) \text{ (the f group keeps the vacuum invariant)} \\ \pi_2(SU(2)/U(1)) = \pi_1(U(1)) = Z. \end{cases} \tag{10.11}$$

This implies that there are infinite possibilities for magnetic monopoles, although only one has been found so far. Therefore, the study of the homotopy group π_2 is very useful in determining whether non-Abelian theories can have finite-energy magnetic monopole solutions.

Of course, there are many applications of homotopy theory in physics, some of which are still not very clear and require further research.

10.3 Topological Solitons in One-Dimensional Space

First, describe Derrick's theorem [7], as

$$\mathcal{L}(x) = \frac{1}{2} \partial_\mu \varphi \partial^\mu \varphi - U(\varphi) \tag{10.12}$$

(Where $U(\varphi) > 0, U(\varphi) = 0$ corresponding to the vacuum state) described scalar field theory, except for the case where the spatial dimension $D = 1$, there is no static (time-independent) non-singular solitary wave solution.

This is because: Not a complete sentence $\varphi(x)$ is a solitary wave solution with energy $H = V_1 + V_2$, where

$$V_1 = \int (\Delta\varphi_s(x))^2 \, d^l x, V_2 = \int U(\varphi_s) \, d^l x,$$

In this way, the configuration of the field $\varphi(x/a)$ has energy:

$$H(a) = a^{D-2}V_1 + a^D V_2$$

However, H must be stable under arbitrary variations of φ. In particular, under a scaling transformation, it must satisfy

$$\left.\frac{\delta H(a)}{\delta a}\right|_{a=1} = (D-2)V_1 + DV_2 = 0$$

Since V_1 and V_2 are both positive, this equation can only have a solution when $D = 1$. Thus, the proof is complete. Clearly, this conclusion can be generalized to situations with more than one scalar field.

For pure Yang-Mills theory (with a compact gauge group), it can be proven that a variant of Derrick's theorem states that except for $D = 4$, in D spatial dimensions, the only static finite energy solutions to the pure gauge field equations are the $A_\mu = 0$ transformations. This was derived by Coleman using arguments similar to the scale-invariance discussions used for scalar fields. The above is known as the "n_0-g_0" theorem in quantum field theory, stating that for pure scalar fields, there are only one-dimensional solitary wave solutions, and for pure Yang-Mills fields, there are only four-dimensional solitary wave solutions, with no other solitary wave solutions existing beyond these cases [9].

It should be noted that the non-topological solitons in dynamics, due to their time dependence, are not constrained by this theorem. Furthermore, this theorem does not impose restrictions on static scalar-gauge field solitons in 2-dimensional and 3-dimensional space. In fact, the vortices and magnetic monopoles discussed in the following sections are examples of such possibilities. Additionally, the pseudoparticles or instantons in $D = 4$ are exceptions to the Coleman and Deser theorem.

Now we study the topological solitons in one-dimensional space [10]. Since topological solitons require vacuum degeneracy, V must have more than one minimum value. Assume that the minimum value of V is zero, as shown in Figure 10-2.

$$V(a) = V(b) = \cdots = 0$$

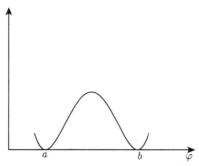

Figure 10-2   The form of $V(\varphi)$

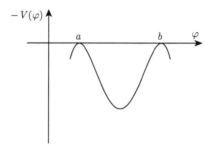

Figure 10-3   The form of $-V(\varphi)$

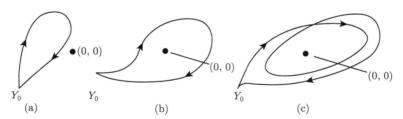

Figure 10-4   Topological soliton solution

From equation (10.12), the equation of motion can be obtained as

$$\frac{\partial^2 \varphi}{\partial x^2} - \frac{dV}{d\varphi} = 0$$

Assume $\varphi = \varphi(x)$ is a real field and is independent of time. Then we get

$$\frac{\partial^2 \varphi}{\partial x^2} - \frac{dV}{d\varphi} = 0$$

Integrating once, we get

$$\frac{1}{2}\left(\frac{d\varphi}{dx}\right)^2 - V(\varphi) = \text{constant} \tag{10.13}$$

The corresponding non-relativistic mechanical model is the motion of a particle with respect to the spatial coordinate φ, time x, and mass 1. equation (10.13) represents the conservation of the particle's energy, with its potential energy being $-V$, as shown in Figure 10-3. The kinetic energy is $\frac{1}{2}\left(\frac{d\varphi}{dx}\right)^2$. Assume that at $x = -0$, the particle is at point a. Give it a slight push to the right, and it will roll down along the curve, reaching point b at $x = +0$. Its energy is obviously finite and does not disperse. The soliton solution is shown in Figure 10-4. Its energy density is confined within a certain range between a and b. Because as $x \to a$ or b, $\frac{d\varphi}{dx} \to 0$ and $V \to 0$, while the boundary conditions at $x = \pm\infty$ are different, it is called a topological soliton. It can be considered as the lowest energy solution that satisfies the boundary conditions $\varphi = a$ as $x \to -\infty$ and $\varphi = b$ as $x \to +\infty$. Therefore, it is stable. If the particle is at point b when $x = -\infty$ and at point a when $x = +\infty$, the resulting soliton is called an anti-soliton solution. Thus, both solitons and anti-solitons exist in classical field theory.

All these soliton solutions can be obtained by setting the constant in equation (10.13) to zero, that is

$$x - x_o = \int_{\varphi_0}^{\varphi} \frac{d\varphi'}{\sqrt{2V(\varphi')}} \tag{10.14}$$

where x is a constant. Moreover, since equation (10.12) is Lorentz invariant, the above soliton solution can be obtained in a moving frame by performing a Lorentz transformation.

The most studied in one-dimensional space are soliton solutions for $U(\varphi) = \frac{\beta}{4}(\varphi^2 - m^2/\beta)^2$ of the φ^4 field theory and the sine-Gordon (SG) equation with $U(\varphi) = \sin\varphi$. Both of these equations have kink soliton solutions, and their equations of motion obey Lorentz invariance. We will study the φ^4 field equation here, while the SG equation will be studied in the next chapter.

For the φ^4 field theory, the Lagrangian function is:

$$L = \int \mathcal{L}\, dx = \int \left[\frac{1}{2}\partial_\alpha\varphi\partial^\alpha\varphi - \frac{\beta}{4}(\varphi^2 - m^2/\beta)^2\right] dx \quad (\alpha = 0, 1) \tag{10.15}$$

Making the transformation: $\varphi' = \sqrt{\dfrac{\beta}{m}}\varphi$, $x' = mx$, then equation (10.15) becomes:

$$L = \frac{m^3}{\beta}\int\left[\frac{1}{2}\partial_\alpha\varphi'\partial_\alpha\varphi' - \frac{1}{4}\left(\varphi'^2 - 1\right)\right]dx \qquad (10.16)$$

The equations of motion are

$$(\partial_x^2 - \partial_t^2)\varphi' + \varphi'(1 - \varphi'^2) = 0 \qquad (10.17)$$

For the static solution, it can be written as

$$\varphi'_{xx} + \varphi'(1 - \varphi'^2) = 0 \qquad (10.18)$$

We are already very familiar with the solution to this equation and can immediately write it down as

$$\varphi' = \pm 1 \text{ (vacuum state)} \qquad (10.19)$$

$$\varphi' = \pm\tanh\left(\frac{x' - x'_0}{\sqrt{2}}\right)\left(\begin{array}{l}\text{`` + ''denotes kink}\\ \text{`` - ''denotes anti-kink}\end{array}\right) \qquad (10.20)$$

that is

$$\varphi = \pm\frac{m}{\beta^{1/2}} \text{ (vacuum state)} \qquad (10.21)$$

$$\varphi = \pm\frac{m}{\sqrt{\beta}}th\left(\frac{m(x - x_0)}{\sqrt{2}}\right)\left(\begin{array}{l}\text{`` + ''denotes kink}\\ \text{`` - ''denotes anti-kink}\end{array}\right) \qquad (10.22)$$

In fact,

$$V(\varphi) = \frac{\beta}{4}\left(\varphi^2 - m^2/\beta\right)^2$$

Substituting it into equation (10.14), we can immediately write down the above solution. Figure 10-5 shows this solution.

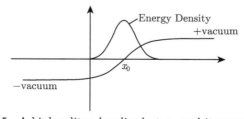

Figure 10-5   A kink soliton localized at x_0 and its energy density

The energy difference between its kink soliton solution and the vacuum state can be obtained as

$$E_{\text{kink}} \longrightarrow E_{\text{vacuum}} = 2\sqrt{2}\,m^3/3\beta \tag{10.23}$$

The non-trivial topological charge K associated with the twist is called the twist number. It is the result of the conservation of the current J, that is

$$J_\mu = \frac{1}{2}\varepsilon_{\mu L}\partial^\nu\varphi \quad (\varepsilon_{01} = \varepsilon_{10} = 1,\ \ \varepsilon_{00} = \varepsilon_{11} = 0)$$

then

$$K = \int_{-\infty}^{+\infty} J_0 dx = \frac{1}{2}\int_{-\infty}^{+\infty}\frac{d\varphi}{dx}dx = \frac{1}{2}\varphi\,(x)\,\Big|_{-\infty}^{+\infty} \tag{10.24}$$

For the kink (anti-kink) $K = 1(-1)$, it is a non-trivial mapping, whereas for the vacuum $K = 0$. Since the topological charge is absolutely conserved, it is generally quite difficult for the field configuration to dissipate into the vacuum. Therefore, the topological effect of the field acts like an infinite barrier, requiring infinite energy to make the kink decay into the vacuum. Hence, this soliton is non-decaying, making the kink stable. This stability of the soliton can be considered as a result of the degenerate vacuum, where these two vacua can transform into each other under the reflection $\varphi \to -\varphi$. The kink is inserted between the two vacua at $x = \pm\infty$ and approaches each vacuum within an infinite range. Only at $x = x_0$, where the energy density is at its maximum, do they differ significantly. In this situation, the homotopy mapping is the field energy $\varphi = \pm1$ mapped to the point $x = \pm\infty$, which is a trivial mapping. Therefore, deforming the kink mapping to any vacuum state requires infinite energy.

As mentioned in the previous section, such kink solitons can be used to describe hadrons. The quark model's SLAC bag [11] model is based on the above theory. In this model, quarks are distributed along the edges of the bag, as shown in Figure 10-6, and cannot escape. Therefore, free quarks cannot be found unless enough energy is provided to release free quarks from the hadron. This explains the problem of "quark confinement."

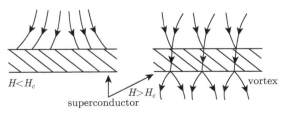

Figure 10-6   The Meissner effect and vortex in superconductor

10.4   Topological Solitons in Two-Dimensional

This type of soliton is similar to the vortex structure (flux line) formed by the quantization of magnetic flux in type-II superconductors. Therefore, it is necessary to first introduce the relevant knowledge in this area.

The superconductivity of matter refers to the phenomenon where some elements, alloys, and compounds suddenly lose their electrical resistance below a certain critical temperature T_c. In addition to complete conductivity, superconductors also exhibit complete diamagnetism. That is, when $T \leqslant T_c$, they expel magnetic flux from their interior, making the magnetic induction inside $B = 0$, which is also known as the Meissner effect. It is now clear that the superconducting state is a long-range coherent state formed when conduction electrons pair up due to the attractive interaction with lattice vibrations, creating Cooper pairs.

Consider a superconducting ring. When a magnetic field is applied at $T > T_c$, the superconducting ring is in the normal state, and the induced current in the ring immediately disappears, allowing magnetic flux to penetrate the ring. Then, if the temperature is lowered below T_c and the magnetic field is removed, the induced current in the now-superconducting ring will disappear, leaving some magnetic flux trapped in the ring. Experiments in 1961 measured this flux to be quantized, as

$$\varphi = \oiint \mathscr{H} \cdot ds = n \frac{h}{2e} = n\Phi_0$$

where

$$\Phi_0 = \frac{h}{2e} = 2.07 \times 10^{-5} Wb$$

referred to as the flux quantum, and n is an integer. This equation indicates that the magnetic flux in a superconductor is quantized, which is a macroscopic quantum effect.

Experiments have shown that is a type-II superconductor with the Ginsburg-Landau parameter $K > \dfrac{1}{\sqrt{2}}$. When the applied magnetic field H exceeds the lower critical field H_{c1} of the superconductor, magnetic flux lines will "infiltrate" the superconductor one by one in units of $\dfrac{h}{2e}$, forming columnar structures of magnetic flux lines within the material. These magnetic flux lines are also called vortex lines, which "embed" into the type-II superconductor, forming a mixed state. These quantized vortex line structures are known as the Abrikosov structure of the superconductor. In an ideal type-II superconductor, in a stable state, they are orderly arranged into a triangular structure. This Abrikosov [3] structure can be derived by solving the nonlinear Ginsburg-Landau (GL) equation that the macroscopic quantum wave function of the superconductor satisfies in an external magnetic field.

Nielsen and Olesen applied these concepts and ideas to field theory. They pointed out that the scalar Higgs field acts similarly to the order parameter in superconductors. Therefore, a relativistic field theory similar to the Abelian Higgs model should also have static vortex solutions. Below, we will study the Abelian vortex soliton solution and the non-Abelian vortex soliton solution separately.

(1) Abelian Vortex Soliton Solution

In the two-dimensional case, the Lagrangian density of an Abelian Higgs model can be written as

$$\mathcal{L} = -\frac{1}{4}F_{\mu\nu}F^{\mu\nu} + \frac{1}{2}(D_\mu\varphi)^* D^\mu\varphi - \frac{1}{4}\beta\left(\varphi\varphi^* - \frac{m^2}{\beta}\right)^2 \tag{10.25}$$

where

$$F_{\mu\nu} = \partial_\mu A_\nu - \partial_\nu A_\mu, \quad D_\nu\varphi = (\partial_\mu - ieA_\mu)\varphi \tag{10.26}$$

It indicates: under the local gauge invariant theory, Not a complete sentence. From the Euler-Lagrange equations, the equations of motion are

$$\partial^\mu F_{\mu\nu} = j_n = -\frac{1}{2}ie(\varphi^*\partial_\mu\varphi - \varphi\partial_\mu\varphi^*) + e^2 A_\mu\varphi\varphi^* \tag{10.27}$$

$$D_n D^\mu\varphi = -\beta\varphi(\varphi\varphi^* - m^2/\beta) \tag{10.28}$$

These are the G-L equations in this case. Nonlinear partial differential equations like these are extremely difficult to solve. We now transform to cylindrical coordinates and seek the asymptotic solution under such an Ansatz [5,13].

Now assume that

$$A_0 = 0, \ A = \hat{\theta}A(r), \ \varphi = f(r)e^{in\theta}, \ r^2 = x^2 + y^2 \tag{10.29}$$

Equation (10.27) and (10.28) can be simplified as

$$-\frac{1}{r}\frac{d}{dr}\left(r\frac{d}{dr}f\right) + \left[\left(\frac{n}{r} - eA\right)^2 + (\beta^2 - m^2/\beta)\right]f = 0 \tag{10.30}$$

$$-\frac{d}{dr}\left(\frac{1}{r}\frac{d}{dr}(rA)\right) + \left(Ae^2 - \frac{ne}{r}\right)f^2 = 0 \tag{10.31}$$

When seeking the asymptotic solution as $r \to 0$, the requirement for finite energy of the vortex line per unit length implies that the vortex field must have the following asymptotic form:

$$f(r) \xrightarrow{r \to \infty} \left(1 - \text{const } e^{[1-r/\xi]}m/\beta^{\frac{1}{2}}\right) \tag{10.32}$$

$$A(r) \xrightarrow{r \to \infty} \frac{n}{er} + \text{const } e^{[-r/\delta]} \tag{10.33}$$

Where the mass of the Higgs scalar field particle is $m_s = \sqrt{2}m$, while the mass of the vector field A_μ is $m_A = m\sqrt{\beta}$. In the case of the scalar field φ^4 with self-interaction terms and a virtual mass, spontaneous symmetry breaking occurs. In this process, the gauge particle—the photon "eats" the Goldstone boson it produces and simultaneously gains mass. The coherence length $\xi = \sqrt{2}/m$ provides the spatial scale of the Higgs field's variation, and the penetration depth $\delta = \dfrac{1}{m_A}$ of the electromagnetic field describes its spatial fluctuation amplitude. Thus, a vortex line soliton solution as shown in Figure 10-7 appears.

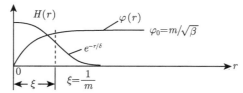

Figure 10-7   The vortex line of $\varphi(r)$ and $H(r)$

Here, K is called the Ginzburg-Landau parameter and is given by

$$K = \delta/\xi = m_s/\sqrt{2}\,m_A = \sqrt{\beta}/e \tag{10.34}$$

This dimensionless parameter divides superconductors into two types: When $K < \dfrac{1}{\sqrt{2}}$, it is a type-I superconductor, and when $K > \dfrac{1}{\sqrt{2}}$, it is a type-II superconductor, where vortices appear. However, for any K, a Nielsen-Olesen vortex solution may also exist.

For the properties of such vortices, refer to the literature [11-13]. Here, we only discuss the topological properties and energy form of the vortex.

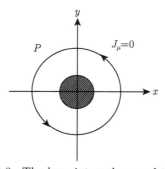

Figure 10-8   The loop integral around the vortex

For a magnetic vortex along the z-direction, we are mainly concerned with F_{12}, which can be calculated as the magnitude of the magnetic flux through the unit area of the (x, y) plane. By parameterizing the Higgs field as $\varphi = |\varphi| e^{i\alpha}$, the flux through the area enclosed by the closed loop P (as shown in Figure 10-8) is given by

$$\varphi = \int F_{12} dx dy = \int_P A_i dx^i = -\frac{1}{\varepsilon} \int_P \partial_i \alpha dx^i$$

Here, the fact that $j_\mu = 0$ along the path P is used. From the requirement of the single-valuedness of $\varphi(x)$, we can obtain:

$$\varphi = \frac{1}{e} \left[\partial(2\pi) - \partial(0) \right] = \frac{2\pi}{e} n = n \varphi_o$$

$$(n = 0, \pm 1, \pm 2, \cdots)$$

(10.35)

Thus, the magnetic flux is quantized. This quantized magnetic flux is the total topological charge of the vortex enclosed by the path P.

To illustrate the topological nature of the flux, we note that the vacuum of this theory is defined by the condition

$$|\varphi| = \varphi_o = \sqrt{\frac{m^2}{2\beta}}.$$

In this model, we only require that the energy of the vortex line per unit length is finite, which implies that the complex scalar field must have the asymptotic form $\varphi(\theta) = e^{in\theta}$, where θ is the polar angle in two dimensions and n is an integer (due to the single-valuedness requirement). This means that φ can be determined up to a phase factor $\alpha = n\theta$. This indicates that on the complex φ-plane parameterized by c, there exists a degenerate vacuum circle. Let P be a circle in the (x, y) plane. As one moves along this circle, the phase factor $\alpha(x, y) = n\theta$ changes from 0 to $2\pi n$. Thus, $n(\theta)$ describes a mapping from the real circle in the (x, y) plane to a circle in the complex φ internal space (see Figure 10-9). This mapping $U(1) \rightarrow S^1$ is characterized by the class of the map

$$\pi_1 \left(U(1) \right) = Z \text{ (integer set)}$$

(10.36)

This equation indicates that there can be an infinite number of discrete vortices with flux $\varphi = np$ (where $n = 0, \pm 1, \pm 2, \ldots$). The integer labeling each homotopy class is called the winding number. It reflects the number of rotations in the φ-plane corresponding to one full rotation of 2π in the (x, y) plane. The net flux of the vortex is proportional to the winding number. The field configurations with non-zero flux have a non-zero winding number as well. Due to the conservation of topological charge, it is impossible to continuously deform a configuration with a winding number into a configuration with zero winding number (i.e., it cannot be deformed into a constant $a(\theta)$). Such a deformation would require infinite energy, which confirms that the vortex solution we obtained is stable. In other words, the shape of the vortex remains invariant under topological transformations (non-trivial mappings).

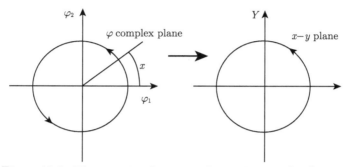

Figure 10-9   The mapping from complex φ-plane to (x, y) plane

The energy per unit length of a vortex with n units of magnetic flux can also be calculated. For a vortex carrying n units of magnetic flux, the energy per unit length is given by [14]

$$\begin{cases} \varepsilon_n > n\pi \mathrm{m}^2 \sqrt{2}\, k/\beta \quad (k > 1/\sqrt{2}) \\ \varepsilon_n = n\pi \mathrm{m}^2/\beta = n\pi \mathrm{m}_A{}^2/e^2 \quad (k = 1/\sqrt{2}) \\ \varepsilon_n > n\pi \mathrm{m}^2/\beta \quad (k > 1/\sqrt{2}) \end{cases} \qquad (10.37)$$

For the interaction between vortex lines, when $\kappa < \dfrac{1}{\sqrt{2}}$, they seem to attract each other, while for $\kappa > \dfrac{1}{\sqrt{2}}$, they seem to repel each other, and for $\kappa = \dfrac{1}{\sqrt{2}}$, the interaction is uncertain and can be either attractive or repulsive. Matricon et al. found numerically that $\epsilon_2 > 2\epsilon_1$ for all $\kappa > \dfrac{1}{\sqrt{2}}$, indicating that in type-II superconductors, a vortex with double quantized flux is energetically unfavorable. For $n \geqslant 2$ unit fluxes, Bogomolnys general analysis of the energy functional also leads to the same conclusion: a vortex with n units of magnetic flux will split into n separated topologically equivalent unit vortices (for $\kappa > \dfrac{1}{\sqrt{2}}$).

(2) Non-Abelian Vortex Soliton Solution

Tze and Ezwand [1] extended the Higgs model described above to the non-Abelian case and obtained the following main results.

For a gauge group G, if we require that there are static, axially symmetric vortex solutions with finite energy per unit length, this implies that as the radius approaches infinity, the Higgs scalar field must be a covariant constant. That is,

$$l_\mu \varphi = (\partial_\mu - iet^\alpha A_\mu^\alpha)\varphi \xrightarrow{r\to\infty} 0 \ (r^2 = x^2 + y^2) \tag{10.38}$$

Among them, t^α is the matrix representation of the group generators acting on the φ space. A_μ^α is the gauge field, which must belong to the adjoint representation of G. Given this condition, we find that the values of the φ field at any two points P_1 and P_2 on the path P (at large r) have the following relationship:

$$\varphi(P_2) \doteq s(P_2, P_1)\varphi(P_1) \tag{10.39}$$

where

$$s(P_2, P_1) = T \exp\left[-ie \int_{P_1}^{P_2} t^\alpha A_\mu^\alpha(x)\, dx^\mu\right]$$

is a non-integrable phase factor. T is the path-ordered exponential of the matrix along the path. For a circular path parameterized by the angle θ, the group to which the phase factor $S(\theta)$ belongs determines the types of vortices allowed in the model. For example, in the Abelian model,

$$S(\theta) = e^{in\theta} \in U(1) \tag{10.40}$$

leads to the result in Equation (10.40) and vortex solutions with allowed flux $\varphi_n = \varphi_0 n \ (n = 0, \pm1, \ldots)$.

For $G = SU(2)$, the situation is different. If the gauge invariance is broken by a doublet Higgs scalar, then

$$S(\theta) = e^{2in\theta\tau_s} \in SU(2) \tag{10.41}$$

The overall $SU(2)$ symmetry means that this theory cannot have vortex solutions. Since $\pi_1(SU(2)) = 0$, it indicates that any simple vortex solution can be continuously deformed into the vacuum, i.e., there is no topological stability. However, if the symmetry is broken by a scalar triplet, it can lead to:

$$S(\theta) = e^{in\theta\tau_s} \in SO(3) \tag{10.42}$$

Homotopy theory tells us

$$\pi_1(SO(3)) = \pi_1(SU(2)/Z_2) = Z_2 \tag{10.43}$$

This means that the theory can have solutions with 0 or ± 1 flux units. Other vortex solutions are not possible. This indicates that Abelian and non-Abelian theories are different, with the latter having only a finite number of vortex solutions.

10.5 Three-Dimensional Magnetic Monopole Solution

A magnetic monopole is the smallest microscopic magnetic element with only one magnetic pole. This concept was first introduced by Dirac in 1931. Now, let's see how this concept arises.

In quantum theory, the wave function $\psi(x_1, x_2, x_3, t)$ multiplied by a phase factor e^{ir} yields another wave function $\Psi = e^{ir}\psi$. If the phase factor r is a function of (x_1, x_2, x_3, t), i.e.,

$$\psi(x_1, x_2, x_3, t) = e^{ir(x_1, x_2, x_3, t)} \cdot \psi(x_1, x_2, x_3, t)$$

then we have:

$$\frac{\partial \Psi}{\partial x_i} = e^{ir} \left(\frac{\partial}{\partial x_i} + i\Pi_i \right) \psi \qquad \left(\Pi_i = \frac{\partial r}{\partial x_i} \right) (i = 1,\, 2,\, 3)$$

Therefore, when r is a function of (x_1, x_2, x_3, t), the operators have the following transformation relationship

$$\frac{\partial}{\partial x_i} \to \frac{\partial}{\partial x_i} + i\Pi_i$$

This is similar to the transformation relationship of the momentum operator of an electron in the electromagnetic field A

$$\hat{P}_i \to \hat{P}_i + eA_i \quad \text{or} \quad \frac{\partial}{\partial x_i} \to \frac{\partial}{\partial x_i} + ieA_i$$

Then this transformation relationship is exactly the same if $\Pi_i = eA_i$. This means that introducing a non-integrable phase factor $r(x_1, x_2, x_3, t)$ is the same as introducing the electromagnetic potential A. When we travel around a closed loop, the total change in phase r is expressed as:

$$(\Delta r)_{\text{loop}} + 2\pi n = \oint_{\text{loop}} \Pi_i dx_i$$

$$= e \oint_{\text{loop}} A_i dx_i$$

$$= e \iint_{\text{surfaces of loop}} \mathscr{H} \cdot dS \, (\mathscr{H} = \nabla \times A),$$

where n is an integer. $\iint \mathcal{H} \cdot dS$ represents the magnetic flux through the surface of the loop, which is closely related to the phase change when going around the loop.

Now we consider the region of ψ equal zero. If $\psi = 0$, then r is completely undetermined. If ψ is close to zero, then a small change in ψ can correspond to a very significant change in r. This condition is usually met along a line, which we call a nodal line. At $\psi = 0$, there can be several nodal lines. Suppose there are some wave functions containing a nodal line, and this nodal line has only one endpoint. In this case, the endpoint of the nodal line is some kind of singularity of the field. If we take a closed surface surrounding this singularity (in this case $(\Delta r)_{\text{loop}} = 0$), then e times the total magnetic flux through this closed surface equals $2\pi n$, i.e.,

$$e \oiint \mathcal{H} \cdot dS = 2\pi n$$

If the magnetic flux through a closed surface is non-zero, it implies that there is a magnetic monopole within the closed surface. If we denote its strength by q_m, we have $\oint B \cdot dS = 4\pi q_m$. This is similar to Gauss's theorem in electrostatics and can be referred to as Gauss's theorem in magnetism. It indicates that the magnetic flux through any closed surface surrounding a magnetic monopole is equal to 4π times the strength q_m of the enclosed magnetic monopole. Among them, $q_m = \dfrac{n}{2} e$. This shows the close relationship between particle charge and magnetic monopole strength. Thus, if there is a magnetic monopole, the charges of all charged particles in nature must be quantized. Moreover, if a magnetic monopole appears in quantum theory, this expression can be derived very definitively from quantum mechanical considerations and is unavoidable. Additionally, if magnetic monopoles exist, Maxwell's electromagnetic equations can be expressed in a symmetric form. After Dirac proposed the magnetic monopole, much theoretical and experimental work was done to find its existence in nature or laboratories. Recently, Blas Cabrera at Stanford University in the United States precisely measured the change in magnetic flux in a superconducting niobium coil. Through 151 days of observation, a sudden increase in magnetic flux was measured in one observation, which precisely matched the magnetic flux caused by a magnetic monopole, leading to the claim that a magnetic monopole was observed.

It should be noted that Dirac's magnetic monopole must be accompanied by a troublesome nodal line or singular string, which is unnatural. In 1974, A.M. Polyakov of the Soviet Union and G. 't Hooft of the Netherlands pointed out that the mass of a magnetic monopole is more than five thousand times the mass of a proton (in the current grand unified theory of electromagnetic, weak, and strong interactions, the mass of a magnetic monopole is considered to be 10^{16} times the mass of a proton). They also pointed out that if the Dirac electromagnetic $U(1)$ gauge group is "embedded" into a non-Abelian compact gauge group, then the magnetic monopole in this theory can be free of singular strings and is also a soliton solution of nonlinear partial differential equations. Next, we will study it.

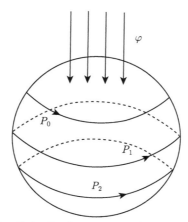

Figure 10-10   The G. 't Hooft structure of compact electrodynamics

Let us consider the G. 't Hooft structure of compact electrodynamics as shown in Figure 10-10, where Φ is the magnetic flux entering the spherical space, and P_0 is a path around the magnetic field line. Along P_0, the potential must be purely gauge. Due to this situation and the fact that all charge fields should be single-valued, we have

$$\Phi = \oint P_0 A_i dx^i = \frac{2\pi n}{e}$$

For the Abelian theory, the flux must flow entirely out of the sphere, so the outer loop P_0 cannot be continuously moved to a constant (i.e., $P_0 \to P_1 \to P_2 \to \ldots \to$ south pole). This means that in an Abelian magnetic monopole scenario, a Dirac string is required. On the other hand, if the electromagnetic $U(1)$ gauge group is embedded into a non-Abelian compact group, the magnetic monopole may not require a singular string. For example, in the $SO(3)$ gauge theory, a 4π rotation can be converted to a constant at the south pole due to the increased degrees of freedom in the gauge transformation. Therefore, in such a theory, a stringless magnetic monopole with magnetic charges $q_m = \frac{n}{2e}(\pm 1, \pm 2, \ldots)$ can exist. To realize this idea, G. 't Hooft and others examined the following Lagrangian function S with $SO(3)$ gauge invariance:

$$
\begin{cases}
\mathscr{L} = -\dfrac{1}{4}\, F_{\mu\nu}^a F^{a\mu\nu} + \dfrac{1}{2}\, D_{\mu\nu}\varphi^a D^a \varphi^a \\[2mm]
\qquad - \dfrac{\beta}{4}\left(\varphi^a \varphi^a - m^2/\beta\right)^2 \\[2mm]
F_{\mu\nu}^a = \partial_\mu A_\nu^a - \partial_\nu A_\mu^a + e\varepsilon^{abc} A_\mu^b A_\nu^c \\[2mm]
D_\mu \varphi^a = \partial_\mu \varphi^a + e\varepsilon^{abc} A_\mu^b \varphi^c \ (a = 1,\, 2,\, 3)
\end{cases} \tag{10.44}
$$

It describes the interaction between the gauge field A_μ and the Higgs isospin vector field φ^μ. At the same time, it is $SO(3)$ gauge invariant, i.e., it is the Georgi-Glashow model of weak-electromagnetic interaction. It describes a massless photon and two massive charged intermediate vector bosons. The latter acquire mass through the Higgs mechanism. Thus, due to the presence of a degenerate vacuum $\varphi_0^2 = \dfrac{m^2}{\beta}$, the $SO(3)$ gauge symmetry is spontaneously broken. In this case, only the $U(1)$ gauge symmetry remains unbroken.

Through the E-L equations, the classical equations of motion derived from Eq. (10.44) are:

$$
\begin{cases}
D_\mu F_{\mu\nu}^a = -e\varepsilon^{abc}\varphi^b D_r \varphi^c \\[2mm]
D^\mu D_\mu \varphi^a = -\beta\varphi^a \left(\varphi^a - m^2/\beta\right)
\end{cases} \tag{10.45}
$$

G. 't Hooft and Polyakov found that this equation has a magnetic monopole solution. For this, they made a static spherically symmetric assumption:

$$
\begin{cases}
A_0^a = 0, \ A_i^a = \varepsilon_{aij} x_j [1 - K(r)] e r^2 \\[2mm]
\varphi^a = -x_a H(r)/e r^2, \ r^2 = x^2 + y^2 + z^2
\end{cases} \tag{10.46}
$$

Reducing Eq. (10.45) to a system of radial equations, i.e.,

$$
\begin{cases}
r^2 K'' = K(K^2 - 1) + KH^2 \\[2mm]
r^2 H'' = 2HK^2 + \beta/e^2 (H^3 - c^2 r^2 H)(c = M_c/\beta^{1/2})
\end{cases} \tag{10.47}
$$

For the exact solution with $K = 0, H = cr$, which corresponds to infinite energy, we are not interested and ignore it. For a solution with the same asymptotic behavior along the radius and approaching a pure gauge at the origin ($F_{\mu\nu}^\alpha = 0$) ($K(0) = \pm 1$), it is of particular significance. As shown in Figure 10-11, this solution is called a magnetic monopole. The energy or mass of a magnetic monopole can be obtained by numerical calculation. Finally, we get:

$$
M = 4z\frac{m}{e\beta} f(\beta/e^2) = \frac{M_\alpha}{\alpha} f(\beta/e^2)
$$

where $a = \dfrac{e^2}{4\pi}$, f is a slowly varying monotonic increasing function with $f(0) = 1$, and M is the mass of the vector boson. Since $1/a = 137$, we estimate $M \approx 50\,\mathrm{GeV}$, indicating that the magnetic monopole has a very large mass.

In 3-D space, the topological invariant is the magnetic charge. To illustrate the nontriviality or stability of the topological soliton solution, G. 't Hooft constructed a gauge-invariant electromagnetic field tensor:

$$\begin{cases} F_{\mu\nu} = \hat{\varphi}^a F_{\mu\nu}^a - \dfrac{1}{e}\,\varepsilon^{abc}\hat{\varphi}^a D_\mu \hat{\varphi}^b D_\nu \hat{\varphi}^c \\ \hat{\varphi}^a = \varphi^a / |\varphi| \end{cases} \tag{10.48}$$

It can also be written as:

$$\begin{cases} F_{\mu\nu} = \partial_\mu B_\nu - \partial_\nu B_\mu - \dfrac{1}{e}\varepsilon^{abc}\hat{\varphi}^a \partial_\mu \hat{\varphi}^b \partial_\nu \hat{\varphi}^c \\ B_\mu = \hat{\varphi}^a A_\mu^a \end{cases} \tag{10.49}$$

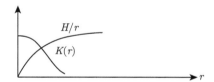

Figure 10-11   The form of magnetic monopole solution

Substituting Equation (10.46) into Equation (10.49), we get:

$$F_{ij} = \varepsilon_{ijk}\chi_k / er^3 \tag{10.50}$$

From a topological point of view [16], it corresponds at least to the magnetic field of a point magnetic monopole with magnetic charge $q_m = \dfrac{1}{2e}$. If we change the sign in Equation (10.46), we obtain an anti-monopole [15]. Now let us consider the topological origin of this charge.

If there is no string singularity in B_μ, the resulting magnetic current is:

$$^*j_\mu = \partial^{\nu\,*}F_{\mu\nu} = \dfrac{1}{2e}\,\varepsilon_{\mu\nu\kappa\beta}\varepsilon^{abc}\partial^\nu(\hat{\varphi}^a \partial^a \hat{\varphi}^b \partial^s \hat{\varphi}^c)$$

It can be proven to be conserved, i.e., $\partial_\mu^* j^\mu = 0$. However, this topological current is not a Noether current. The associated charge does not generate a symmetry of the

Lagrangian. The magnetic flux or magnetic charge is given by:

$$\Phi = 4\,zq_m$$

$$= \int d^3x^* j_0 = \frac{1}{2\,e}\oint_{s_k^2} \varepsilon_{ijk}\hat{\varphi}^a \partial j \hat{\varphi}^b \partial_k \hat{\varphi}^c (d^2\sigma)_i \qquad (10.51)$$

where s_k^2 is a sphere of radius R (in the limit $R \to 0$). Since a sphere can be parameterized by two coordinates $\xi^\alpha (\alpha = 1, 2)$, the above equation can be written as:

$$\begin{cases} 4zq_m = \dfrac{1}{e}\displaystyle\int_{s_k^2} d^2\xi \,\dfrac{1}{2}\, \epsilon_{\alpha\beta}\epsilon^{abc}\hat{\varphi}^a \partial_\alpha \hat{\varphi}^b \partial_\beta \hat{\varphi}^c = \dfrac{1}{e}\displaystyle\int d^2\xi \,\sqrt{g} \\ g = \det(\partial_a \hat{\varphi}^a \partial_\beta \hat{\varphi}^a) \end{cases} \qquad (10.52)$$

We know that this integral is four times the Kronecker index of the mapping $s_k^2 \to s_\varphi^2$. The Kronecker index must be an integer, so we obtain:

$$q_m = \frac{n}{2e}$$

The emergence of a magnetic monopole soliton leads to the following results:

(1) By introducing an electric field through the consistent hypothesis of Julia and Zee: $A_0^a = \dfrac{\chi_a J(r)}{er^2}$, a dyon solution appears [17]. This dyon has finite energy and continuous electric and magnetic charges $q_m = \dfrac{1}{2e}$. In quantum theory, the allowed values of charge become discrete: $q = ne$ [18].

(2) This theory can be extended to higher rank gauge groups, such as $SU(3)$. In this case, new magnetic monopoles with various charges can be formed.

(3) In the limit $\beta \to 0$ (keeping the condition $H(r) \xrightarrow[r \to \infty]{} Cr$), Prasad and Sommerfield [19] obtained an exact solution for Equation (10.47):

$$K(r) = \frac{Cr}{\sinh(Cr)}, \quad H(r) = (Cr)_{\coth}(Cr) - 1$$

(4) Hasenfratz and G. 't. Hooft, as well as Jackiw and Rebbi, pointed out that if the isospin doublet of Lorentz scalars is added to the monopole model using $SU(2)$, there is a state with a magnetic monopole plus isospin. Its angular momentum is $\dfrac{1}{2}$. This composite system clearly obeys Dirac-Fermi statistics, implying that the spin originates from isospin.

(5) Since the original $SU(2)$ model has a magnetic monopole $q_m = \dfrac{1}{2e}$, is there a finite energy solution with multiple magnetic charges? It is now believed that

for suitably defined "spherical symmetry," there is no $SU(2)$ spherically symmetric magnetic monopole with $|q_m| > \dfrac{1}{2e}$. For $SU(3)$, there exist spherically symmetric monopoles with multiple magnetic charges [22].

(6) It is impossible to form a topologically stable static finite energy solution without long-range fields (due to the Higgs mechanism making all gauge fields quite heavy) in 3D. Because in three-dimensional space, an infinite barrier requires long-range gauge fields, this infinite barrier is used to provide topological stability, preventing the soliton from decaying into the trivial vacuum.

10.6 Topological Solitons in Four-Dimensional Space—Instantons

Now we study the topological soliton solutions in four-dimensional Euclidean space. This problem was first studied by Belavin, Polyakov, Schwarz, and Tyupkin (BPST). They found source-free solutions of the non-Abelian gauge field equations under $SU(2)$ in four-dimensional real space. These solutions are called instantons. In this case, the gauge potentials are analytic throughout Euclidean space, but the field strengths are localized in certain regions of spacetime. The Euclidean energy and momentum are zero. Below we briefly describe the work of BPST and others. For detailed derivations, please refer to the original papers.

In 4D, the Lagrangian density of the gauge field, as expressed by BPST, is:

$$\mathscr{L} = -\frac{1}{4}\, F^a_{\mu\nu} F^a_{\mu\nu} \ (\mu,\, \nu = 1,\, 2,\, 3,\, 4) \tag{10.53}$$

where

$$F^a_{\mu\nu} = \partial_\mu A^a_\nu - \partial_\nu A^a_u + g\, C^{abc} A^b_\mu A^c_\nu \ \left(C^{abc} = \text{structure constants}\right)$$

Now we consider the gauge group $SU(2)$, and use the matrix representation:

$$A_\mu = A^a_\mu \tau^a/2, \quad F_{\mu\nu} = F^a_{\mu\nu}\tau^2/2$$

where τ^a are the 2×2 Pauli matrices. When moving these quantities, we use the following relation:

$$\begin{cases} \left[\tau^a,\, \tau^b\right] = 2\,i\,\epsilon^{abc}\tau^c, \quad Tr\tau^a\tau^b = 2\,\delta^{ab} \\ Tr\tau^a\tau^b\tau^c = 2i\varepsilon^{abc} \ (Tr = \text{trace}) \end{cases} \tag{10.54}$$

Using Eq. (10.53), from the Euler-Lagrange equation, we obtain the field equation:

$$D_\mu F_{n\nu} = \partial_\mu F_{\mu\nu} - ig[A_\mu,\, F_{\mu\nu}] = 0 \tag{10.55}$$

We can see that in this theory, there is only one topological invariant, namely the Pontryagin index, or the second Chern number. It corresponds to $\pi_3(SU(2)) = \mathbb{Z}$. In this case, the topological charge Q is defined as:

$$Q = \frac{g^2}{32\pi^2} \int dx^4 Tr \left(\varepsilon_{\mu\nu\alpha\beta} F_{\alpha\beta} F_{\mu,\nu} \right) = n \tag{10.56}$$

where $n = 0, \pm1, \pm2, \ldots$, and $Q = n$ is the winding number of the topological mapping of $SU(2)$, denoted by the homotopy class of $g(x)$, which is a matrix of the gauge group. This topological charge in the gauge $A_0 = 0$ simply equals the change in the winding number between $t = -\infty$ and $t = \infty$.

For such a physical system, we also introduce an inequality:

$$\int d^4x \, Tr \left(F_{\mu\nu} - \frac{1}{2} \varepsilon_{\mu\nu\alpha\beta} F_{\alpha\beta} \right)^2 \geqslant 0 \tag{10.57}$$

This condition is equivalent to imposing a constraint on the Euclidean energy E:

$$E = \frac{1}{2} \int d^4x Tr(F_{\mu\nu} F_{\mu\nu}) \geqslant 8\pi^2 |Q| \tag{10.58}$$

If A_μ has no singularities, i.e.,

$$\partial_\mu \partial_\nu A_\lambda = \partial_\nu \partial_\mu A_\lambda,$$

then

$$Tr \left(\frac{1}{2} \varepsilon_{\mu\nu\alpha\beta} F_{\alpha\beta} F_{\mu\nu} \right) = \frac{\partial}{\partial_\mu} \left(2\varepsilon_{\mu\nu\alpha\beta} Tr \cdot \left(A_\nu \partial_\alpha A_\beta + \frac{2g}{3i} A_\gamma A_\alpha A_\beta \right) \right) \tag{10.59}$$

Here the right-hand side of equation (10.6.7) is gauge-independent. But

$$2\varepsilon_{\mu\nu\alpha\beta} Tr \left(A_r \partial_\alpha A_\beta + \frac{2g}{3i} A_\nu A_\alpha A_\beta \right)$$

is gauge-dependent. If equation (10.55) is integrable and $g = 1$, Q can be expressed as

$$Q = \frac{1}{16\pi^2} \oint_{s^3} \left[2\varepsilon_{\mu\nu a\beta} Tr \left(A_\nu \partial_a A_\beta + \frac{2g}{3i} A_\nu A_a A_\beta \right) \right] d^3\sigma^\mu \tag{10.60}$$

In order to make $F_{\mu\nu}$ asymptotically zero, we require that A_μ is a pure gauge when $r \to \infty$, i.e.,

$$A_\mu(x) = ig^{-1}(x)\partial_\mu g(x), \quad x \in s^3 \tag{10.61}$$

The constraint on E given by equation (10.58) indicates that the energy of soliton solutions with non-trivial charge has a lower bound. If

$$F_{\mu\nu} = \frac{1}{2}\,\varepsilon_{\mu\nu\alpha\beta}F_{\alpha\beta} \tag{10.62}$$

exists, we should take the lower bound of the energy in equation (10.58). When this condition is satisfied, the field equation is automatically fulfilled. Because:

$$D_\mu F_{\mu\nu} = D_\mu\left(\frac{1}{2}\,\varepsilon_{\mu\nu\alpha\beta}F_{\alpha\beta}\right) = 0$$

(for non-singular fields, this is an identity).

Using the spherical symmetry assumption of the gauge field, we can write:

$$\begin{cases} A = if(r)g^{-1}(x)\partial_\mu g(x), \quad \gamma^2 = x_1^2 + x_2^2 + x_3^2 + x_4^2 \\[2mm] g(x) = \dfrac{x_4 - ix_a\tau^a}{r} \end{cases} \tag{10.63}$$

Substituting it into equation (10.62) and rearranging, we find the self-dual condition:

$$rf' \mp 2i(1-f) \tag{10.64}$$

Solving this equation, we get a non-trivial solution:

$$f(r) = \frac{\gamma^2}{\gamma^2 + \beta^2} \tag{10.65}$$

This solution is called an instanton, where β is an arbitrary scale with the dimension of length, representing the size of the instanton. This instanton can be localized anywhere and has any volume scale. From this solution and equation (10.63), we can obtain:

$$\begin{cases} A_\mu = \dfrac{i\gamma^2}{\gamma^2 + \beta^2}\,g^{-1}\partial_\mu g \\[3mm] F_{\mu\nu} = \dfrac{4\beta^2}{(\gamma^2 + \beta^2)^2}\,\sigma_{\mu\nu} \end{cases} \tag{10.66}$$

where

$$\sigma_{ij} = \frac{1}{4i}[\tau^i, \tau^i]\,, \sigma_{i4} = \frac{1}{2}\tau^i = -\sigma_{4i}$$

For this instanton, the Pontryagin index $Q = 1$, since $F_{\mu\nu} \to 0$ as $\gamma \to \infty$, and A_μ approaches a pure gauge. By performing a translation transformation in equation (10.66): $x_\mu \to x_\mu - a_\mu$, another instanton solution can be obtained, located at $x = a$.

Similarly, there is an anti-instanton solution, which is obtained by replacing β with $-\beta$ in the first equation of (10.66). This solution satisfies [24].

$$F_{\mu\nu} = -\left(\frac{1}{2}\varepsilon_{\mu\nu\alpha\beta}F_{\alpha\beta}\right) = \frac{4\beta^2}{(\gamma^2+\beta^2)^2}\bar{\sigma}_{\mu\nu}$$

where

$$\bar{\sigma}_{ij} = \sigma_{ij}, \quad \bar{\sigma}_{j4} = -\sigma_{i4}$$

It corresponds to the topological charge $Q = -1$. For

$$F_{\mu\nu} = \pm\frac{1}{2}\varepsilon_{\mu\nu\alpha\beta}F_{\alpha\beta}$$

and $|Q| = 1$, there seems to be no other solution except for the gauge transformation mentioned above. For a single instanton, five parameters are commonly used to represent it. Among them, four determine its position and one determines its size.

It should be pointed out that instantons are not physical particles. They are solutions to the field equations in four-dimensional Euclidean space (imaginary time) rather than in four-dimensional physical space. That is, instanton processes occur in imaginary time (very similar to the tunneling effect in quantum mechanics). They can be regarded as classical solutions (or paths) that connect the gauge potentials of two classical vacua with $\Delta n = 1$ in imaginary time, as

$$i\,|\rangle \quad \xleftarrow{\;\;anti-instanton\;\;} \quad \xrightarrow{\;\;instanton\;\;} \quad |\,i+1\rangle$$

Thus, they are called instantons. There is no interaction between instantons themselves; interaction only exists between instantons and anti-instantons when quantum fluctuations are taken into account. Despite this, they exhibit many properties similar to particles in Euclidean space and have observable physical effects. For example, the instanton tunneling effect can cause observable effects physically. These effects can also be explained using perturbation methods in natural vacua. For this reason, they are also called pseudoparticles. Additionally, instantons are closely related to the theory of strong interactions (QCD), making instanton physics an important topic of current research in gauge theory.

Recently, it has also been pointed out that based on the similarity between the gauge field strength $F_{\mu\nu\lambda}$ and the metric tensor $R_{\mu\nu\rho\sigma}$ of Riemann space, the concept of instantons has also been introduced in general relativity. These are called gravitational instantons. The significance of gravitational instantons is that they are inserted between different vacua $(R_{\mu\nu\rho\sigma} = 0)$ in Minkowski space. It is very likely that in the so-called super gauge theory, there are also super gravitational instantons.

10.7 Nontopological Solitons

For non-topological solitons, they differ from topological solitons in that they do not require the existence of a degenerate vacuum. Their boundary conditions at infinity and the field equations without soliton solutions are the same, and they also require the existence of an additive conservation law such as "charge" and a scalar field. Therefore, the simplest way to produce a non-topological soliton is to introduce a complex field [28]:

$$\varphi = \varphi_1 + i\varphi_2, \quad \varphi = \varphi_1 - i\varphi_2 \tag{10.67}$$

where φ_1 and φ_2 are Ermi fields. Here, we only discuss the one-dimensional case. The corresponding Lagrangian density can be expressed as

$$\mathcal{L} = \frac{1}{2} \frac{\partial \varphi^*}{\partial x_\mu} \frac{\partial \varphi}{\partial x_\mu} - U(\varphi^*\varphi) \tag{10.68}$$

Using the Euler-Lagrange equation, we can obtain the equation of motion from Eq. (10.68):

$$\frac{\partial^2 \varphi}{\partial x_\mu^2} - \varphi \frac{\partial}{\partial(\varphi^*\varphi)} U(\varphi^*\varphi) = 0 \tag{10.69}$$

Using the above equation, we can directly prove:

$$N = i \int (\varphi^*\dot{\varphi} - \dot{\varphi}^*\varphi) dx \tag{10.70}$$

is a conserved quantity (assuming $\varphi^* \dfrac{\partial \varphi}{\partial x} - \varphi \dfrac{\partial \varphi^*}{\partial x}$ is zero at $x = \pm\infty$).

In this system, the reason for the existence of such a conserved quantity is that $\mathcal{L}$ is invariable under the transformation $\varphi \to \varphi e^{i\theta}$, then the Hamiltonian function remains unchanged. Therefore, $\mathcal{H}$ is independent of θ, we can be taken N as the conjugate momentum of θ. Using the Hamiltonian equation, we obtain: $\dot{N} = \dfrac{\partial \mathcal{H}}{\partial \theta} = 0$, since N is a conserved quantity. In classical field theory, N can be any real number because θ is a phase variable. When $\theta \to \theta + 2\pi$, $\varphi \to \varphi$. Hence, in quantum field theory, N must be an integer, like angular momentum. (This reasoning can be reversed: if N is a conserved quantity, $\mathcal{H}$ must be independent of its conjugate coordinate. In quantum field theory, if N is an integer, then θ must be a cyclic variable, so θ can be taken as a phase variable. This explains why we introduced the complex field initially).

When $N \neq 0$, φ must be time-dependent. It is easy to prove that when we fix N, the solution with the minimum energy as a function of time should follow the harmonic oscillator relation:

$$\varphi = \sigma(x) e^{-i\omega t} \tag{10.71}$$

Substituting φ from Eq. (10.71) into Eq. (10.69), we get:

$$\frac{d^2\sigma}{dx^2} + \omega^2\sigma - \sigma\frac{d}{d\sigma^2}U = 0$$

After integrating the above equation, we obtain

$$\frac{1}{2}\left(\frac{d\sigma}{dx}\right)^2 - V(\sigma) = \text{constant} \tag{10.72}$$

where

$$V = \frac{1}{2}\left(U - \omega^2\sigma^2\right), U = U(\sigma^2)$$

Since non-topological solitons do not require vacuum degeneracy, we can set $U(0) = 0$. At this point, U can take the form shown in Fig. 10-12.

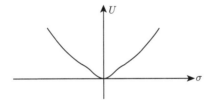

Figure 10-12   The potential energy curve

To obtain the solution for non-topological solitons, we assume

$$V = \frac{1}{2}\left(U - \omega^2\sigma^2\right)$$

which has the form shown in Fig.10-13, that is

$$U(\varphi^*\varphi) \to \omega^2\varphi^*\varphi = 0$$

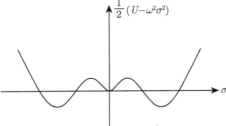

Figure 10-13   The curve of $\frac{1}{2}(U - \omega^2\sigma^2)$

In addition to the solution $\varphi = 0$, it also has solutions where $\varphi \neq 0$.

In the non-relativistic mechanical simulation case, the potential energy of the particle should be $-V$. Its form is as shown in Fig.10-14. When $x = -\infty$, placing the particle at point O, if we gently push it, it will roll along the curve to point A, then roll back from point A. When $x = +\infty$ again, it will roll back to point O. This motion pattern is easy to obtain. The general solution from equation (10.72) is:

$$x - x_0 = \int_A^\sigma \frac{e}{\sqrt{2V(\sigma)}} \quad (x = x_0 \text{ when } \sigma = A,) \tag{10.73}$$

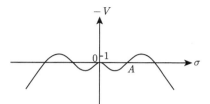

Figure 10-14   The form of $V(\sigma)$

In this case, the field energy is confined to a finite region in space and does not disperse, so it is a soliton. Since the solution approaches zero at $x = \pm\infty$, it is a non-topological soliton, as shown in Fig.10-15.

From Fig.10-12, we see that when $\sigma \to 0$, $U \to m^2\omega^2$, where m^2 is a constant. If we take V as shown in Fig.10-14, we can prove

$$\omega < m \tag{10.74}$$

Now, if we take

$$U = \frac{m^2\varphi^*\varphi}{1 + \varepsilon^2} \left[(1 - g^2\varphi^*\varphi)^2 + s^2 \right]$$

then the solution is obtained by equation of motion, as

$$\varphi = \frac{1}{g} \left[\frac{a}{1 + \sqrt{1 - a}\cosh y} \right]^{\frac{1}{2}} e^{-iwt} \tag{10.75}$$

where

$$a = (1 + \varepsilon^2)(m^2 - \omega^2)/m^2, y = 2\sqrt{m^2 - \omega^2}(x - x_0)$$

In Eq.(10.75), when we let $|x| \to \infty$, we get the asymptotic form of this solution:

$$\varphi \propto \frac{1}{g}\sqrt{m^2 - \omega^2}\, e^{-\sqrt{m^2-\omega^2}|x|} \tag{10.76}$$

This is a decaying solution. When $|x| \to \infty$, it decays to zero, consistent with the definition of a non-topological soliton. Therefore, the asymptotic form in Eq.10.76 has generality, and we can strictly prove it from other aspects as well. Next, we explain the stability of this soliton.

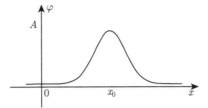

Figure 10-15   The soliton solution φ

The stability of this soliton is determined entirely by comparing its energy with the energy of the plane wave solution under the same conserved quantity. First, we discuss the one-dimensional case.

We know that any nonlinear field equation has a plane wave solution of the form

$$\varphi = \sqrt{\frac{N}{2\omega Q}}\, e^{iK \cdot x - i\omega t} \qquad (\omega = \sqrt{m^2 + K^2}) \tag{10.77}$$

This is because when the system volume $Q \to 0$, the amplitude of the above equation becomes infinitely small, so the higher-order terms in the Lagrangian function can be ignored. By retaining the first-order term, this solution can be obtained. For such a plane wave solution, the energy and the conserved quantity (e.g., N) have the following linear relationship:

$$E_{\text{plan}} = N\omega \geqslant Nm \tag{10.78}$$

For the soliton solution, the energy is a nonlinear function of N. From Eq. (10.74), we know that $\omega < m$. Therefore, the non-topological soliton solution can be considered an analytic continuation of the plane wave, extending from $\omega \geqslant m$ to $0 < \omega < m$. For any conserved quantity N and any coupling constant g, the lowest energy corresponding to the non-topological soliton solution is

$$E_{\text{soliton}} < Nm \tag{10.79}$$

From the previous sections, we know that the conserved quantity N and the phase angle $\theta = \omega t$ of the complex field are conjugate variables. Thus, we have the relationship: $\dot{N} = -\dfrac{\partial \mathcal{H}}{\partial \theta}$ and $\dot{\theta} = -\dfrac{\partial \mathcal{H}}{\partial N}$. Since $\dot{\theta} = \omega$ and the eigenvalue of $\mathcal{H}$ on any solution is the energy E, the above equation can be written as:

$$\omega = \frac{dE}{dN} \tag{10.80}$$

For plane waves, ω is independent of N, so $E = N\omega$. But for non-topological solitons, this relationship does not hold because we need to consider the relationship between ω and N. If we consider the limit $\omega \to m^-$, from Eq. (10.76), we know that when $\omega \to m^*$,

$$N \to 2m \int |\varphi|^2 \, dx \sim \sqrt{m^2 - \omega^2} \to 0$$

From the integral of Eq. (10.80) and since $\omega < m$, we have

$$E_{\text{soliton}} = \int_0^N \omega dN < m \int_0^N dN = mN \tag{10.81}$$

Therefore, $E_{\text{soliton}} < E_{\text{plan}}$. That is, in one-dimensional space, if a non-topological soliton solution exists, for any conserved quantity N, the lowest energy state is not the plane wave but the soliton. Moreover, this soliton solution is always stable. For the two-dimensional case, it can also be proved that under certain conditions, the soliton is the lowest energy state and is stable.

The above is the static soliton solution obtained in one-dimensional space. When extended to (3+1) dimensions, due to Derrick's theorem, there is no static soliton solution for a pure scalar field. However, if the nonlinear scalar field has an internal symmetry group G, there may exist soliton solutions of the form $\{e^{i\alpha_i(t)T_i}\}\varphi_c(x)$, where T_i are the generators of the group G in the representation to which the scalar field φ belongs, and $\alpha_i(t)$ are the group parameters dependent on time t. When G is an Abelian group, it was studied by Lee et al. [25-27]. When G is a non-Abelian group, Zhou et al.[28] studied the case where $G = SU(2)$ is the isospin group, and proved that only the non-topological solitons $I_3 = \pm I_{3\text{max}}$ components with non-zero in $\varphi_c(x)$ are stable.

10.8   Quantization of Solitons

In current field theory research, certain nonlinear equations are often regarded as classical approximate equations of localized quantum field equations. Soliton solutions are obtained, and then quantized. Quantized solitons better reflect their particle nature, i.e., wave-particle duality. This is beneficial for studying the properties of solitons and their interactions with each other (or with other particles). However, two questions arise: Can solitons of nonlinear field equations be quantized? If so, how can they be quantized? These issues are complex and unresolved. Many papers have been published; here we briefly introduce these problems.

We all know that the equations like SG and φ^4 field equations introduced in the first chapter have soliton solutions, which are called classical solitons. Quantization is discussed in this context. Can they be quantized? Let's first consider the following facts.

In any field theory, the system we consider has many different fields, which we collectively denote as φ. These fields can have many different coupling constants among them. For simplicity, a general coupling constant g can be defined using the Lagrangian function in Equation (10.12). The classical soliton solution in classical field theory can be expressed as:

$$\varphi_{\text{classical}} = \frac{1}{g}\,\sigma$$

then

$$\mathscr{L} = \frac{1}{g^2}\mathscr{L}_\sigma$$

where

$$\mathscr{L}_\sigma = -\frac{1}{2}\left(\frac{\partial\sigma}{\partial x_\mu}\right)^2 - V(\sigma)$$

$\mathscr{L}_\sigma$ is independent of g. Since the classical solution is determined by the extremum of the action integral, and the action is proportional to $\int \mathscr{L}_\sigma d^4x$. Therefore, $\mathscr{L}_\sigma$ is independent of g, implying that the corresponding soliton solution σ must also be independent of g.

In the quantum theory, the action is equal to $\hbar^{-1}$ times the classical action, and the latter depends linearly on $\mathscr{L}$, i.e., $g^{-1}\mathscr{L}_\sigma$. Hence, we can expect that g^2 plays a similar role to $\hbar$. We can understand it this way: if there are two points A and B in the generalized coordinate space, considering two different paths, each with an action phase:

$$\frac{1}{\hbar}\int \mathscr{L}d^4x = \frac{1}{hg^2}\int \mathscr{L}_\sigma d^4x$$

Summing over all different paths of $e^{\frac{i}{hg^2}\int \mathscr{L}_\sigma d^4x}$ is the amplitude in quantum mechanics. Thus, $g^2 \to 0$ and $\hbar \to 0$ have similar effects, both transitioning quantum mechanical solutions to classical mechanical solutions. Therefore, once a classical bosonic field soliton solution exists, it can be proven that at least in the case of weak coupling, a corresponding quantum soliton solution exists. If we expand the quantum soliton solution in powers of g^2, the expansion form is very similar to the expansion in powers of $\hbar$, with the first term identical to the WKB approximation. Therefore, if we expand the energy E of the quantum soliton solution in powers of g^2, i.e.,

$$E_{\text{quantum}} = O\left(\frac{1}{g^2}\right) + O(1) + O(g^2) + \cdots \qquad (10.82)$$

where the first term is the energy of the classical solution $E_{\text{classical}}$. It is $O\left(\frac{1}{g^2}\right)$ since $\varphi_{\text{classical}} \propto \frac{1}{g}$. If it is renormalization field theory, the latter term $O(1), O(g^2)$ can be proven to be finite (the main reason being that $\varphi_{\text{classical}}$ has no singularities in space x).

This discussion shows that whether a field or system has a quantized soliton solution depends on whether the field or system has a classical soliton solution. Therefore, T. D. Lee et al. believe that in a renormalizable bosonic field, for every classical soliton solution, there is a corresponding quantized solution. When the coupling constant $g \to 0$, its mass approaches the mass of the classical soliton, and it also has a shape factor. Renormalizable means that in the S-matrix of the interaction matrix elements between particles, parameters like charge and mass can be replaced by their physical values (experimental values), and introducing renormalization factors makes S converge. For spin-zero fields, when the interaction Lagrangian or Hamiltonian between them does not exceed the fourth power of the field quantity φ, quantum electrodynamics has proven it to be renormalizable. For such fields, there exists a quantized soliton solution whose shape factor is measurable and its shape exactly matches the classical soliton. Similarly, for Fermi fields, a similar definition can be made.

Then how to quantize a classical soliton? We all know that for a general bound state, when $\hbar \to 0$, the solution does not exist. However, in quantized bound states, as discussed earlier, when $\hbar \to 0$, the solution exists. Moreover, since the soliton solution is related to nonlinear fields, the usual perturbation expansion method is not suitable because when the coupling constant g is very small, the field quantity φ may be very large. However, for general soliton systems such as the nonlinear Schrödinger equation, SG equation, and φ^4 field equation, they are all completely integrable canonical systems. Each soliton in them has certain energy, momentum, charge, mass, and other basic physical properties. Therefore, using the commonly used canonical quantization method for quantization is very appropriate [24-26,29-31]. Of course, there are also current methods using path integrals. Below we briefly introduce the quantum expansion method proposed by T. D. Lee et al. [24-26,29]. For details and other methods [32,33], readers can refer to the relevant original papers.

This quantization method expands its quantum solution to the classical solution, with the expansion coefficients being the creation and annihilation operators of particles. They should satisfy the required commutation relations. In this quantum expansion of the soliton solution, we mainly seek the corresponding classical soliton solution and then determine the expansion method without spending effort on finding its commutation relations. As the study in the previous chapters shows, for a nonlinear diffusion-type system, there is always a classical soliton solution. If so, then multi-soliton or anti-soliton solutions related to time can always be established through appropriate asymptotic states at $t \to -\infty$. According to the definition of quantum solutions, we can use this classical solution as a basis to expand the desired quantum solution in small parameters g. It can be seen that the lowest order term in this expansion (i.e. $O\left(\dfrac{1}{g^2}\right)$) is the classical solution. Below, we illustrate this problem with the quantum expansion of solitons in (1+1)-dimensional systems.

For a (1+1)-dimensional system, the Lagrangian is represented by Eq. (10.12). The corresponding equation of motion is

$$\frac{\partial^2 \varphi}{\partial x^2} - g^{-1} V'(g\varphi) = 0 \tag{10.83}$$

For soliton solutions to exist, the absolute minimum of V must be degenerate. Without loss of generality, we can assume that the absolute minimum of $V(g\varphi)$ is zero, and there are several $g\varphi_i$ such that $V(g\varphi_i) = 0$.

Let $\varphi = g^{-1}\sigma(x)$, then from Eq. (10.8.2), we get:

$$\frac{d^2 \sigma}{dx^2} - V'(\sigma) = 0 \tag{10.84}$$

This means

$$\frac{1}{2}\left(\frac{d\sigma}{dx}\right)^2 - V(\sigma) = \text{constant}$$

The classical soliton solution σ is determined by

$$\int_0^\sigma [2V(\sigma)]^{-\frac{1}{2}} d\sigma = x \tag{10.85}$$

According to the quantum expansion method defined earlier, its quantum solution can be expressed as

$$\varphi = g^{-1}\sigma(x - z) + q_n(t)\psi_n(x - z) \tag{10.86}$$

Summing from $n = 2$ to ∞, ψ_n satisfies the orthogonal condition

$$\int_{-\infty}^\infty \psi_n(x - z)\psi_{n'}(x - z)dx = \delta_{nn'} \tag{10.87}$$

The constraint condition is

$$\int_{-\infty}^{\infty} \frac{\partial \sigma(x-z)}{\partial z} \psi_n(x-z) dx = 0 \qquad (10.88)$$

And ψ_n satisfies the eigenvalue equation

$$\left[-\frac{d\psi_n(x-z)}{dx^2} + V''(\sigma) \right] \psi_n(x-x) = \omega^2 \psi(x-x) \qquad (10.89)$$

If we introduce canonical momentum π_n and canonical coordinates q_n, the Hamiltonian function of the system can be expressed as

$$H = \frac{1}{2} \sum_{n=2}^{\infty} (\pi_n^2 + \omega_n^2 q_n^2) + m \qquad (10.90)$$

The corresponding energy spectrum is

$$E = m_r + \sum_n N_n \omega_n$$

where m and m_r are the unrenormalized and renormalized masses of the soliton. Ignoring radiative corrections, we have

$$m = m_r = g^{-2} \int_{-\infty}^{\infty} \left[\frac{\partial \sigma(x)}{\partial x} \right]^2 dx$$

Solving Equations (10.85) and (10.87)– (10.90) completely determines the quantum expansion in Equation (10.86). For example, for the SG equation

$$\frac{\partial^2 \varphi}{\partial t^2} - \frac{\partial^2 \varphi}{\partial x^2} + \frac{\mu_-^2}{g^2} \sin(\varphi g) = 0$$

with the soliton-scattering solution, the quantum solution is:

$$\varphi(x) = \frac{\psi}{g} \tan^{-1} \left[\frac{\mu \text{sh}(r_\mu(x-z)\mu)}{e^\xi} \right] + q_n \psi_n(x-z, \xi)$$

$$\xi = \ln(\text{ch}(r_\mu z \mu)) \ (-\infty \leqslant \xi \leqslant \infty)$$

and ψ_n satisfy

$$\int_{-\infty}^{\infty} \frac{\delta \sigma(Y, \xi)}{\partial y} \psi_n(Y, \xi) dy = 0$$

$$\int_{-\infty}^{\infty} \frac{\delta \sigma(y, \xi)}{\partial y} \psi_n(Y, \xi) dy = 0$$

References

[1] 江泽涵. 拓扑学引论 [M]. 上海科学技术出版社 1978.

[2] Hu S. Homotopy theory[M]. Academic press, 1959.

[3] Husemöller D, Husemöller D. Fibre bundles[M]. New York: McGraw-Hill, 1966.

[4] Shankar R. More SO (3) monopoles[J]. Physical Review D, 1976, 14(4): 1107.

[5] Nielsen H B, Olesen P. Vortex-line models for dual strings[J]. Nuclear Physics B, 1973, 61: 45-61.

[6] Finkelstein D, Rubinstein J. Connection between spin, statistics, and kinks[J]. Journal of Mathematical Physics, 1968, 9(11): 1762-1779.

[7] t Hooft G. Magnetic monopoles in unified theories[J]. Nucl. Phys. B, 1974, 79(CERN-TH-1876): 276-284.

[8] Derrick G H. Comments on nonlinear wave equations as models for elementary particles[J]. Journal of Mathematical Physics, 1964, 5(9): 1252.

[9] Deser S. Absence of static solutions in source-free Yang-Mills theory[J]. Physics Letters B, 1976, 64(4): 463-464.

[10] Friedberg R, Lee T D, Sirlin A. Class of scalar-field soliton solutions in three space dimensions[J]. Physical Review D, 1976, 13(10): 2739.

[11] Bardeen W A, Chanowitz M S, Drell S D, et al. Heavy quarks and strong binding: A field theory of hadron structure[J]. Physical Review D, 1975, 11(5): 1094.

[12] Parks R D. SUPERCONDUCTIVITY. VOLUMES 1 AND 2[J]. 1969.

[13] Belavin A A, Polyakov A M, Schwartz A S, et al. Pseudoparticle solutions of the Yang-Mills equations[J]. Physics Letters B, 1975, 59(1): 85-87.

[14] De Vega H J, Schaposnik F A. Classical vortex solution of the Abelian Higgs model[J]. Physical Review D, 1976, 14(4): 1100.

[15] Arafune J, Freund P G O, Goebel C J. Topology of Higgs fields[J]. Journal of Mathematical Physics, 1975, 16(2): 433-437.

[16] Boulware D G, Brown L S, Cahn R N, et al. Scattering on magnetic charge[J]. Physical Review D, 1976, 14(10): 2708.

[17] Julia B, Zee A. Poles with both magnetic and electric charges in non-Abelian gauge theory[J]. Physical Review D, 1975, 11(8): 2227.

[18] Tomboulis E, Woo G. Soliton quantization in gauge theories. Abelian model, Higgs scalar triplet, collective coordinates, perturbation expansion[J]. Nucl. Phys., B;(Netherlands), 1976, 107(2).

[19] Prasad M K, Sommerfield C M. Exact classical solution for the't Hooft monopole and the Julia-Zee dyon[J]. Physical Review Letters, 1975, 35(12): 760.

[20] Hasenfratz P. Fermion-boson puzzle in a gauge theory[J]. Physical Review Letters, 1976, 36(19): 1119.

[21] Jackiw R, Rebbi C. Spin from isospin in a gauge theory[J]. Physical Review Letters, 1976, 36(19): 1116.

[22] Weinberg E J, Guth A H. Nonexistence of spherically symmetric monopoles with multiple magnetic charge[J]. Physical Review D, 1976, 14(6): 1660.

[23] Patrascioiu A. Extended particles and magnetic charges[J]. Physical Review D, 1975, 12(2): 523.

[24] Lee T D. IV. Examples of four-dimensional soliton solutions and abnormal nuclear states[J]. NON-LINEAR AND COLLECTIVE PHENOMENA IN QUANTUM PHYSICS, 1976: 21.

[25] Friedberg R, Lee T D, Sirlin A. Class of scalar-field soliton solutions in three space dimensions[J]. Physical Review D, 1976, 13(10): 2739.

[26] Christ N H, Lee T D. Quantum expansion of soliton solutions[J]. Physical Review D, 1975, 12(6): 1606.

[27] 周光召, 朱重远, 戴元本, 等具有非 Abel 内部对称性的非拓扑性孤立子 [J]. 中国科学, 1979(11): 1057-1071.

[28] Lee T D. Nontopological solitons and applications to hadrons[J]. Physica Scripta, 1979, 20(3-4): 440.

[29] Friedberg R, Lee T D, Sirlin A. Gauge-field non-topological solitons in three space-dimensions. I[J]. Nucl. Phys., B;(Netherlands), 1976, 115(1).

[30] Christ N H. IX. Quantum expansion of soliton solutions[J]. Physics Reports, 1976, 23(3): 294-300.

[31] Jackiw R. VII. The quantum theory of solitons and other non-linear classical waves[J]. NON-LINEAR AND COLLECTIVE PHENOMENA IN QUANTUM PHYSICS, 1976: 40.

[32] Enz U. A new type of soliton with particle properties[J]. Journal of Mathematical Physics, 1977, 18(3): 347-353

[33] Faddeev L D, Korepin V E. Quantum theory of solitons[J]. Physics Reports, 1978, 42(1): 1-87.

Chapter 11
Solitons in Condensed Matter Physics

For the "soliton problem", a current focus of research is to find out what kind of matter in nature exhibits soliton motion. What are its properties and characteristics? In the first chapter, we provided many examples of specific physical processes to illustrate that a large number of soliton motions exist in nature. But how do we derive the form of soliton motion from a specific process? In the previous chapter, we studied the soliton problem in basic particle physics. In fact, this form of matter motion also exists extensively in condensed matter physics, plasma physics, fluid mechanics, astrophysics, oceanography, molecular biology, materials science, and statistical physics. This is especially evident in condensed matter physics such as superconductors, nonlinear optics, solids, lattice vibrations, ferromagnets, and ferroelectrics. This is because condensed matter is composed of a large number of atoms, ions, and molecules gathered together by a strong cohesive force. Under appropriate conditions (such as high pressure, high density, and low temperature), the nonlinear effects in the system become very significant, and particles within the system can couple through certain interactions into a special "quasiparticle" state and exhibit soliton motion. When solitons move in condensed matter in the form of "quasiparticles", it is accompanied by physical phenomena such as charge density waves, spin density waves, and non-Ohmic electrical transport phenomena. For example, in one-dimensional chain-like crystals (whose crystal structure resembles a chain, with only weak interactions between chains, such as the artificially synthesized organic conductor TTF-TCNQ), if the lattice displacement is used as a variable, the φ^4 field equation can describe such chain-like crystals. The domain walls within are found to be soliton solutions of the φ field equation, and their energy and the resulting charge density waves align well with experimental results.

In particular, the phenomenon of soliton motion in superconductors (including superfluid helium) should be highlighted, as the superconducting state is a coherent state formed after the system undergoes spontaneous symmetry breaking. In this state, the nonlinear interactions between electrons play a crucial role. It's no wonder that Josephson, who won the Nobel Prize for discovering the superconducting tunneling effect, emphasized the importance of spontaneous symmetry breaking in superconductors in his award speech. The Higgs field is also closely related to spontaneous symmetry breaking. Therefore, it is imaginable that solitons hold a significant position in superconducting research. In this chapter, we will use several specific physical processes and phenomena, such as superconductors, ferroelectrics, solids, and lattice vibrations, as research objects to establish their motion equations. We will then use the obtained soliton solutions to explain some discovered physical phenomena. Finally, we will also study the statistical laws of systems with N solitons. The purpose of this is to make it clear how to establish the soliton motion equation through a specific physical process.

11.1 Soliton Motion in Superconductors

In superconducting phenomena, the wave function of bound electron pairs coupled through lattice vibrations (phonons) (macroscopic quantum wave function or order parameter) is often expressed as

$$\varphi(r) = \varphi_A(r) e^{i\theta(x)} \cdot \varphi_A(r) \propto \sqrt{h_s(r)} \tag{11.1}$$

where $h_s(r)$ is the density of superconducting electron pairs. In 1935, Landau expressed the free energy density of a superconducting system in the presence of an electromagnetic field $\mathbf{A}$ as

$$
\begin{aligned}
f_{SH} &= f_{n0} + \alpha |\varphi(r)|^2 + \frac{1}{2}\beta |\varphi(r)|^2 \\
&+ \left| \frac{1}{4m}\left(-i\hbar\nabla - \frac{2e}{c}\mathbf{A}(r) \right)\varphi(r) \right|^2 + \frac{h^2(r)}{8\pi}
\end{aligned}
\tag{11.2}
$$

Here, f_{n0} is the free energy density of the normal state, $\mathbf{A}(r)$ is the vector potential of the applied electromagnetic field, $\mathbf{h}(r) = \nabla \times \mathbf{A}(r)$ is the magnetic field at point r in the superconductor, and α and β are constants related to T. The free energy of the superconductor can be expressed as

$$
\begin{aligned}
F_{SH} &= \int f_{SH} dr = F_{n0} + \int \left\{ \alpha |\varphi|_1^2 + \frac{\beta}{4}|\varphi|^4 \right. \\
&\left. + \frac{1}{4m}\left| -i\hbar\nabla - \frac{2e}{c}\mathbf{A}(r)\varphi(r) \right|^2 + \frac{h^2(r)}{8\pi} \right\} dr
\end{aligned}
\tag{11.3}
$$

From the condition that F_s is minimized, i.e., the variational condition $\delta F_s = 0$, the famous Ginzburg-Landau (GL) equations for superconductors is [1]:

$$\begin{cases} \dfrac{1}{4m}\left(-i\hbar\nabla - \dfrac{2e}{c}\mathbf{A}\right)^2 \varphi + 2\varphi + \beta\,|\varphi|^2\varphi = 0 \\[2ex] \mathbf{h}\cdot\left(-i\hbar\nabla - \dfrac{2e}{c}\mathbf{A}\right)\varphi = 0 \\[2ex] j_s(r) = \dfrac{-ie\hbar}{2m}\left(\varphi^*(r)\nabla\varphi(r) - \varphi(r)\nabla\varphi^*(r)\right) \\[2ex] \qquad\quad - \dfrac{2e^2}{m_e}\varphi^*(r)\varphi(r)A(r) \end{cases} \tag{11.4}$$

These equations can also be expressed as

$$\begin{cases} \xi^2(T)\left(\nabla - i\dfrac{2e}{\hbar c}\right)\mathbf{A}(r)^2\varphi(r) + \varphi(r) \\[2ex] \qquad\quad - \dfrac{1}{\varphi_0^2}|\,\varphi(r)\,|^2\varphi(r) = 0 \\[2ex] \mathbf{j}_s(r) = \dfrac{-c}{4\pi\lambda^2(T)\varphi_0^2}\big[|\varphi(r)|^2\mathbf{A}(r) \\[2ex] \qquad\quad + \dfrac{i\hbar c}{4e}\varphi^*(r)\nabla\varphi(r) - \varphi(r)\nabla\varphi^*(r)\big] \end{cases} \tag{11.5}$$

where

$$\varphi_0(r) = \varphi_A(r)|_{h=n_0} = \sqrt{h_0}$$

$\mathbf{h}_0$ is the density of superconducting electron pairs when there are no flux lines. Here,

$$\nabla \times h = \frac{4\pi}{c}\mathbf{j}_s(r), \quad \nabla \times A = h(r)$$

$\xi(T)$ is the coherence length of the superconducting state, $\xi^2(T) = \dfrac{b}{4m|a(T)|}$, $\lambda(T)$ is the penetration depth, K is the GL parameter and $K = \dfrac{\lambda(T)}{\xi(T)}$.

For

$$H > H_{cl}(T)\left(= \frac{1}{\sqrt{2}K}(\ln K)H_c(T)\right)$$

H_c is the thermodynamic critical field and the lower critical field in the mixed state. At thermodynamic equilibrium, it can be solved directly. When $H \lesssim H_{c2}(T)$ (where $H_{c2} = \sqrt{2}KH_c(T)$ is the upper critical field), $|\varphi_1|$ is very small, $\mathbf{A}(r) \approx H_{c2}X\hat{Y}$, Eq. (11.5) can be approximately written as the linear equation [2]:

$$\xi^2(T)\left(\nabla - i\frac{2e}{\hbar c}H_{cl}X\hat{Y}\right)^2\varphi(r) + \varphi(r) = 0 \tag{11.6}$$

Its solution is

$$\varphi(X,Y) = \epsilon^{iKY} e - \frac{1}{2\xi^2(T)} \left(X - \xi^2(T) K \right)^2$$

where K is an arbitrary constant. The linear combination of the above equation gives the general solution of Eq. (11.6). That is,

$$\varphi_L(X_9 Y) = \varphi_0 \sum_{n=-\infty}^{\infty} C_n e^{iK_n Y} e^{-\frac{1}{2\xi^2(T)}(X-\xi^2(T)K)^2} \tag{11.7}$$

In an ideal type-II superconductor, using the existing physical conditions, some constants in Equation (11.7) can be determined, and Equation (11.7) can be expressed as:

$$\varphi_L(X, Y) = C_0\varphi_0 e^{-\frac{X^2}{2\xi^2(T)}} \sum_{n-\infty}^{\infty} e^{u\sqrt{\sqrt{3}\mu} \cdot \frac{X+iY}{\xi(T)}} e\mu^{2(\frac{\sqrt{3}}{2}+i\frac{1}{2})\pi} \tag{11.8}$$

where $n_s(X,Y) = |\varphi_L(X,Y)|^2$, Obviously, it has triangular symmetry, the lattice constant is:

$$a = \frac{2\pi}{\sqrt{\sqrt{3}n}} \xi(T)$$

where l, m are arbitrary integers.

From this solution, it can be concluded that in an ideal type-II superconductor, the flux lines or vortex lines are arranged in a triangular periodic manner. This structure is known as the Abrikosov structure [3]. This structure has been experimentally confirmed. In a non-ideal type-II superconductor, from an overall perspective, the vortex lines are not distributed in a triangular shape, but locally they still have this distribution. We can prove that these flux lines (vortex lines) are very stable, and the energy of each flux line is:

$$\epsilon = \frac{\varphi_0 H_c(T) l_u K}{4\sqrt{2\pi K}}$$

From equations (11.4) or (11.5), it can be seen that the GL equations are nonlinear partial differential equations, and theoretically, they should have soliton solutions. For example, in equations (11.4) or (11.5), when $A = 0$, the nonlinear Schrödinger equation satisfied by electron pairs with mass $2m$ and charge $2e$ is obtained. It is well known that this equation has a soliton solution in the form of a wave packet (bell-shaped). This solution was obtained by de Gennes [1]. However, when $A \neq 0$, it is very difficult to find the soliton solution of equations (11.4) or (11.5). Recently, Jacobs and others [4] used the Abel-Higgs field method to study this problem.

They expressed the free energy of the superconductor as:

$$F = \frac{C_0}{e} \int \left[\frac{1}{2} \left| \left(\partial_i - \frac{ie}{\hbar c} A_i \right) \varphi \right|^2 + \frac{1}{4} F_{ij} F^{ij} \right. $$
$$\left. + \frac{1}{8} \lambda^2 (|\varphi|^2 - 1)^2 \right] d^3 x \tag{11.9}$$

where

$$F_{ij} = \partial_i A_j - \partial_j A_i$$
$$i, j = 1, 2, 3, \ \partial_i = \frac{\partial}{\partial x_i} \ (i = 1, 2, 3)$$

A is the gauge electromagnetic potential, and λ is the coupling constant.

Using the Euler-Lagrange method, with the above equation, we have:

$$\begin{cases} \left(\partial - i\frac{e}{\hbar c} A \right) \left(\bar{\partial} - i\frac{e}{\hbar c} \overline{A} \right) \varphi + \left(\bar{\partial} - i\frac{ie}{\hbar c} \overline{A} \right) \\ \qquad \times \left(\delta - \frac{ie}{\hbar c} A \right) \varphi - \frac{1}{4} \lambda^2 \varphi(\varphi\bar{\varphi} - 1) = 0 \\ 4\partial\bar{\partial}A - 4\partial^2 A - i\bar{\varphi}2\varphi + i\varphi\partial\bar{\varphi} - \partial A\varphi\bar{\varphi} = 0 \end{cases} \tag{11.10}$$

Note that Jacobs and others only studied the field configuration invariant under translation along a specific third axis, at which point the field only depends on x_1, x_2, and $A_3 = 0$. Therefore, it can be assumed that

$$z = x_1 + ix_2, \quad \bar{z} = x_1 - ix_2$$
$$A = \frac{1}{2} (A_1 - iA_2), \quad \overline{A} = \frac{1}{2} (A_1 + iA_2)$$
$$\partial = \frac{\partial}{\partial z} = \frac{1}{2} \left(\frac{\partial}{\partial x_1} - i\frac{\partial}{\partial x_2} \right)$$
$$\bar{\partial} = \frac{\partial}{\partial \bar{z}} = \frac{1}{2} \left(\frac{\partial}{\partial x_1} + i\frac{\partial}{\partial x_2} \right)$$
$$\bar{x}_i = \frac{\hbar c}{e \mid \varphi \mid_0} \bar{x}_i, \quad \tilde{\varphi} = \mid \varphi_0 \mid \varphi$$
$$\tilde{A}_i = \mid \varphi \mid_0 A, \quad \mid \varphi \mid_0 = C_o$$

They further assumed that:

$$\varphi = e^{i\theta n} f(r) \quad A = -(ni/2\varepsilon)a(r)$$
$$r = |z|, \quad f(\infty) = a(\infty) = 1$$

Then Eq. (11.10) can be transformed into:

$$\begin{cases} \dfrac{d^2 f}{dr^2} + \dfrac{1}{r}\dfrac{df}{dr} - \dfrac{n^2(a-1)}{r^2}f - \dfrac{1}{2}\lambda f(f^2 - 1) = 0 \\[3mm] \dfrac{d^2 a}{dr^2} - \dfrac{1}{r}\dfrac{da}{dr} - (a-1)f^2 = 0 \end{cases} \tag{11.11}$$

where n is related to the magnetic flux, that is:

$$\Phi(B) = \frac{-i}{e}\int(\partial\overline{A} - \bar\partial d)dzd\bar z = \frac{2\pi n}{e}$$

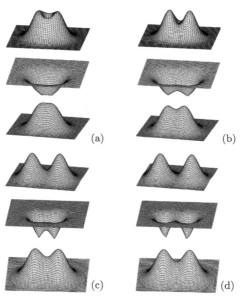

(a)   (b)

(c)   (d)

Figure 11-1 For $\lambda = 1$ and distances $d = 1, 2, 3, 4$ (corresponding to a, b, c, d respectively), the distributions of the energy density of the two vortices (top), the magnitude of the matter field φ (middle), and the magnetic field $\mathbf{h}$ (bottom) are shown

They assumed that:

$$f(r) = 1 + e^{-\lambda p}\sum_{l=0}^{n}(f_l p^l / l!)$$
$$a(r) = 1 + e^{-p}\sum_{l=0}^{n}(a_i p^l / l!)$$

Substituting into equation (11.11), Jacobs et al. used the variational method to obtain the relationship between the interaction energy between two flux lines and the distance d between the two flux lines and the coupling constant λ between the φ field and the electromagnetic field A. The calculation results show that when $\lambda < 1$, the two flux lines attract each other; when $\lambda > 1$, the two flux lines repel each other; and when $\lambda = 1$, there is no interaction. They also used numerical integration to obtain the energy density ε, field φ, and magnetic field $\mathbf{h}$ distributions of the two flux lines when $\lambda = 1$ and $d = 1, 2, 3, 4$. The results are shown in Figure 11-1.

The above describes the movement of superconducting electron pairs in a bulk homogeneous superconductor. Next, we study the movement of superconducting electron pairs through a weakly connected superconducting junction with phase change.

For this problem, in 1962, Josephson theoretically predicted that if the oxide layer of a superconductor-oxide-superconductor junction (Josephson junction) is very thin ($\sim 10 - 30$ Å), the superconductors on both sides of the oxide layer are still weakly connected, and the electron pairs can still pass through the oxide layer, forming a direct current Josephson effect. This was experimentally confirmed in 1964.

Phenomena that appear when the junction interacts with an electromagnetic field are called alternating current Josephson effects. If a voltage V is applied across the oxide layer of the junction, electron pairs will pass through the oxide layer from the side with higher chemical potential to the side with lower chemical potential. This process is energy lossless. This results in an additional energy of $2eV$. According to the law of conservation of energy, this part of the energy is emitted in the form of electromagnetic waves, with a frequency satisfying $2eV = h\nu$. Additionally, if microwaves are irradiated onto the Josephson junction from the outside and a bias current source is used, a series of equal voltage steps will appear on the I-V (current-voltage) characteristic curve. The voltage of the n-th step V satisfy $V_n = nh\nu/2e$ or $2eV_n = nh\nu$. These phenomena have been observed multiple times in experiments. Studies indicate that they are directly related to the phase change $\Delta\theta = \varphi$ caused by Cooper pairs of superconducting electrons passing through the Josephson junction, leading to the appearance of the Josephson current and the change of magnetic flux lines, which can propagate along the superconducting junction. Let us now study the laws governing this motion.

The well-known Josephson relation existing in the superconducting junction is [5]:

$$\begin{cases} j = j_c \sin\varphi \, , \, \hbar\dfrac{\partial\varphi}{\partial t} = 2eV \\[3mm] \hbar\dfrac{\partial\varphi}{\partial x} = \dfrac{2ed}{c} H_y, \, \hbar\dfrac{\partial\varphi}{\partial y} = \dfrac{2ed}{c} H_x \quad (\varphi = \theta_1 - \theta_2) \end{cases} \tag{11.12}$$

Equation (11.12) is not a closed set of equations because V and H are unknown and usually need to be solved in conjunction with Maxwell's equations:

$$\nabla \times H = \frac{4\pi}{c}\mathbf{j}$$

Under the condition $\mathbf{H} = (H_x, H_y, 0)$, we have:

$$\frac{\partial}{\partial x} H_y(x, y, t) - \frac{\partial}{\partial y} H_x(x, y, t) = \frac{4\pi}{c} \mathbf{j}(x, y, t) \tag{11.13}$$

Since

$$\mathbf{j} = \mathbf{j}_s(x, y, t) + \mathbf{j}_n(x, y, t) + \mathbf{j}_\alpha(x, y, t) + I_0 \tag{11.14}$$

Here, $\mathbf{j}_n$ is the normal current density in the junction. If there is a resistance $R(V)$ across the junction and a voltage on both sides, then $\mathbf{j}_n = \dfrac{V}{R(V)}$. $\mathbf{j}_\alpha$ is called the displacement current and can be expressed as $\mathbf{j}_\alpha = c\dfrac{dV(t)}{dt}$, I_0 is the usual current. By combining Equations (11.12) —(11.14), we can obtain:

$$\begin{aligned}
\frac{\hbar c}{e^* d}\left(\frac{\partial^2}{\partial x^2}\varphi + \frac{\partial^2}{\partial y^2}\varphi\right) &= \frac{4\pi}{c}\left(\frac{\hbar c}{e^*}\frac{d^2\varphi}{dt^2}\right. \\
&\left. + \frac{\hbar}{e^* R}\frac{d\varphi}{dt} + j_c \sin\varphi + I_0\right)
\end{aligned} \tag{11.15}$$

Assuming

$$V_0 = \left(\frac{c^2}{4\pi cd}\right)^{1/2} \qquad v_0 = \frac{1}{RC}$$

$$\lambda_J = \left(\frac{c^2\hbar}{4\pi de^*}\right)^{1/2} \qquad j_0 = \frac{4 I_0 \pi e^* d}{\hbar c^2}$$

Equation (11.15) becomes:

$$\nabla^2\varphi - \frac{1}{V^2}\left(\frac{\partial^2}{\partial t^2}\varphi + \gamma_0 \frac{\partial}{\partial t}\varphi\right) = \frac{1}{\lambda_J^2}\sin\varphi + j_0 \tag{11.16}$$

This is a complete equation that the phase difference φ of the superconducting electron wave function in the superconducting junction satisfies [2,5].

When there is no Josephson current $j_s = 0$ and $j_0 = 0$ in the junction, Equation (11.16) reduces to:

$$\nabla^2\varphi - \frac{1}{V^2}\frac{\partial^2}{\partial t^2}\varphi - \frac{r_0}{V^2}\frac{\partial}{\partial t}\varphi = 0$$

If we take partial derivatives of the above equation with respect to x, y, and t respectively, using Equation (11.12) and

$$E_z = \frac{1}{d} V$$

we obtain:

$$
\begin{cases}
\nabla^2 E_z - \dfrac{1}{V_0^2} \dfrac{\partial^2}{\partial t^2} E_z - \dfrac{\gamma_0}{V_0^2} \dfrac{\partial}{\partial t} E_z = 0 \\[3mm]
\nabla^2 H - \dfrac{1}{V_0^2} \dfrac{\partial^2}{\partial t^2} H - \dfrac{\gamma_0}{V_0^2} \dfrac{\partial}{\partial t} H = 0
\end{cases}
$$

This is a set of wave equations for the electromagnetic field, whose solution represents electromagnetic waves propagating along the junction plane with a damping coefficient γ_0, and the propagation speed is V_0.

Generally, in a superconductor, $\mathbf{j}_n$ is very small, i.e., the junction resistance R of a single electron is very large, so we can consider $j_m \to 0$. Then, Equation (11.16) reduces to:

$$\nabla^2 \varphi - \frac{1}{V^2} \frac{\partial^2 \varphi}{\partial t^2} = \frac{1}{\lambda_J^2} \sin \varphi \tag{11.17}$$

In the equilibrium state (i.e., independent of t), Equation (11.17) becomes:

$$\nabla^2 \varphi = \frac{1}{\lambda_J^2} \sin \varphi \tag{11.18}$$

Under one-dimensional and weak magnetic field conditions, Equation (1.18) reduces to:

$$\nabla^2 \varphi = \frac{1}{\lambda_J^2} \varphi$$

The solution is

$$
\begin{cases}
\varphi(x) = -\dfrac{2e^* d \lambda_J}{\hbar c} H_0 e^{-x/\lambda_J} \\[3mm]
H(x) = H_0 e^{-x/\lambda_J}
\end{cases}
\tag{11.19}
$$

In two dimensions,

$$
\begin{cases}
\varphi(x, y) = -\dfrac{4e^* d \lambda_J}{\hbar c} H_0 e^{-(x+y)/\lambda_J} \\[3mm]
H(x, y) = H_0 e^{-(x+y)/\lambda_J}
\end{cases}
\tag{11.20}
$$

This is precisely the Meissner effect occurring in superconductors. λ_J is the Josephson penetration depth.

In one dimension, the solution of equation (11.18) is:

$$\varphi = 4\tan^{-1} e^{\frac{x-x_n}{\lambda_J}}$$

This is a static kink soliton.

The system described by Equation (11.15) is a system of spontaneous symmetry breaking, which has infinitely many degenerate ground states. Therefore, the appearance of soliton solutions in this case is not an issue. The kink soliton solution of Equation (11.17) for the (1+1)-dimensional system is well known to us, namely:

$$\varphi_s = \pm 4\tan^{-1} e^{\pm\frac{x-ut}{\lambda_J - \sqrt{1-u^2}}}$$

$$\begin{cases} \text{``$\pm\,\pm$''denote the positive kink soliton} \\ \text{``$-\,+$''denote the negative kink soliton} \end{cases} \tag{11.21}$$

This indicates that the phase field φ is performing soliton motion, which is accompanied by the generation of the Josephson effect. The mass of the soliton for equation (11.21) is

$$M^{cl} = \frac{8m'^3}{\beta''}\left(\frac{m^4}{\beta''} = \frac{1}{\lambda_L^2}\right)$$

It can also be proven that such solitons are stable.

The above study concerns the solution of the SG equation for a Josephson junction without boundary conditions, but it is not universal. In fact, general superconducting junctions have definite boundary conditions or are connected to appliances. In such cases, the SG equation is difficult to solve. Recently, Scott and Constabile et al.[7] have conducted extensive research, and due to its complexity, only the results are listed below. For definite boundary conditions such as:

$$\varphi_x(0,t) = \varphi_x(L,t) = 0 \tag{11.22}$$

Then, what is the solution to the sine-Gordon equation? Lamb has obtained the solution to the SG equation under this condition as:

$$\varphi(x,\,t) = 4\tan^{-1}[\,h(x)\,g(x)\,] \tag{11.23}$$

where h and g are general Jacobian elliptic functions, satisfying:

$$(h')^2 = ah^4 + (1+\ b)h^2 - c \text{ and } g'^2 = cg^4 + bg^2 - a$$

$(a, b, c$ are arbitrary constants). Constabile et al.[7] have given three basic solutions for oscillating soliton waves:

(1) Plasma oscillation solution, in the form of:

$$\varphi = 4\tan^{-1}\left[A\mathrm{cn}\left(\frac{\beta x}{\lambda_J\sqrt{1-\mu^2}}, K_f\right), \mathrm{cn}(Q_t; K_g)\right]$$

here

$$K_f^2 = \frac{A^2[\beta^2(1+A^2)+1]}{\beta^2(1+A^2)^2}$$

$$K_g^2 = \frac{A^2[\Omega^2(1+A^2)+1]}{\Omega^2(1+A^2)^2}$$

where Ω, β and A obey the nonlinear dispersion equation:

$$\Omega^2 - \beta^2 = \frac{1-A^2}{1+A^2}$$

From the periodicity required by boundary conditions, we obtain:

$$\beta_N = \frac{ZN}{L}\mathbb{K}(K_1)$$

Here, $N = 1, 2, \ldots$ is the node number of the stable wave. $\mathbb{K}(K_1)$ is the first kind of complete elliptic integral.

(2) Breather oscillation solution: it can take two forms, (a) localized vortex-antivortex bound state in the center of the Josephson junction, (b) virtual antivortex state constrained by the vortex at the Josephson end. The solution is:

$$\varphi = 4\tan^{-1}\left\{A\mathrm{dn}\left(\frac{\beta(x-x_0)}{\lambda_j(1-u^2)^{1/2}}; \ K_f\right), \mathrm{sn}(\Omega_t; K_g)\right\}$$

$$K_f^2 = 1 - \left(\frac{1-\beta^2(1+A^2)/A^2}{\beta^2(1+A^2)}\right)$$

$$K_s^2 = \frac{A^2[1-\Omega^2(1+A^2)]}{\Omega^2(1+A^2)}$$

Its nonlinear dispersion relation is: $\beta = \Omega A$. Required by open boundary conditions:

$$\beta_n = \left(\frac{n}{L}\right)\mathbb{K}(K_f)$$

(3) Vortex line oscillation solution, in the form of:

$$\varphi = 4\tan^{-1}\left\{A\mathrm{dn}\left(\frac{\beta x}{\lambda_J(1-u^2)^{1/2}}; \ K_f\right), \mathrm{tn}(Q_t; K_g)\right\}$$

$$K_f^2 = 1 - \left(\frac{(\beta^2/A^2)\,(A^2-1)-1}{\beta^2(A^2-1)} \right)$$

$$K_g^2 = 1 - \left(\frac{A^2 Q^2 (A^2-1)-1}{Q^2(A^2-1)} \right)$$

Its nonlinear dispersion relation is the same as above.

For the general Equation (11.16) with dissipative structure, a stable soliton solution can only be obtained when j_0 and $\dfrac{\gamma_0}{V}\dfrac{\partial \varphi}{\partial t}$ are very small and cancel each other out. The solution remains Equation (11.23). Since the sine-Gordon equation has Lorentz invariance, it remains invariant under

$$x \to x' = (x - ut)/(1 - u^2)^{1/2}$$

$$t \to t' = (t - u^{-1}x)/(1 - u^2)^{1/2}$$

Then the solution is

$$\varphi(x,\,t) = 4\tan^{-1}\left[f\left(\frac{x - ut}{(1 - u^2)^{1/2}} \right) g\left(\frac{t - x/u}{(1 - u^2)^{1/2}} \right) \right] \tag{11.24}$$

This wobbling fluxon vortex line can generate radiation. If one end of the junction is connected to a resistor, equation (11.22) becomes:

$$\begin{cases} \varphi_x(0,\,t) = 0 \\ -\varphi_t(L,\,t)/\varphi_x(L,\,t) = R_0 \end{cases} \tag{11.25}$$

In this case, transverse waves propagate at speed $u = -\dfrac{\varphi_t}{\varphi_x}$. When g is a constant, the solution of equation (11.24) is a simple transverse wave. When g is not constant, the solution of equation (11.24) cannot satisfy the boundary condition of equation (11.25). We expect the resistor to terminate at one end, where the vortex line oscillation solution gains energy and converts to plasma and breather vortex oscillation solutions.

If the boundary condition changes to:

$$\begin{cases} \varphi_x(0,t) - \alpha\varphi_t(0,t) = 0 \\ \varphi_x(L,\,t) + \beta\varphi_t(L,\,t) = 0 \\ \varphi(x,\,0) = F(x,\,0,\,n) \\ \varphi_t(0,\,x) = F_t(x,\,0,\,n) \end{cases} \tag{11.26}$$

Then the solution to the corresponding sine-Gordon equation is:

$$F(x,\, t,\, n) = 4\tan^{-1}\exp\left[\frac{x - u_0 t - x_0}{\lambda_J\sqrt{1 - u^2}}\right]$$
$$+ (n-1)4\tan^{-1}\exp\left[\frac{x - u_1 t - x_1}{\lambda_J\sqrt{1 - u^2}}\right] \tag{11.27}$$

where the first term represents the vortex line with initial position and speed x_0 and u_0, respectively. When $n = 2$, the second term represents the movement of the second vortex line with initial speed u_1 and position x_1, respectively.

From the above, it can be seen that the transmission process of vortex lines along the superconducting junction is a very complex physical process. Under different physical conditions and environments, the transmission equations of vortex lines are very different, and the properties and characteristics exhibited are also different. This process makes us deeply understand that the motion of vortex lines as soliton states is due to the action of a "nonlinear potential field". Therefore, a series of nonequilibrium phenomena appear on the superconducting junction.

11.2   Soliton Motion in Ferroelectrics

We already know that there are two types of domain wall configurations in ferromagnets: one is the Ising domain wall (for materials with larger uniaxial anisotropy) and the other corresponds to the Bloch domain wall. This change in configuration is a two-dimensional phase transition, with the order parameter having spatial helical characteristics. As mentioned in the first chapter, the motion of such Bloch walls obeys the sine-Gordon equation, which has kink soliton solutions. Is there a similar phenomenon occurring in ferroelectrics? Current research shows that in ferroelectrics, there are also two types of configuration states: the ferroelectric phase and the antiferroelectric phase. What are the structure and motion of ferroelectric domain walls formed by the competition between these two ordered states? This is worth studying.

When in the ordered states of ferroelectric and antiferroelectric phases, Landau theory is often used to describe them [8,12]. Similar to the ferromagnet case, an order parameter needs to be introduced. If P_1 and P_2 are the polarization intensities of two sublattices, then the free energy of the system can be expressed as:

$$F = \frac{c}{2}[(\Delta P_1)^2 + (\nabla P_2)^2] + \frac{a}{2}(P_1^2 + P_2^2)$$
$$+ b\,P_1 P_2 + \frac{d}{4}(P_1^2 + P_2^2)^2\ (d > 0) \tag{11.28}$$

Now we introduce two variables corresponding to ferroelectricity and antiferroelectricity:

$$B = \frac{1}{\sqrt{2}}(P_1 + P_2),\ \ A = \frac{1}{\sqrt{2}}(P_1 - P_2) \tag{11.29}$$

Then the free energy can be expressed as:

$$F = \frac{c}{2}[(\nabla B)^2 + (\nabla A)^2] + \frac{a+b}{2}B^2 + \frac{a-b}{2}A^2 + \frac{d}{4}(B^2 + A^2)^2 \qquad (11.30)$$

When $A = 0$ and $B = \pm\sqrt{-\dfrac{a+b}{d}}$, it is in the ferroelectric phase, and when

$B = 0$ and $A = \pm\sqrt{b - \dfrac{a}{d}}$, it is in the antiferroelectric phase. The phase diagram is shown in Figure 11-2, where regions B, A, and P represent the ferroelectric phase, antiferroelectric phase, and paraelectric phase, respectively. Region I indicates the domain wall formed in the ferroelectric phase with antiferroelectric characteristics. Region II indicates the domain wall formed in the antiferroelectric phase with ferroelectric characteristics. Thus, if b changes, a first-order phase transition from ferroelectric to antiferroelectric occurs at $b = 0$. This phenomenon has been observed in crystals such as $\mathrm{Pb(Zr_{1-x}Ti_x)O_3}$, confirming the correctness of the above theory.

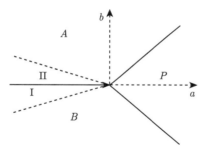

Figure 11-2   The phase diagram of ferroelectric

As seen above, the ferroelectric and antiferroelectric phases are completely symmetrical. Therefore, we only study the ferroelectric domain wall. From the Euler-Lagrange equation, we get:

$$\begin{cases} c\nabla^2 B - (a+b)B - d(B^2 + A^2)B = 0 \\ c\nabla^2 A + (b-a)A - d(B^2 + A^2)A = 0 \end{cases} \qquad (11.31)$$

When $A = 0$, the similar φ^4-type field equation for B is obtained:

$$c\nabla^2 B - (a+b)B - dB^3 = 0 \qquad (11.32)$$

The soliton solution for this equation, representing the ferroelectric domain wall, is easily obtained:

$$B_0(x) = \sqrt{-\frac{a+b}{d}} \tanh\left(x\sqrt{-\frac{a+b}{2c}}\right) \qquad (11.33)$$

Now we study the stability of this domain wall, i.e., small excitations around the above solution $B_0(x)$. If we take the effective density of the ferroelectric and antiferroelectric phases as ρ_1 and ρ_A (for simplicity, set $\rho_1 = \rho_A = \rho$), and express B and A as:

$$B = B_0(x) + B_1(x,t), \quad A = A_1(x,t) \tag{11.34}$$

Substituting Equation (11.34) into Equation (11.31), we get the eigenvalue equation in the first-order approximation, as

$$
\begin{cases}
\rho\omega^2 B_1 = -c\nabla^2 B_1 + (a+b)B_1 \\
\qquad\quad - 3(a+b)\, B_1 \tanh^2\left(x\sqrt{-\dfrac{a+b}{2c}}\right) \\
\rho\omega^2 A_1 = -c\nabla^2 A_1 + (a-b)A_1 \\
\qquad\quad - (a+b)\, A_1 \tanh^2\left(x\sqrt{-\dfrac{a+b}{2c}}\right)
\end{cases}
\tag{11.35}
$$

It is easy to find the eigenvalues of this Schrödinger-like equation. If these eigenvalues are positive, the system is stable. For the first equation, all eigenvalues are positive. For the second equation, we can write the lowest eigenvalue as $\rho\omega_0^2 = \dfrac{a-3b}{2}$.

Thus, if $a > 3b$, the antiferroelectric fluctuations are suppressed, making the domain wall stable. If $3b > a$, the system is unstable due to these fluctuations. Near the threshold $3b = a$, the system exhibits soft mode behavior for ω_0. Therefore, in the cooperative process, the domain wall changes configuration.

Regarding the nonlinear equations (11.31), Sarker et al.[9] have conducted research. We list two sets of solutions here:

$$(a.)\; 3b - a < 0,\, B = \sqrt{-\frac{a+b}{d}}\,\tanh\left(x\sqrt{-\frac{a+b}{2c}}\right),\; A = 0$$

$$(b.)\; 3b - a > 0,\;
\begin{cases}
B = \sqrt{-\dfrac{a+b}{d}}\,\tanh\left(x\sqrt{-\dfrac{2b}{c}}\right) \\
A = \sqrt{\dfrac{3b-a}{d}}\cdot\dfrac{1}{\mathrm{ch}(x\sqrt{-2b/c})}
\end{cases}
$$

Thus, the amplitude of the antiferroelectric component is precisely the order parameter in our problem. Therefore, the domain wall exhibits an antiferroelectric component.

Interestingly, as b approaches zero, near the ferroelectric-antiferroelectric phase transition, the width of the antiferroelectric component diverges. This indicates that the nucleation process of the new phase has occurred within the domain wall during the first-order phase transition.

The same situation occurs in the antiferroelectric phase, where an antiferroelectric domain wall contains a ferroelectric component. When an additional gradient field is applied, this structure is very sensitive. Comparing this with the ferroelectric case, we get an effective field:

$$\mathbf{E}_{eff} = (\mathrm{grad}\mathbf{E}) \times \delta$$

where δ is the width of the domain wall, given by $\sqrt{\dfrac{c}{-2b}}$. The effective field is generally very small. We can expect the same situation to occur in the ferromagnetic-antiferromagnetic phase transition [10].

11.3 Solitons of Coupled Systems in Solids

In solids, a wide variety of phenomena occur. As pointed out in the first chapter, considering the anharmonicity of atomic vibrations leading to acoustic and optical modes, soliton motion of the wave-packet type can accompany these modes [13]. These occur in nonlinear systems of a single mode. But what happens if there exists a system with coupled modes in solids? Moskalenko et al.[14] studied systems with exciton-electromagnetic mode coupling and two-level atom-electromagnetic mode coupling, finding wave-packet soliton motion within. This study inspired Nelson to investigate coupled systems of optical modes and electromagnetic modes (where the linear sinusoidal excitation is called a polariton) and optical modes and acoustic modes [10]. The research indicated soliton motion also existed in these systems. Why does soliton motion appear in such physical systems? To clarify its physical origin, let's consider an optical mode. Its potential energy includes quadratic and quartic terms of the optical mode amplitude. We assume each term can generate a restoring force, though the restoring force corresponding to the quadratic term is small. In other words, we consider the soft mode behavior above the phase transition temperature. If this transition is a ferroelectric-paraelectric transition, we study the coupling of an electromagnetic wave and an optical mode. If this transition is a ferroelastic-paraelastic transition, we study the coupling of a sound wave and an optical mode. In both cases, the coupling term for the traveling wave can lead to a negative quadratic term, turning it into an anti-restoring force. Therefore, the effective potential energy of the optical mode has two minima under interaction. It is well known that the existence of such a double-well potential energy curve, similar to the degenerate vacuum in topological solitons, induces soliton motion. Below, we conduct a theoretical study of such a coupled system.

First, we study the coupled system of optical and electromagnetic modes. We construct the nonlinear interaction of the optical mode and the applied electromagnetic mode in this system. It can be described by the scalar amplitude E of an electric field's eigenmode:

$$\frac{\partial^2 E}{\partial z^2} - \frac{K_h}{c^2} \frac{\partial^2 E}{\partial t^2} = \frac{q}{\varepsilon_0 c^2} \frac{\partial^2 y}{\partial t^2} \tag{11.36}$$

and the nonlinear force equation of the optical mode's normal coordinate y:

$$m \frac{\partial^2 y}{\partial t^2} = -A_{20}y - A_{40}y^3 + qE \tag{11.37}$$

Here, K_h is the dielectric constant in the frequency range above the optical mode's resonance frequency. q and m are the charge and mass density associated with the optical mode, A_{20} and A_{40} are the characteristic constants corresponding to the linear and nonlinear restoring force terms of the optical mode. We still assume $A_{20} > 0, A_{40} > 0$, where the potential energy of the optical mode has a single minimum at $y = 0$. The non-perturbed state of the crystal has $y = 0$ and $E = 0$.

Now let's find its stable quasi-pulse solution. To do this, make the coordinate transformation $\xi = z - Vt$ in Eq. (11.36), then it becomes:

$$\frac{\partial^2 E}{\partial \xi^2} = \frac{V^2 q}{\varepsilon_0} \frac{\partial^2 y}{\partial \xi^2} \cdot \frac{1}{(c^2 - K_h V^2)}$$

Integrating twice, we get:

$$E = V^2 qy / \varepsilon_0 (c^2 - K_h V^2) \tag{11.38}$$

To satisfy the boundary condition:

$$E = y = \frac{\partial E}{\partial \xi} = \frac{\partial y}{\partial \xi} = 0 \quad (\xi = \pm\infty) \tag{11.39}$$

Thus, the two integration constants in Eq. (11.38) are both taken as zero.

The equation of motion for the optical mode (11.37) now becomes:

$$m^2 V^2 \frac{\partial^2 y}{\partial \xi^2} = -A_{20}y - A_{40}y^3 \tag{11.40}$$

Here, we have substituted Eq. (11.38) into Eq. (11.37). Where

$$A'_{20} = A_{20}V_h^2(V_l^2 - V^2)/V_l^2 \, (V_h^2 \longrightarrow V^2)$$

$$V_h^2 = c^2/K_h, \quad V_l^2 = c^2/K_l$$

$$K_l = K_h + q^2/\varepsilon_0 A_{20}$$

K_l is the dielectric constant in the frequency range below the optical mode's resonance frequency. For a soliton solution to exist, we must have

$$A'_{20} < 0, \text{ i.e. } V_l < V < V_h \tag{11.41}$$

This indicates that the allowable range of soliton velocity coincides with the forbidden range of polariton phase velocity.

Under this condition, integrating twice using the boundary condition, we get the wave-packet soliton solution:

$$\begin{cases} y = y_0 \text{sech}[K(\xi - \xi_0)] \\ E = \dfrac{\varepsilon y_0}{\varepsilon_0 (V^{-2} - V_h^{-2})} \text{sech}[K(\xi - \xi_0)] \end{cases} \tag{11.42}$$

where the amplitude of optical mode $y_0 = \dfrac{AK\nu}{\pi}$, and the quasi-wave vector is:

$$K = \left(\frac{\omega_L}{V}\right)(V^2 - V_h)^{1/2}/(V_{11}^2 - V^2)^{1/2} \tag{11.43}$$

where ξ_0 is integrate constant, A is the area constant,as

$$A = \int_{-\infty}^{\infty} y dt = \pi \left(\frac{\partial m}{A_{40}}\right)^{1/2}$$

where ω_L and ω_T are longitudinal and transverse frequency. We have

$$\begin{cases} \left(\dfrac{\omega_L}{\omega_T}\right)^2 = \left(\dfrac{V_h}{V_l}\right)^2 = K_l/K_b \\ \omega_T = (A_{20}/m)^{1/2} \end{cases} \tag{11.44}$$

Eqs. (11.42) and (11.43) are called the Lyddane-Sachs-Teller (LST) relations. Generally, V_h and V_l are considered as limiting velocities that can be measured from the high and low frequency ranges of the optical mode resonance, respectively. Examining the soliton solution (11.42), (11.43), we find another interpretation of the LST relation: V_h is the velocity of the maximum amplitude, the narrowest soliton, and V_l is the velocity of the minimum amplitude, the widest soliton.

Eq. (11.43) is the dispersion relation of the soliton, showing the relationship between the quasi-wave vector K and the velocity V. This relationship depends only on the linear characteristics of the medium, meaning the dispersion relation of an intrinsic nonlinear wave does not include nonlinear features. Defining a quasi-frequency $\omega = KV$, the dispersion relation becomes more explicit. Eq. (11.43) becomes:

$$K^2 = (\omega/V_h)^2(\omega_L^2 + \omega^2)/(\omega_T^2 + \omega^2) \tag{11.45}$$

Comparing this with the dispersion relation of the monochromatic plane wave solution (polaritons) without nonlinear parameters (i.e., $A_{40} = 0$):

$$K_p^2 = (Q_p/V_h)^2(\omega_L^2 - Q_p^2)/(\omega_T^2 - Q_p^2) \tag{11.46}$$

(Here, K_p and Q_p have their usual meaning for polaritons, i.e., $E \sim \exp[i(K_p z - \Omega_p t)]$), we see that replacing K_p with iK and Ω_p with $i\omega$ gives the soliton's dispersion relation. Therefore, the soliton's dispersion relation corresponds to the dispersion relation of a polariton with imaginary wave number and frequency. Figure 11-3 shows the relationship between them.

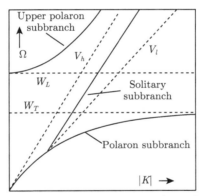

Figure 11-3　The dispersion relation of coupled system

Now, let's discuss the coupling system between optical modes and acoustic modes. This interaction system can be described by the acoustic wave equation

$$\rho\frac{\partial^2 u}{\partial t^2} = A_{02}\frac{\partial^2 u}{\partial X^2} + A_{11}\frac{\partial y}{\partial X} \tag{11.47}$$

and the nonlinear force equation of the optical mode

$$m\frac{\partial^2 y}{\partial t^2} = -A_{20}y - A_{40}y^3 \longrightarrow A_{11}\frac{\partial u}{\partial X} \tag{11.48}$$

Here, u is the scalar amplitude of an acoustic eigenmode, X is the particle coordinate in the propagation direction, ρ is the mass density of the medium, A_{02} is the elastic stiffness produced by various sources other than the normal coordinate y of the optical mode, and A_{11} is the coupling constant between y and the displacement gradient $\dfrac{\partial u}{\partial X}$. The meanings of the other quantities are the same as before.

As in the previous method, we can immediately write out the soliton solutions of the above two equations in the form of wave packets:

$$y = y_0 \text{sech}[K(\xi - \xi_0)]$$

$$\frac{\partial u}{\partial X} = \frac{A_{11}y_0}{\rho} \frac{}{(V^2 - V_h^2)} \text{sech}[K(\xi - \xi_0)]$$

where $\xi = X - Vt$. The soliton existence conditions (11.41), (11.43), and (11.44) also apply here. But now,

$$V_h^2 = A_{02}/\rho, \ \ V_l^2 = (A_{02} - A_{11}^2/A_{20})\rho, \ \ \omega_L = (A_{20}/m)^{1/2}$$

where $\left(\dfrac{A_{20}}{m}\right)$ is called the longitudinal optical frequency ω_L, not the transverse optical frequency ω_T, because the elastic compliance (which is the reciprocal of elastic stiffness) is similar to the dielectric constant. Since the speed of the linear wave is inversely proportional to the square root of the relevant quantity in each case, the frequency in the above equation corresponds to the condition where the compliance is zero. Therefore, $\left(\dfrac{A_{20}}{m}\right)^{1/2}$ is referred to as the longitudinal optical frequency.

There are many instances of mode coupling phenomena that occur in solids. We can study other coupling systems similarly.

11.4 Statistical Mechanics of Toda Lattice Solitons

Previously, we studied the dynamics of a single soliton. What about a thermodynamic system composed of N solitons?

For the statistical mechanics and thermodynamic quantities of a mechanical system of N sine-Gordon and φ^4 field kink solitons, at low temperatures, such quasi-solitons can often be treated like a non-interacting phonon gas [15-19]. However, for non-topological Toda lattice soliton systems, the situation is somewhat different [21].

We know that the Toda lattice is a linear chain with nearest-neighbor interactions. Its Hamilton is

$$H = \sum_{i=1}^{N} \left(\frac{P_i^2}{2m} + V(x_{i+1} - x_i) \right) \tag{11.49}$$

The Toda lattice potential is given by:

$$V(r) = \frac{a}{b}(e^{-br} - 1) + ar$$

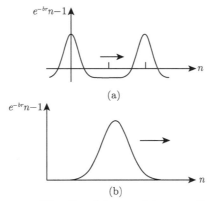

Figure 11-4   The Cnoidal wave (a) and soliton (b)

For small amplitude systems, it can almost be considered harmonic oscillations, whose periodic solutions are phonons. For high amplitude systems, the periodic solutions are cnoidal waves, as shown in Figure 11-4. Cnoidal waves can be considered as phonons deformed due to anharmonicity. At low temperatures, they can be simplified to a single phonon. If the cnoidal wave with $\lambda \to \infty$ becomes a soliton, as shown in Figure 11-4, then the soliton is a more severely deformed phonon, with its energy and velocity determined by the parameter a. i.e [20].

$$E_{kink}(\alpha) = 2\frac{a}{b}\mathrm{sh}\alpha \left(\mathrm{ch}\alpha \; - \; \frac{\mathrm{sh}\alpha}{\alpha} \right)$$

$$E_{pot}(a) = \; 2\frac{a}{b}\alpha \left\{ \left(\frac{\mathrm{sh}\alpha}{a} \right)^2 - 1 \right\}$$

then

$$\begin{cases} E(\alpha) = E_{kink} + E_{pot} = 2\dfrac{a}{b}\,(\mathrm{sh}\alpha\mathrm{ch}\alpha - \alpha) \\[3mm] V(\alpha) = \sqrt{\dfrac{ab}{m}}\dfrac{\mathrm{sh}\alpha}{\alpha} \end{cases} \qquad (11.50)$$

In principle, a single soliton can form an N-soliton system. However, the energy spectrum and velocity of Toda lattice solitons are different from those of the φ^4 field and sine-Gordon systems, as shown in Figure 11-5. In the sine-Gordon lattice system, the soliton can be compared to a relativistic particle, with the sound speed C_s acting as the speed of light. We need energy to excite a soliton and more energy to make it move. In contrast, in the Toda lattice, the sine-Gordon phonons are different from kink excitations. Therefore, in general, solitons and phonons must be used together to describe sine-Gordon excitations. However, the Toda lattice is not like this.

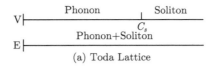

(a) Toda Lattice

(b) Sine-Gordon

Figure 11-5   Energy spectrum

The Hamiltonian of the Toda lattice system can be approximately expressed as the sum of Hamiltonians corresponding to elementary excitations with one degree of freedom:

$$H = H_1(P_1, Q_1) + H_2(P_2, Q_2) + \cdots + H_N(P_N, Q_N) \tag{11.51}$$

Then, the classical partition function of the system is:

$$Z = Z_1 \cdot Z_2 \cdots \cdots Z_N, \text{ while } Z_i = \int e^{-\beta H_i(P_i, Q_i)} dP_i dQ_i \tag{11.52}$$

Replacing the integral over P with an integral over energy E, we have:

$$Z = \int \frac{\partial P(E, Q)}{\partial E} e^{-\beta E} dE dQ \tag{11.53}$$

Using the Hamiltonian equations:

$$\dot{Q} = \frac{\partial H}{\partial P}, \dot{P} = -\frac{\partial H}{\partial Q}, \frac{\partial P(E, Q)}{\partial E} = \frac{1}{\dot{Q}} \tag{11.54}$$

making t as the new variable, from Eq. (11.53), we have

$$Z = \int dE \int_{H(P,Q)=E} e^{-\beta E} dt = \int T(E) e^{-\beta E} dE \tag{11.55}$$

Here, $T(E)$ is the period of the excitation passing through the phase space orbit $H(P, Q) = E$. For a phonon with wavelength λ, we have:

$$T(E) = \frac{2\pi}{\omega(\lambda)}$$

Then

$$Z = \int_0^\infty \frac{2\pi}{\omega(\lambda)} e^{-\beta E} dE = \frac{2\pi}{\omega(\lambda)\beta}$$

For a soliton,

$$T(E) = \frac{2N}{V(E)}$$

Here, N is the number of solitons on the chain, and "$V \cdot t$" is the number of solitons passing through the chain in time t. The factor 2 accounts for the two possible velocities, V and $-V$, for a given energy E. For this soliton system, the partition function Z simplifies to:

$$Z_{sol} = \int_0^\infty \frac{2N}{V(E)} e^{-\beta E} dE \tag{11.56}$$

due to

$$Z = \sum_n \varepsilon^{-\beta E_a} \xrightarrow{h \to 0} \int \rho(E) e^{-\beta E} dE = \frac{1}{2\pi\hbar} \int e^{-\beta E} dP dq \tag{11.57}$$

Comparing with (11.55), we see that $T(E)$ is proportional to the classical limit of the quantum mechanical density of states $\rho(E)$.

The velocity of its traveling wave is V, then the equation of its position is:

$$\dot{Q} = \frac{\partial H}{\partial P} = V = \text{const} \tag{11.58}$$

If we assume the chain is uniform, H is independent of Q, and we have:

$$\dot{P} = -\frac{\partial H}{\partial Q} = 0$$

Eq. (11.58) can be written as

$$\frac{dP(E)}{dE} = \pm \frac{1}{V(E)} \tag{11.59}$$

then Eq. (11.54) transforms into:

$$Z = \int dQ \cdot 2 \cdot \int_0^\infty \frac{1}{V(E)} e^{-\beta E} dE = \int_0^\infty \frac{2N}{V(E)} e^{-\beta E} dE \tag{11.60}$$

This is equivalent to Eq. (11.56).

The action J for a soliton is proportional to $P(a)$:

$$J = \sum_i \oint P_t dq_i = \sum_i \left[\int_0^T P_i(t) \dot{q}_i(t) d(t) \right] \tag{11.61}$$

If a soliton exists, then P_i and q_i are the momentum and coordinate of the soliton. According to our definition of soliton's coordinate and momentum, we have:

$$J = \oint P dQ = NP \tag{11.62}$$

This can be seen directly from the solution of equation (11.61). That is,

$$J = \int_0^{\mathrm{T}} E_{kin}(t) dt = T E_{kin} = \frac{2N}{V(\alpha)} E_{kin}(\alpha)$$

$$= 4N \frac{a}{b} \sqrt{\frac{m}{ab}} \left(2\mathrm{ch}\alpha - \mathrm{sh}\alpha\right) = NP(\alpha)$$

At low temperatures, since solitons are not the lowest energy state of the Toda lattice, but phonons are, the partition function of phonons is:

$$Z_{p\lambda} = \prod_{n=1}^{N/2} \left(\int \frac{2\pi}{\omega_n} e^{-\beta E} dE \right)^2 = \prod_{n=1}^{N/2} \left(\frac{2\pi}{\beta \omega_n} \right)^2 \tag{11.63}$$

At extremely low temperatures, for a system of non-interacting solitons, we have:

$$Z_{sol} = \frac{1}{N!} \left(\int \frac{2N}{C_s} e^{-\beta E} dE \right)^N = \left(\frac{e}{N} \frac{2N}{C_\varepsilon \cdot \beta} \right)^N \tag{11.64}$$

where $N!$ arises due to the indistinguishability of N solitons. For the longest wavelength phonons, we have:

$$\frac{2\pi}{\omega_n} = \frac{\lambda_n}{V_n} \rightarrow \frac{2N}{C_s} \tag{11.65}$$

Therefore, solitons can be considered as long-wavelength phonons. For general phonons:

$$\omega_n = 2 C_s \sin \frac{\pi n}{N}$$

then, the partition function of various phonons is

$$Z_{p\lambda} = \left(\frac{\pi}{C_s \cdot \beta} \right)^N \prod_{n=1}^{N/2} \frac{1}{\sin \frac{2\pi}{N}} = \left(\frac{\pi}{C_s \cdot \beta} \right)^N \cdot Z^N = \left(\frac{\pi}{e} \right)^N Z_{sol} \tag{11.66}$$

Hence, Z_{ph} and the partition function Z_{sol} of interacting solitons are different.

Using the above partition functions, we can derive the free energy of the soliton system by $F = -\frac{1}{\beta} \ln Z_s$. Subsequently, all thermodynamic quantities can be also derived, such as specific heat. The method is similar to the familiar statistical physics methods, so it is not introduced here.

References

[1]   Landau L D, Ginzburg V L. On the theory of superconductivity[J]. Zh. Eksp. Teor. Fiz., 1950, 20: 1064.

[2]   吴杭生, 管惟炎, 李宏成. 超导电性 [M]. 科学出版社,1979

[3]   Abrikosov A A. On the magnetic properties of superconductors of the second group[J]. Soviet Physics-JETP, 1957, 5: 1174-1182.

[4]   Jacobs L, Rebbi C. Interaction energy of superconducting vortices[J]. Physical review B, 1979, 19(9): 4486.

[5]   Josephson B D. Supercurrents through barriers[J]. Advances in Physics, 1965, 14(56): 419-451.

[6]   GL LAMB J R. Analytical descriptions of ultrashort optical pulse propagation in a resonant medium[J]. Reviews of Modern Physics, 1971, 43(2): 99.

[7]   Costabile G, Parmentier R D, Savo B, et al. Exact solutions of the sine-Gordon equation describing oscillations in a long (but finite) Josephson junction[J]. Applied Physics Letters, 1978, 32(9): 587-589.

[8]   Kittel C. Theory of antiferroelectric crystals[J]. Physical Review, 1951, 82(5): 729.

[9]   Sarker S, Trullinger S E, Bishop A R. Solitary-wave solution for a complex one-dimensional field.[Crossover behavior, anisotropy][J]. Phys. Lett., A;(Netherlands), 1976, 59(4).

[10] Lajzerowicz J, Niez J J, Bishop A R, et al. Solitons and condensed matter physics[C]//Proc. Symp. on Nonlinear (Soliton) Structure and Dynamics in Condensed Matter.(Oxford, UK, 27-29 June). 1978: 195-8.

[11] De Gennes P G. Superconductivity of metals and alloys[M]. CRC press, 2018.

[12] M. E. Lines and A. M. Glass, Principles and Application of Ferro electrics and Related Materials, Clarendon Press, Oxford, 1977.

[13] Scott A C, Chu F Y F, McLaughlin D W. The soliton: a new concept in applied science[J]. Proceedings of the IEEE, 1973, 61(10): 1443-1483.

[14] Moskalenko S A, Sinyak V A, Khadzhi P I. Propagation of coherent excitons and photons in a crystal[J]. Soviet Journal of Quantum Electronics, 1976, 6(4): 464.

[15] Krumhansl J A, Schrieffer J R. Dynamics and statistical mechanics of a one-dimensional model Hamiltonian for structural phase transitions[J]. Physical Review B, 1975, 11(9): 3535.

[16] Currie J F, Krumhansl J A, Bishop A R, et al. Statistical mechanics of one-dimensional solitary-wave-bearing scalar fields: Exact results and ideal-gas phenomenology[J]. Physical Review B, 1980, 22(2): 477.

[17] Trullinger S E, Miller M D, Guyer R A, et al. Brownian motion of coupled nonlinear oscillators: Thermalized solitons and nonlinear response to external forces[J]. Physical Review Letters, 1978, 40(4): 206.

[18] Currie J F, Fogel M B, Palmer F L. Thermodynamics of the sine-Gordon field[J]. Physical Review A, 1977, 16(2): 796.

[19] Gupta N, Sutherland B. Investigation of a class of one-dimensional nonlinear fields[J]. Physical Review A, 1976, 14(5): 1790.

[20] Toda M. Waves in nonlinear lattice[J]. Progress of Theoretical Physics Supplement, 1970, 45: 174-200.

[21] Bolterauer H, Opper M. Solitons in the statistical mechanics of the Toda lattice[J]. Zeitschrift für Physik B Condensed Matter, 1981, 42: 155-161.

Chapter 12
Rogue Wave and Wave Turbulence

12.1 Rogue Wave

Rogue waves, also known as freak waves, extreme waves, or monster waves, refer to extreme wave phenomena with unusual characteristics that genuinely exist in nature. Due to the fact that the phenomenon of rogue waves was first discovered in the ocean and caused numerous devastating maritime disasters in maritime history, so it is also called as monster waves or killer waves. In oceanography, the current authoritative theory is that rogue waves are some peculiar lar-amplitude waves that suddenly appear in the ocean, with the basic characteristic of "coming and going without a trace"[1,2].

Thanks to the unremitting efforts of a group of pioneering scientists, the phenomenon of rogue waves has been proven to be an extreme natural phenomenon caused by nonlinear effects[3]. Through research on controllable experimental systems such as nonlinear optical fibers, Bose-Einstein condensates(BEC), and plasmas, scientists have discovered that the dynamic characteristics of localized waves on a plane wave background can accurately describe the "rogue wave phenomenon" that actually exists in nature. Nowadays, the concept of rogue waves has been extended to fields such as nonlinear optics, BEC, atmospheric science, superfluidity, and finance[1,2,4,5].

12.2 Formation of Rogue Wave

Initially, random disturbances in an unstable state trigger modulation instability (MI), resulting in the formation of rogue waves. Here is an example to illustrate this phenomenon. Considering the following NLS equation

$$i\psi_t - \psi + \psi_{xx} + |\psi|^2\psi = 0 \tag{12.1}$$

The simplest condensation solution $\psi = 1$ is unstable. Because the solution of (12.1) is

$$\psi = 1 + k e^{ikx + i\Omega t}, \quad |k| \ll 1 \tag{12.2}$$

the linearization equation yields $\Omega^2 = k^4 - 2k^2$. Therefore, $k \in (-2, 2)$ is unstable.

Figure 12-1   Rogue wave (originating from Google)

Consider the initial condition for equation (12.1) as

$$\psi|_{t=0} = 1 + \epsilon(x), \quad |\epsilon(x)| \ll 1 \tag{12.3}$$

here, $\epsilon(x)$ is white noise.

The first characteristic of turbulence is defined as the wave action spectrum

$$I_k(t) = <|\psi_k(t)|^2> \tag{12.4}$$

where $< \dots >$ represents the arithmetic mean of the initial ensemble,

$$\psi_k(t) = \mathcal{F}[\psi(x,t)] = \frac{1}{L} \int_{-L/2}^{L/2} \psi(x,t)e^{-ikx} \, dx$$

$$\psi(x,t) = \mathcal{F}^{-1}[\psi_k(t)] = \sum_k \psi_k(t)e^{ikx}$$

For the wave action spectrum with $\psi = 1$, it is concentrated at $k = 0$

$$I_k = \begin{cases} 1, & k = 0 \\ 0, & k \neq 0 \end{cases}$$

and the power form of the spectrum at $k = 0$ can be obtained

$$I(k) \approx k^{-\alpha}, |k| \leqslant 0.15, \ \alpha \approx \frac{2}{3} \tag{12.5}$$

Another important characteristic of turbulence is the singularity of spatial correlation

$$g(x,t) = \left\langle \frac{1}{L} \int_{-L/2}^{L/2} \psi(y,t)\psi^*(y-x,t)\mathrm{d}y \right\rangle \tag{12.6}$$

$$g(x,t) = \mathcal{F}^{-1}\left[I_k(t)\right] \tag{12.7}$$

It can be discovered that

$$g(x,t) \sim \frac{1}{|x|}, \quad |x| \leqslant 2$$

in the MI stage. Numerical methods indicate that for the equation (12.1), by employing spatial differencing and the 4th-order Runge-Kutta method for t

$$\mathrm{i}\frac{\mathrm{d}\psi_n}{\mathrm{d}t} + \frac{\psi_{n+1} - 2\psi_n + \psi_{n-1}}{h^2} - \psi_n + |\psi_n|^2 \frac{\psi_{n+1} + \psi_{n-1}}{2} = 0 \tag{12.8}$$

is obtained, where $-M/2 \leqslant n \leqslant M/2 - 1$.

Take

$$\epsilon(x) = A_0 \left(\frac{\sqrt{8\pi}}{\theta L}\right)^{\frac{1}{2}} \mathcal{F}^{-1}\left[10^{-v_k} \times \mathrm{e}^{-k^2/\theta^2 + i\xi_k}\right] \tag{12.9}$$

here, A_0 is the amplitude of white noise, θ is the width of white noise in k space, and ξ_k is the phase. For $t \in [0, 1000]$, $x \in [-L/2, L/2]$, $L = 1024\pi$, direct calculations gives $|\psi| > 2.8$, i.e. $|\psi|^2 > 8$, it forms a rogue wave. In the nonlinear stage of MI, define

$$Q = \frac{|< H_4 >|}{|< H_d >|}$$

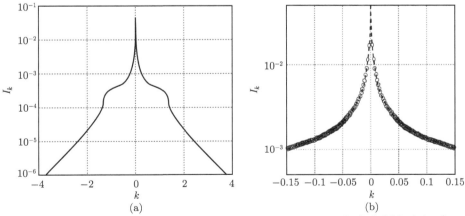

Figure 12-2   (a) Image of asymptotic wave action spectrum I_k (solid black line); (b)The image of asymptotic wave action spectrum I_k near $k = 0$ (black circle) and its fitting with function $f(k) = b\,|k|^{-\alpha}$, $\alpha \approx 0.659$, $b \approx 2.97 \mathrm{x} 10^{-4}$

where H_d represents kinetic energy, H_4 represents potential energy

$$H_d = \frac{1}{L}\int_{-L/2}^{L/2} |\psi_x|^2 \,\mathrm{d}x, \quad H_4 = -\frac{1}{2L}\int_{-L/2}^{L/2} |\psi|^4 \,\mathrm{d}x$$

When $|Q| \ll 1$, it indicates weak turbulence, while $Q = 2$ indicates strong turbulence, as shown in Fig.12-2-Fig.12-4.

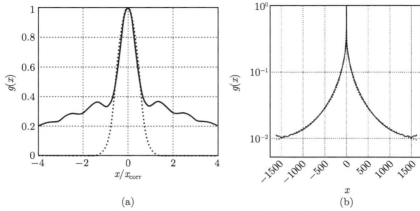

(a)   (b)

Figure 12-3   (a) The graph of the asymptotic spatial correlation function $g(x)$(solid black line) and Gaussian distribution $g(x) \approx \exp\left[-4\ln 2\left(\dfrac{x}{x_{corr}}\right)^2\right]$ (dashed red line) with respect to x/x_{corr}, here $x_{\text{corr}} \approx 4.016$ is the full width at half maximum of $g(x)$; (b)The asymptotic spatial correlation function $g(x)$(solid black line) and its fit to the functions $f(x) = b_1/\left(|x| + b_2\right), b_1 \approx 16.1, b_2 \approx 82.7$ (dashed red line)

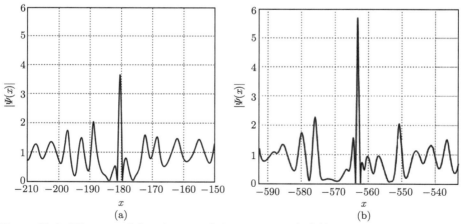

(a)   (b)

Figure 12-4   The spatial distribution of the amplitude $|\psi(x)|$ of the rogue wave. (a) A typical rogue wave generated near the local minimum of $|<H_4>|$. The maximum amplitude, $\max|\psi| \approx 3.6$, is achieved at $t \approx 19.8$. That is, it approaches the second local minimum of $|<H_4>|$ at $t \approx 19.6$, lasting for a duration of $\Delta T \sim 1$; (b) An extremely large rogue wave is generated at $t \approx 715.1$, with a maximum amplitude of $\max|\psi| \approx 5.7$. The duration of this time is $\Delta T \sim 0.5$

12.3   Wave Turbulence

Wave turbulence occurs in nonlinear dispersion systems and exists in many common natural physical phenomena, such as capillary waves, magnetic fluids with or without magnetic fields in plasmas, superfluid helium, Bose-Einstein condensation, nonlinear optics, sonic waves, etc. Wave turbulence is also important in oceanography and cosmology, including ocean surface waves, internal waves, Rossby waves, and Rossby, internal gravity waves in the ocean and atmosphere.

The nonlinearity in wave turbulence is small and the wave interaction is weak, therefore, it can be described by several kinetic equations. Among them, nonlinear structures were introduced, which results in the emergence of solitons and quasi solitons. Not a complete sentence. After colliding with each other, solitons are elastic and stable, but in many practical problems, soliton collisions are not elastic. The so-called quasi soliton is unstable. In nuclear physics, quasi soliton turbulence is generated from unstable particles. High amplitude quasi solitons are unstable and lead to singularity. The rogue wave or singular waves in the ocean originate from coherent structures.

In the following, consider the weak wave turbulence in the Majda- McLaughlin-Tabak (MMT) mathematical model. That is to consider the following dynamic equation:

$$i\psi_t = \left|\frac{\partial}{\partial x}\right|^{\alpha}\psi + \lambda\left|\frac{\partial}{\partial x}\right|^{\frac{\beta}{4}}\left(\left|\left|\frac{\partial}{\partial x}\right|^{\frac{\beta}{4}}\psi\right|^2\left|\frac{\partial}{\partial x}\right|^{\frac{\beta}{4}}\psi\right) \tag{12.10}$$

where $\psi(x,t)$ represents the complex wave function, $\lambda = \pm 1$. Real number α controls dispersion, parameter β controls nonlinearity, and $\left|\frac{\partial}{\partial x}\right|^{\alpha}$ represents the fractional derivative. The fractional order integral $D^{-\nu}\psi(x)$ is defined as

$$D^{-\nu}\psi(x) = \frac{1}{\Gamma(\nu)}\int_0^x (x-\xi)^{\nu-1}\psi(\xi)d\xi, \quad \nu > 0$$

with $\alpha > 0$ fractional order derivatives

$$D^{\alpha}\psi(x) = \frac{d}{dx^m}\left[D^{-(m-\alpha)}\psi(x)\right]$$

where m is an integer, $m \geqslant [\alpha]$. Its Fourier transform is

$$F\left[D^{\alpha}\psi(x)\right] = (ik)^{\alpha}\widehat{\psi}(k)$$

For NLS equation, $\alpha = 2$, $\beta = 0$, $\left|\frac{\partial}{\partial x}\right|^2 = -\frac{\partial^2}{\partial x^2}$, $\left|\frac{\partial}{\partial x}\right|^{\alpha}\psi = |k|^{\alpha}\widehat{\psi}$, the system of equations (12.10) can be written in Hamiltonian form, where the Hamiltonian quantity is

$$H = E + H_{NL} = \int \left\| \left| \frac{\partial}{\partial x} \right|^{\frac{\alpha}{2}} \psi \right|^2 dx + \left\| \frac{1}{2}\lambda \int \left\| \frac{\partial}{\partial x} \right|^{\frac{\beta}{4}} \psi \right|^4 dx \quad (12.11)$$

and there are two conserved quantities

$$N = \int |\psi|^2 dx, \quad M = \frac{1}{2}i \int \psi \frac{\partial \psi^*}{\partial x} - \frac{\partial \psi}{\partial x} \psi^* dx$$

To perform the Fourier transformation on equation (12.10), there is

$$i\frac{\partial \widehat{\psi}_k}{\partial t} = w(k)\widehat{\psi}_k + \lambda \int T_{123k}\widehat{\psi}_1 \widehat{\psi}_2 \widehat{\psi}_3^* \delta\left(k_1 + k_2 - k_3 - k\right) dk_1 dk_2 dk_3 \quad (12.12)$$

where $\widehat{\psi}_k(t) = \dfrac{1}{2\pi} \displaystyle\int_{-\infty}^{\infty} \psi(x,t)e^{-ikx}dx, k \in \mathbb{R}$, dispersion relation $w(k) = |k|^\alpha, \alpha > 0$, simple interaction coefficient

$$T_{123k} = T\left(k_1, k_2, k_3, k\right) = |k_1 k_2 k_3 k|^{\frac{\beta}{4}}$$

In the $\mathcal{F}$ space, Hamilton quantity is

$$H = \int w(k) \left|\widehat{\psi}_k\right|^2 dk + \frac{1}{2}\lambda \int T_{123k}\widehat{\psi}_1 \widehat{\psi}_2 \widehat{\psi}_3^* \widehat{\psi}_k^* \delta\left(k_1 + k_2 - k_3 - k\right) dk_1 dk_2 dk_3 dk.$$

(12.12) can be written as

$$i\frac{\partial \widehat{\psi}_k}{\partial t} = \frac{\delta H}{\delta \widehat{\psi}_k^*}$$

$$N = \int |\psi_k|^2 dk, \quad M = \int k \, |\psi_k|^2 dk$$

It can be inferred that $T_{123k} = T_{213k} = T_{3k12}$. In equation (12.12), $\widehat{\psi}_k(t) = \left|\widehat{\psi}_k(t)\right| e^{i\varphi(k,t)}$, and $\varphi(k,t)$ is a phase function that is random.

Let $n(k,t)$ be the spectral density of the wave function ψ

$$\int |\psi(x,t)|^2 dx = \int n(k,t)dk = N$$

In frequency space

$$N(w,t) = n(k(w),t)\frac{dk}{dw}, \quad E(w,t) = wN(w,t)$$

then

$$N = \int N(w)dw, \quad E = \int E(w)dw$$

Introduce the four-wave correlation function

$$\left\langle \widehat{\psi}_{k_1}(t)\widehat{\psi}_{k_2}(t)\psi_{k_3}^*(t)\psi_k^*(t) \right\rangle = J_{123k}\delta\left(k_1 + k_2 - k_3 - k\right)$$

For the equation (12.12) satisfied by the original $n_k(t)$, consider it to have a random phase. The Kinetic equation is

$$\frac{\partial n_k}{\partial t} = 2\lambda \int \mathrm{Im}\, J_{123k}\delta\left(k_1 + k_2 - k_3 - k\right) dk_1 dk_2 dk_3$$

From approximation of quasi Gaussian random phase

$$\mathrm{Re}\, J_{123k} \approx n_1 n_2 \left[\delta\left(k_1 - k_3\right) + \delta\left(k_1 - k\right)\right]$$

the imaginary part of J_{123k} can be approximated by the solution of the correlation function

$$\mathrm{Im}\, J_{123k} \approx 2\pi\lambda T_{123k}\delta\left(w_1 + w_2 - w_3 - w\right)$$

$$\cdot \left(n_1 n_2 n_3 + n_1 n_2 n_k - n_1 n_3 n_k - n_2 n_3 n_k\right)$$

So the dynamic wave equation is obtained

$$\frac{\partial n_k}{\partial t} = 4\pi \int T_{123k}^2 \left(n_1 n_2 n_3 + n_1 n_2 n_k - n_1 n_3 n_k - n_2 n_3 n_k\right)$$

$$\times \delta\left(w_1 + w_2 - w_3 - w_4\right)\delta\left(k_1 + k_2 + k_3 - k\right) dk_1 dk_2 dk_3$$

Consider the average of wave numbers

$$\frac{\partial N(w)}{\partial t} = \frac{4\pi}{\alpha^4} \int \left(w_1 w_2 w_3 w\right)^{\frac{\beta/2 - \alpha + 1}{\alpha}} \left(n_1 n_2 n_3 + n_1 n_2 n_w - n_1 n_3 n_w - n_2 n_3 n_w\right)$$

$$\times \delta\left(w_1 + w_2 - w_3 - w\right)\left[\delta\left(w_1^{1/2} + w_2^{1/\alpha} - w_3^{1/\alpha} + w^{1/\alpha}\right)\right.$$

$$+ \delta\left(w_1^{1/\alpha} + w_2^{1/\alpha} + w_3^{1/\alpha} - w^{1/\alpha}\right) + \delta\left(w_1^{1/\alpha} - w_2^{1/\alpha} - w_3^{1/\alpha} - w^{1/\alpha}\right)$$

$$\left. + \delta\left(-w_1^{1/\alpha} + w_2^{1/\alpha} - w_3^{1/\alpha} - w^{1/\alpha}\right)\right] dw_1 dw_2 dw_3, \quad w_i > 0 \qquad (12.13)$$

where $n_w = n(k(w))$.

Let $n(w) = w^{-\alpha}$, then the dynamic equation is

$$\frac{\partial N(w)}{\partial t} \approx w^{-\alpha-1} I(\alpha, \beta, \gamma)$$

where

$$
I(\alpha, \beta, \gamma) = \frac{4\pi}{\alpha^4} \int_\Delta (\xi_1 \xi_2 \xi_3)^{\beta/2\alpha + 1/\alpha - 1 - \gamma} (1 + \xi_3^\gamma - \xi_1^\gamma - \xi_2^\gamma) \cdot \delta (1 + \xi_3 - \xi_1 - \xi_2)
$$

$$
\cdot \delta \left(\xi_1^{1/\alpha_1} + \xi_2^{1/\alpha} + \xi_3^{1/\alpha} - 1 \right) (1 + \xi_3^y - \xi_1^y - \xi_2^y) \, d\xi_1 d\xi_2 d\xi_3 \qquad (12.14)
$$

where $\Delta = \{0 < \xi_1 < 1, 0 < \xi_2 < 1, \xi_1 + \xi_2 > 1\}$, $y = 3\gamma + 1 - \dfrac{2\beta + 3}{\alpha}$.

According to the convergence of integral (12.13), it can be divided into low-frequency and high-frequency parts. For the low-frequency part, it is easy to obtain in (12.14) that

$$
2\gamma < -1 + \frac{\beta + 4}{\alpha}
$$

For the high-frequency part, substituting $n(w) \approx w^{-\alpha}$ into (12.14) yields

$$
\gamma > \frac{\beta + \alpha - 1}{\alpha}
$$

Together, $\beta \leqslant 3(2 - \alpha)$. We can discuss the case of $\alpha = 1/2$.

12.4   Soliton and Quasi Soliton

We are now discussing solitons and quasi solitons in coherent structures. Consider the following form of solution for equation (12.12)

$$
\widehat{\psi}_k(t) = e^{i(\Omega - kv)t} \widehat{\phi}_k
$$

where Ω and v are constants. In physical space,

$$
\psi(x, t) = e^{i\Omega t} \xi(x - vt)
$$

where $\xi(\cdot)$ is the inverse $\mathcal{F}$ transformation of $\widehat{\phi}_k$, and v is the velocity of the soliton or quasi soliton. Therefore, $|\psi(x, t)| = |\xi(x - vt)|$, and $\widehat{\phi}_k$ satisfies the integral equation

$$
\widehat{\phi}_k = -\frac{\lambda}{\Omega - kv + \omega(k)} \int T_{123k} \widehat{\phi}_1 \widehat{\phi}_2 \widehat{\phi}_3^* \delta (k_1 + k_2 - k_3 - k) \, dk_1 dk_2 dk_3 \qquad (12.15)
$$

Introducing functional

$$
T(k) = \lambda \int T_{123k} \widehat{\phi}_1 \widehat{\phi}_2 \widehat{\phi}_3^* \delta (k_1 + k_2 - k_3 - k) \, dk_1 dk_2 dk_3
$$

and

$$
F = -\Omega + kv - \omega(k) = -\Omega + kv - |k|^\alpha
$$

we can obtain $\hat{\phi}_k = \dfrac{T(k)}{F}$.

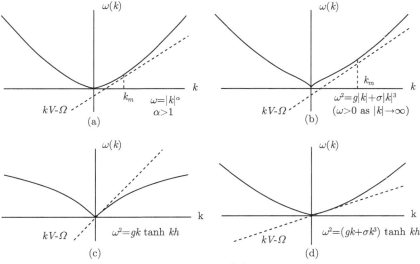

Figure 12-5   Soliton

Figure 12-5 shows a typical soliton. In Figure (a), for any $k \in \mathbb{R}$, constant $\Omega > 0$, function $F < 0$, the dispersion relation is $\omega = |k|^\alpha$. Specifically, when $\alpha = 2$, it is the dispersion relation of the NLS equation. In Figure (b), the dispersion relation is $\omega^2 = g|k| + \sigma|k|^3$, where g is the gravitational acceleration and σ is the coefficient of surface tension. This soliton corresponds to a stable envelope soliton of capillary gravity waves. In Figure (c), the dispersion relation is $\omega^2 = gk$ tanh kh. This soliton is an soliton of gravity waves in shallow water. In Figure (d), the dispersion relationship is $\omega^2 = (gk + \sigma k^3)$ tanh kh, where $\sigma > gh^2/3$. This soliton corresponds to capillary gravity waves in very shallow water.

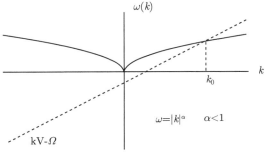

Figure 12-6   An example where the "true" soliton does not exist, where $\omega = |k|^\alpha (\alpha < 1)$

In these three examples, the straight line always passes through the curve $\omega = \omega(k)$. Therefore, in equation (12.15), the denominator $\Omega - kV + \omega(k)$ has one or more zero points. The dispersion relationship in Figure 12-7(a) is $\omega = |k|^\alpha (\alpha < 1)$. The dispersion relation in Figure (b) is $\omega^2 = g|k| + \sigma|k|^3$, corresponding to capillary gravity waves. The dispersion relation of Figure (c) is $\omega^2 = (gk + \sigma k^3) \tanh kh$, where $\sigma < gh^2/3$, which corresponds to capillary gravity waves on the surface of a finite depth h flow layer.

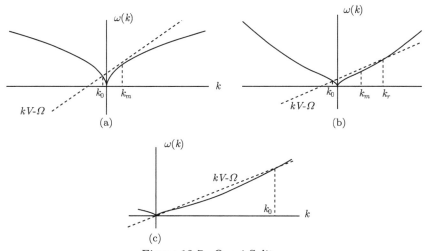

Figure 12-7   Quasi Soliton

12.4.1   The Instability and Blow-up of Solitons

Next, we consider the instability and blow-up of solitons in the focusing MMT model. In equation (12.12), let

$$\widehat{\psi}_k(t) = e^{i\Omega t}\widehat{\phi}_k$$

where $\Omega > 0$. In physical space,

$$\psi(x,t) = e^{i\Omega t}\xi(x)$$

where $\xi(x)$ is the inverse F transformation of $\widehat{\phi}_k$. $\widehat{\phi}_k$ satisfies the integral equation

$$(\Omega + |k|^\alpha)\,\widehat{\phi}_k = \int |k_1 k_2 k_3 k|^{\frac{\beta}{4}}\,\widehat{\phi}_1\widehat{\phi}_2\widehat{\phi_3^*}\delta\,(k_1 + k_2 - k_3 - k)\,dk_1 dk_2 dk_3$$

The free parameter Ω can be removed from scaling

$$\widehat{\phi}_k = \Omega^{-\frac{\beta}{2\alpha}+\frac{1}{2}-\frac{1}{\alpha}}\chi(K), \quad K = \Omega^{-\frac{1}{\alpha}}K$$

where $\chi(K)$ satisfies the following equation

$$(1 + |K|^\alpha)\chi(K) = \int |K_1 K_2 K_3 K|^{\frac{\beta}{4}} \chi_1 \chi_2 \chi_3^* \delta(K_1 + K_2 - K_3 - K) dK_1 dK_2 dK_3$$

The overall effect in soliton is

$$N = \int \left|\widehat{\phi}_k\right|^2 dk = \Omega^{-\frac{\beta}{\alpha}+1-\frac{1}{\alpha}} N_0$$

here $N_0 = \int |\chi|^2 dK$.

If $\dfrac{\partial N}{\partial \Omega} > 0$, it is stable. Here

$$\frac{\partial N}{\partial \Omega} = -\left(\frac{\beta - \alpha + 1}{2}\right)\frac{N}{\Omega}$$

Therefore, if $\beta < \alpha - 1$, soliton is stable, otherwise it is unstable. When $\alpha = \dfrac{1}{2}$, $\beta > -\dfrac{1}{2}$, soliton is unstable.

12.4.2   The Case of Quasi-Solitons

We are now considering the case of quasi-solitons in the defocusing MMT model. Here, we consider that $\Omega - kv + w(k) = 0$ has a zero point at $k = k_0$. $\widehat{\phi}_k$ is localized near $k = k_m$. $\widehat{\phi}_k$ has the width of q at $k = k_m$.

$$T(k_0) \approx e^{-c|k_m - k_0|/q} T(k_m)$$

where c is constants. In other words, $\widehat{\phi}_k$ has a pole at $k = k_0$, and the residue of this pole is exponentially small, indicating that $\psi(x, t) = e^{i\Omega t}\xi(x - vt)$ is non-localized. When $x \to -\infty$, it tends to be a monochromatic wave with a very small amplitude wave number of $k = k_0$ in the backward direction.

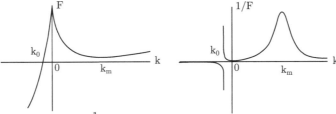

Figure 12-8   Function F and $\dfrac{1}{F}$ are qualitative behaviors. The expression for F is shown in equation (12.16)

If $q/k_m << 1$, then radiation is very slow. From

$$\left.\frac{\partial F}{\partial k}\right|_{k=k_m} = 0$$

the velocity of the quasi soliton v can be obtained $v = \alpha k_m^{\alpha-1}$.

For quasi solitons, it is very narrow in the $\mathcal{F}$ space. Ω is

$$\Omega = (\alpha - 1)k_m^\alpha \left(1 + \frac{1}{2}\alpha(q/m)^2\right), \quad q/k_m << 1$$

then

$$F = k_m^\alpha - |k|^\alpha + \alpha k_m^{\alpha-1}(k - k_m) + \frac{1}{2}\alpha(1 - \alpha)k_m^{\alpha-2}q^2 \tag{12.16}$$

If $\alpha < 1$, F has zero point at $k = k_0 < 0$, $\forall k_m$ has zero point. Therefore, $\frac{1}{F}$ always has a pole on the negative real axis, and the soliton of (12.15) cannot be real. Let $q << k_m$, $\kappa = |k - k_m|$, which can be approximated as

$$F \approx \frac{1}{2}\alpha(1 - \alpha)k_m^{\alpha-2}\left(\mathfrak{H}^2 + q^2\right) \tag{12.17}$$

From this, it can be concluded that the width of the maximum value of $\frac{1}{F}$ is $\kappa \approx q$.

If $\kappa << k_0$, a quasi soliton can be constructed on the branch of $\mathcal{F}$ near k_m. Generally, $|k_0| \approx k_m$, for instance, $\alpha = \frac{1}{2}$ and $q = 0$ gives

$$k_0 = -(\sqrt{2} - 1)^2 k_m$$

The quasi soliton moves to the right, with a velocity of $v(k_m)$, radiating monochromatic waves moving backward, with a wave number of k_0. When $q \to 0$, the expression for the quasi soliton can be found, as $\kappa << k_m$ has an approximation

$$T(k) \approx k_m^\beta \int \widehat{\phi}_1\widehat{\phi}_2\widehat{\phi_3^*}\delta(k_1 + k_2 - k_3 - k_4)\,dk_1dk_2dk_3$$

Considering (12.12) and (12.17), it can be concluded that

$$\frac{1}{2}\alpha(1 - \alpha)k_m^{\alpha-2}\left(\kappa^2 + q^2\right)\widehat{\phi}_k = k_m^\beta \int \widehat{\phi}_1\widehat{\phi}_2\widehat{\phi}_3^*(k_1 + k_2 - k_3 - k_4)\,dk_1dk_2dk_3$$

By inverse F transform, the above equation is transformed into a steady NLS equation

$$\frac{1}{2}\alpha(1 - \alpha)k_m^{\alpha-2}\left(-\frac{\partial^2\phi}{\partial x^2} + q^2\phi\right) = k_m^\beta|\phi|^2\phi$$

which has soliton solution

$$\phi(x) = \sqrt{\frac{\alpha(1-\alpha)}{k_m^{\beta-\alpha+2}}} \frac{q}{\cosh qx}.$$

For $\lambda = 1$, the following approximate quasi soliton solution can be obtained

$$\begin{cases} \psi(x,t) = \phi(x - vt)e^{i\Omega t}e^{ik_m(x-vt)} \\ \Omega = -(1-\alpha)k_m^\alpha - \frac{1}{2}\alpha(1-\alpha)k_m^{\alpha-2}q^2 \\ v = \alpha k_m^{\alpha-1} \end{cases} \qquad (12.18)$$

The quasi soliton (12.18) is an envelope soliton.

　　Seeking formal solutions from equation (12.10)

$$\psi(x,t) = U(x,t)e^{-i(1-\alpha)k_m^\alpha t}e^{ik_m(x-vt)}$$

Using Taylor expansion, the first term U satisfies the unsteady NLS equation

$$i\left(\frac{\partial U}{\partial t} + v\frac{\partial U}{\partial x}\right) = \frac{1}{2}\alpha(1-\alpha)k_m^{\alpha-2}\frac{\partial U}{\partial x^2} + k_m^\beta|U|^2U$$

and has soliton sloution

$$U(x,t) = \phi(x - vt)\exp\left(-\frac{1}{2}i\alpha(1-\alpha)k_m^{\alpha-2}q^2t\right)$$

　　In fact, the parameter $\dfrac{q}{k_m}$ is crucial for quasi soliton solutions. When it is small, the quasi soliton solution approaches the real soliton, and the amplitude of the quasi soliton is proportional to $\dfrac{q}{k_m}$. The small amplitude of the quasi soliton satisfies the integrable NLS equation and is stable. It is not obvious for finite amplitude. For $\beta > 0$, there exists a critical value of quasi soliton amplitude, which is unstable and forms singularity. Numerically, it is speculated that $\beta = 3$.

References

[1] Akhmediev N, Soto-Crespo J M, Ankiewicz A. Extreme waves that appear from nowhere: on the nature of rogue waves[J]. Physics Letters A, 2009, 373(25): 2137-2145.

[2] Akhmediev N, Ankiewicz A, Taki M. Waves that appear from nowhere and disappear without a trace[J]. Physics Letters A, 2009, 373(6): 675-678.

[3] Zakharov V, Gelash A. Freak waves as a result of modulation instability[J]. Procedia IUTAM, 2013, 9: 165-175.

[4] Dudley J M, Dias F, Erkintalo M, et al. Instabilities, breathers and rogue waves in optics[J]. Nature Photonics, 2014, 8(10): 755-764.

[5] Yan Z Y. Vector financial rogue waves[J]. Physics Letters A, 2011, 375(48): 4274-4279.